Lecture Notes in Artificial Intellig

Edited by J. G. Carbonell and J. Siekmann

Subseries of Lecture Notes in Computer Science

Rokia Missaoui   Jürg Schmid (Eds.)

Formal
Concept Analysis

4th International Conference, ICFCA 2006
Dresden, Germany, February 13-17, 2006
Proceedings

 Springer

Series Editors

Jaime G. Carbonell, Carnegie Mellon University, Pittsburgh, PA, USA
Jörg Siekmann, University of Saarland, Saarbrücken, Germany

Volume Editors

Rokia Missaoui
Université du Québec en Outaouais
Département d'informatique et d'ingénierie
C.P. 1250, succursale B, Gatineau, Canada, J8X 3X7
E-mail: Rokia.Missaoui@uqo.ca

Jürg Schmid
Universität Bern
Mathematisches Institut
Sidlerstr. 5, 3012 Bern, Switzerland
E-mail: juerg.schmid@math-stat.unibe.ch

Library of Congress Control Number: 2006920554

CR Subject Classification (1998): I.2, G.2.1-2, F.4.1-2, D.2.4, H.3

LNCS Sublibrary: SL 7 – Artificial Intelligence

ISSN   0302-9743
ISBN-10   3-540-32203-5 Springer Berlin Heidelberg New York
ISBN-13   978-3-540-32203-0 Springer Berlin Heidelberg New York

Springer is a part of Springer Science+Business Media

springer.com

© Springer-Verlag Berlin Heidelberg 2006
Printed in Germany

Typesetting: Camera-ready by author, data conversion by Scientific Publishing Services, Chennai, India
Printed on acid-free paper   SPIN: 11671404   06/3142   5 4 3 2 1 0

Preface

This volume contains selected papers of ICFCA 2006, the 4th International Conference on Formal Concept Analysis. The ICFCA conference series aims to be the prime forum for the publication of advances in applied lattice and order theory and in particular scientific advances related to formal concept analysis.

Formal concept analysis is a field of applied mathematics with its mathematical root in order theory, in particular the theory of complete lattices. Researchers had long been aware of the fact that these fields have many potential applications. Formal concept analysis emerged in the 1980s from efforts to restructure lattice theory to promote better communication between lattice theorists and potential users of lattice theory. The key theme was the mathematical formalization of *concept* and *conceptual hierarchy*. Since then, the field has developed into a growing research area in its own right with a thriving theoretical community and an increasing number of applications in data and knowledge processing including data visualization, information retrieval, machine learning, data analysis and knowledge management.

ICFCA 2006 reflected both practical benefits and progress in the foundational theory of formal concept analysis. This volume contains four lecture notes from invited speakers and 17 regular papers, among them one position paper. All regular papers appearing in these proceedings were refereed by at least two, in most cases three independent reviewers. The final decision to accept the papers was arbitrated by the Program Chairs based on the referee reports.

We wish to thank the Program Committee and the Editorial Board as well as the additional referees for their involvement, which ensured the scientific quality of these proceedings. Our special thanks go to Léonard Kwuida, who on the behalf of the Program Chairs meticulously managed information flow and data organization of the submission and editing process. We also express our gratitude to Ganaël Jatteau for his valuable assistance in managing the electronic submission system.

The ICFCA 2006 conference was hosted by Technische Universität Dresden, Germany. The Conference Chair was Bernhard Ganter. The conference was funded by the German Research Foundation (Deutsche Forschungsgemeinschaft).

February 2006

Rokia Missaoui
Jürg Schmid

Organization

The International Conference on Formal Concept Analysis (ICFCA) is the annual conference and principal research forum in the theory and practice of formal concept analysis. The inaugural International Conference on Formal Concept Analysis was held at the Technische Universität Darmstadt, Germany, in 2003. Succeeding ICFCA conferences were held at the University of New South Wales in Sydney, Australia, 2004, at the Université d'Artois, Lens, France, 2005 and at the Institut für Algebra, Technische Universität Dresden, Germany, 2006.

Conference Chair

Bernhard Ganter   Technische Universität Dresden, Germany

Program Chairs

Rokia Missaoui   Université du Québec en Outaouais, Canada
Jürg Schmid   Math. Institut, Universität Bern, Switzerland

Editorial Board

Peter Eklund	University of Wollongong, Australia
Bernhard Ganter	Technische Universität Dresden, Germany
Robert Godin	Université du Québec à Montréal, Canada
Sergei Kuznetsov	VINITI and RSUH Moscow, Russia
Uta Priss	Napier University, Edinburgh, UK
Gregor Snelting	Universität Passau, Germany
Gerd Stumme	Universität Kassel, Germany
Rudolf Wille	Technische Universität Darmstadt, Germany
Karl Erich Wolff	Fachhochschule Darmstadt, Germany

Program Committee

Radim Bělohlávek	Palacky University of Olomouc, Czech Republic
Claudio Carpineto	Fondazione Ugo Bordini, Rome, Italy
Richard J. Cole	University of Queensland, Brisbane, Australia
Paul Compton	University of New South Wales, Sydney, Australia

Frithjof Dau	Technische Universität Dresden, Germany
Brian Davey	La Trobe University, Melbourne, Australia
Vincent Duquenne	Université Pierre et Marie Curie, Paris, France
Wolfgang Hesse	Universität Marburg, Germany
Marzena Kryszkiewicz	Warsaw University of Technology, Poland
Léonard Kwuida	Math. Institut, Universität Bern, Switzerland
Wilfried Lex	Universität Clausthal, Germany
Christian Lindig	Saarland University, Saarbrücken, Germany
Engelbert Mephu Nguifo	IUT de Lens - Université d'Artois, France
Rokia Missaoui	Université du Québec en Outaouais, Canada
Lhouari Nourine	LIMOS, Université Blaise Pascal, Clermont-Ferrand, France
Alex Pogel	New Mexico State University, Las Cruces, USA
Sergei Obiedkov	University of Pretoria, South Africa
Sandor Radeleczki	University of Miskolc, Hungary
Jürg Schmid	Math. Institut, Universität Bern, Switzerland
Stefan Schmidt	Technische Universität Dresden, Germany
Bernd Schröder	Louisiana Tech University, Ruston, USA
Selma Strahringer	Fachhochschule Köln, Germany
Petko Valtchev	Université de Montréal, Canada
Mohammed J. Zaki	Rensselaer Polytechnic Institute, NewYork, USA

Additional Referees

Joachim Hereth Correia	Technische Universität Dresden, Germany
Pascal Hitzler	AIFB - Universität Karlsruhe, Germany
Ian Horrocks	University of Manchester, UK
Andreas Hotho	Universität Kassel, Germany
Wiebke Petersen	Institut für Sprache und Information, Heinrich-Heine-Universität Düsseldorf, Germany
Reinhard Pöschel	Technische Universität Dresden, Germany
Thomas Studer	Institute of Computer Science, University of Bern, Switzerland
Friedrich Wehrung	Laboratoire LMNO, Université de Caen, France

Table of Contents

Methods of Conceptual Knowledge Processing

Rudolf Wille

Technische Universität Darmstadt,
Fachbereich Mathematik,
Schloßgartenstr. 7, D–64289 Darmstadt
wille@mathematik.tu-darmstadt.de

Abstract. The offered *methods of Conceptual Knowledge Processing* are procedures which are well-planed to mean and purpose and therewith lead to skills for solving practical tasks. The used means and skills have been mainly created as translations of mathematical means and skills of *Formal Concept Analysis*. Those transdisciplinary translations may be understood as transformations from mathematical thinking, dealing with potential realities, to logical thinking, dealing with actual realities. Each of the 38 presented methods is discussed in a general language of logical nature, while citations give links to the underlying mathematical background. Applications of the methods are demonstrated by concrete examples mostly taken from the literature to which explicit references are given.

Contents

1 Conceptual Knowledge Processing

Conceptual Knowledge Processing is considered to be an applied discipline dealing with ambitious knowledge which is constituted by conscious reflexion, dis-

R. Missaoui and J. Schmid (Eds.): ICFCA 2006, LNAI 3874, pp. 1–29, 2006.

cursive argumentation and human communication on the basis of cultural background, social conventions and personal experiences. Its main aim is to develop and maintain methods and instruments for processing information and knowledge which support rational thought, judgment and action of human beings and therewith promote the critical discourse (cf. [Wi94], [Wi97b], [Wi00b]).

The adjective *"Conceptual"* in the name "Conceptual Knowledge Processing" underlines the constitutive role of the thinking, arguing and communicating human being for knowledge and its processing. The term *"Processing"* refers to the process in which something is gained which may be knowledge or something approximating knowledge such as a forecast, an opinion, a casual reason etc. To process knowledge, formal elements of language and procedures must be activated. This pre-supposes formal representations of knowledge and, in turn, knowledge must be constituted from such representations by humans.

To understand this process, the basic relation between *form* and *content* must be clarified for Conceptual Knowledge Processing. A branch of philosophy which makes basic statements on this is *pragmatic philosophy* which was initiated by Ch. S. Peirce [Pe35] and is presently continued among others in the discourse philosophy of K.-O. Apel [Ap76] and J. Habermas [Ha81]. According to pragmatic philosophy, knowledge is formed in an unbounded process of human thinking, arguing and communicating; in this connection, reflection on the effects of thought is significant and real experiences stimulate re-thinking time and again. In this process, form and content are related so closely that they may not be separated without loss.

Theoretically, Conceptual Knowledge Processing is mainly founded upon a mathematization of traditional philosophical logic with its doctrines of concept, judgment, and conclusion. The core of the mathematical basis of Conceptual Knowledge Processing is *Formal Concept Analysis* [GW99a] which has been developed as a mathematical theory of concepts and concept hierarchies during the last 25 years. Although Conceptual Knowledge Processing deals with actual realities, it obtains its basic forms of thinking from mathematics that, according to Peirce ([Pe92]; p.121), has the aim to uncover a "great Cosmos of Forms, a world of potential being". Above all, Formal Concept Analysis as applied mathematics provides Conceptual Knowledge Processing with a rich amount of mathematical forms of thinking; this has been proven useful in a large number of applications. For such a success it is essential that conceptual representations of knowledge can be materialized so that they appropriately merge form and content of the processed knowledge.

As mathematical theory, Formal Concept Analysis with its notions and statements is strictly based on the common *set-theoretical semantics* which is grounded on abstract sets and their abstract elements. For the explanation of the mathematical notions, statements, and procedures in this paper, the reader is referred to the monograph *"Formal Concept Analysis: Mathematical Foundations"* [GW99a]. The notions and statements discussed in the framework of Conceptual Knowledge Processing shall be understood with respect to the semantics of their specific field of application. If they refer to different fields of application, their semantics has to be more abstract (eventually up to the philosophical semantics). To make the connections between Formal Concept Analysis and Conceptual Knowledge Processing clear, notions and statements of Formal Concept Analysis have to be transformed to suitable notions and statements of Conceptual Knowledge Processing and vice versa. For basic notions such transformation is given by the following list of correspondences (cf. [Wi05b], p.28f.)

Formal Concept Analysis	$\leftrightarrow$	**Conceptual Knowledge Processing**
formal context	$\leftrightarrow$	(logical) context
(formal) many-valued context	$\leftrightarrow$	many-valued context
(formal) object	$\leftrightarrow$	object
(formal) attribute	$\leftrightarrow$	attribute
many-valued attribute	$\leftrightarrow$	many-valued attribute
(formal) attribute value	$\leftrightarrow$	attribute value
formal concept	$\leftrightarrow$	concept
extent	$\leftrightarrow$	extension
object extent	$\leftrightarrow$	object extension
intent	$\leftrightarrow$	intension
attribute intent	$\leftrightarrow$	attribute intension
(formal) object concept	$\leftrightarrow$	object concept
(formal) attribute concept	$\leftrightarrow$	attribute concept
(formal) subconcept	$\leftrightarrow$	subconcept
(formal) superconcept	$\leftrightarrow$	superconcept
infimum of formal concepts	$\leftrightarrow$	largest common subconcept of concepts
supremum of formal concepts	$\leftrightarrow$	smallest common superconcept of concepts
concept lattice	$\leftrightarrow$	concept hierarchy

Based on such correspondences, this paper aims to show how Formal Concept Analysis gives rise to a spectrum of methods of Conceptual Knowledge Processing applicable for gaining knowledge for a broad variety of reasons and purposes.

2   Methods

In [Lo84], scientific methods are characterized in general as follows:

> A *method* is a procedure which is well-planned according to mean and purpose and therewith leads to skills for solving theoretical and practical tasks.

In the case of Conceptual Knowledge Processing, basic means and skills for its methods are mainly translations of mathematically defined means and skills of Formal Concept Analysis. Those translations interpret the mathematical means and skills with respect to actual realities so that they become understandable for common users in their specific semantics. In the sense of Peirce [Pe92], the *transdisciplinary translations* may be understood as transformations from mathematical thinking, dealing with potential realities, to logical thinking, dealing with actual realities (cf. [Wi01], [Wi05b]).

2.1 Conceptual Knowledge Representation

The mathematization of conceptual knowledge by Formal Concept Analysis is based on the understanding of *concepts* constituted by their extension and intension, respectively. For a concept, its *extension* contains all objects falling under the concept and its *intension* comprises all attributes (properties, meanings) common to all those objects. Thus, the representation of conceptual knowledge can be grounded on a *context* consisting of a collection of objects, a collection of attributes, and a relation indicating which object has which attribute. A context corresponds to a formal context which, in Formal Concept Analysis, is usually materialized by a cross table. Therefore, Formal Concept Analysis suggests the following elementary representation method of Conceptual Knowledge Processing:

M1.1 Representing a Context by a Cross Table: A context can be represented by a *cross table*, i.e., a rectangular table the rows of which are headed by the object names and the columns headed by the attribute names; a cross in row g and column m means that the object g has the attribute m. An example is given in [GW99a], p.18.

Conceptual knowledge is often represented by 0-1-tables. Then it is necessary to make explicit in which way the zeros and ones shall give rise to concepts. In the case that they lead exactly to the same concepts as the cross table in which the crosses are at the same places as the ones, such table is called a *one-valued context* (cf. M4.2) and considered as equivalent to the corresponding cross table.

M1.2 Clarifying a Context: *Object Clarification* of a context means to remove all objects except one in each class of objects having the same attributes. Dually, *Attribute Clarification* of a context means to remove all attributes except one in each class of attributes applying to the same objects. *Clarifying a Context* means to apply to a context both: Object Clarifying and Attribute Clarifying (cf. [GW99a], p.24).

A cross table of a context resulting from a clarification may be completed by inserting the name of each removed object g in front of the name of that object having the same attributes as g and inserting the name of each removed attribute m above the name of that attribute applying to the same objects as m. The completed cross table is often a considerably smaller and better readable representation of the original not clarified context than the cross table described in M1.1.

M1.3 Reducing a Context: *Object Reduction* of a finite logical context means first to apply Object Clarification to the context and then to remove each remaining object the object concept of which is the smallest common superconcept of proper subconcepts of that object concept. Dually, *Attribute Reducing* of a finite context means first to apply Attribute Clarification to the context and then to remove each remaining attribute the attribute concept of which is the largest common subconcept of proper superconcepts of that attribute concept. *Reducing a Context* means to apply to a context both: Object Reducing and Attribute Reducing (cf. [GW99a], p.24).

A finite context resulting from the reduction of a context is (up to isomorphism) the smallest context the concept hierarchy of which has the same hierarchical structure as the concept hierarchy of the original context. Thus, the finite reduced contexts are structurally the smallest (implicit) representations of finite concept hierarchies.

M1.4 Representation of a Concept Hierarchy by a Line Diagram: The concept hierarchy of a finite context can be visualized by a line diagram as follows: The concepts of the hierarchy are represented by small circles in such a way that upward leading line segments between those circles can indicate the subconcept-superconcept relation. Every circle representing a concept generated by an object/attribute has attached from below/above the name of that object/attribute (cf. M2.1). Those attachments of object and attribute names allow to read off the extension and intension of each concept from the representing line diagram: the extension/intension of a concept consists of all those objects/attributes the names of which are attached to a circle belonging to a downward/upward path of line segments starting from the circle of that concept (cf. [GW99a], p.23).

Unfortunately, up to now, no universal method is known for drawing well-readable line diagrams representing concept hierarchies. For smaller concept hierarchies, the method of *Drawing an Additive Line Diagram* (see [GW99a], p.75) often leads to well-structured line diagrams. This is the reason that quite a number of computer programs for drawing concept hierarchies use that method (e.g. ANACONDA, *Cernato, Concept Explorer, Elba*).

M1.5 Checking a Line Diagram of a Concept Hierarchy: A line diagram represents the concept hierarchy of a given finite context correctly if and only if the line diagram satisfies the following conditions: (1) each circle being the start of exactly one downward line segment must have attached an object name; (2) each circle being the start of exactly one upward line segment must have attached an attribute name; (3) an object g has an attribute m in the given context if and only if the names of g and m are attached to the same circle or there is an upward path of line segments from the circle with the name of g to the circle with the name of m; (4) the line diagram represents a concept hierarchy (cf. [GW99a], p.20, The Basic Theorem on Concept Lattices).

For line diagrams representing less than 50 concepts, it is quite easy to check the conditions (1), (2), and (3). Checking condition (4) by inspection is usually more costly because (4) mathematically means that the line diagram must represent a lattice. Nevertheless, experiences with many realistic data contexts have shown that a failure of condition (4) is usually accompanied by a failure of at least one of the conditions (1), (2), (3).

M1.6 Dualizing a Concept Hierarchy: *Dualizing a Context* means to interchange the roles of objects and attributes, i.e., objects become attributes and attributes become objects, while the context relation turns to its inverse. If one considers objects as instances of Firstness and attributes as instances of Secondness in the sense of Peirce's universal categories, dualizing a context can be understood as interchanging the roles of Firstness and Secondness. *Dualizing a Concept Hierarchy* means to interchange extension and intension in each concept, i.e., each concept becomes a concept of the dualized context, while each subconcept-superconcept-relationship turns to its inverse (cf. [GW99a], p.22). Therewith a line diagram of the dualized concept hierarchy can be obtained by turning a line diagram of the given concept hierarchy upside down.

There are contexts for which its dual context is meaningful, i.e., it is also interesting to view the attributes as objects and the objects as attributes. An example for that is the context in [Wi05a] on p.11 having the tones of the *diatonic scale* as objects and the major and minor triads as attributes; moreover, a tone as object has a triad as attribute if the tone belongs to the triad. This context might be understood as an answer to the questions: Which triads contain a given tone x in the diatonic scale? The dual context, in which the triads are the objects and the tones are the attributes, might be viewed as an answer to the questions: Which tones characterize a given triad y in the diatonic scale?

2.2   Determination of Concepts and Contexts

The basic mean for generating concepts of a given context are the two *derivations* assigning to each collection of objects the collection of all attributes which apply to those objects and assigning to each collection of attributes the collection of all objects which have those attributes. For determining concepts, sometimes even a context with its objects and attributes has to be determined from more general ideas.

M2.1 Generating Concepts: In a context, each collection of objects *generates a concept* the intension of which is the derivation of the given object collection and the extension of which is the derivation of that intension; dually, each collection of attributes *generates a concept* the extension of which is the derivation of the given attribute collection and the intension of which is the derivation of that extension (cf. [GW99a], p.18f.). An *object concept* is a concept generated by one object and an *attribute concept* is a concept generated by one attribute.

The extension of the concept generated by a given object collection is the *smallest concept extension* containing the generating object collection in the underlying context; this has as consequence that the intersection of concept extensions is an extension again. Dually, the intension of the concept generated by a given attribute collection is the *smallest concept intension* containing the generating attribute collection in the underlying context; this has as consequence that the intersection of concept intensions is an intension again.

M2.2 Generating All Concepts Within a Line Diagram: For smaller contexts the following procedure has been proven a success: First, represent the concept having the full object set of the given context as extension by a small circle and attach (from above) to that circle the names of all attributes which apply to all objects of the context. Secondly, choose from the left attributes all those the extension of which are maximal, draw for each of them a circle below the first circle, link them to the first circle by a line segment, and attach (from

above) to them the corresponding attribute names. Then determine all intersections of the extensions of the already represented concepts and represent the
concepts generated by those intersections by small circles with their respective
line segments representing the subconcept-superconcept-relationships. Perform
analogously the next steps until all attributes are treated. Finally, attach each
object name (from below) to that circle from which upward paths of line segments lead exactly to those circles with attached names of attributes applying
to the respective object (cf. [GW99a], p.64ff.).

After finishing the procedure, the user is recommended to check the representation by method M1.5. Quite often one has not represented all concepts;
but usually it is not difficult to insert the missing circles and respective line
segments. Finally, one should try to improve the drawn line diagram to obtain
a better readable diagram.

M2.3 Determining All Concepts of a Context: A fast procedure for determining all concepts of a finite context is given by the so-called *Ganter Algorithm*. This algorithm is based on a lexicographic order on all collections of
objects of the present context. For establishing this order, we assume a linear
order $g_1, g_2, \ldots, g_n$ on all objects. Then an object collection A is defined to be
lectically smaller than an object collection B if B contains the object which
has the smallest index under all objects distinguishing A and B. The algorithm
starts with the smallest concept extension, i.e., the derivation of the collection
of all attributes, and continues by determining always the lectically next concept extension A^+ after the just determined extension A. The extension A^+ is
generated by g_i and the object collection $\underline{A}$ consisting of all $g_1, \ldots, g_{i-1}$ contained also in A where i is the largest index for which g_i is not in A and the
extension generated by $\underline{A}$ and g_i contains the same objects out of $g_1, \ldots, g_{i-1}$ as
the extension A. The algorithm stops when it reaches the extension consisting of
all objects (cf. [GW99a], p.66ff.). Finally, the constructed extensions are turned
into concepts by M2.1. The subconcept-superconcept-relation can now be easily
determined because it agrees with the containment relation between the concept
extensions.

There are several *implementations* of the Ganter Algorithm (e.g. *ConImp*
[Bu00], ANACONDA [Na96], *ConExp* [Ye00]) which allow to compute even large
concept hierarchies and yield the input for drawing programs too. For drawing
well-readable line diagrams of concept hierarchies by hand, a concept list is useful
which indicates for each concept its upper neighbours in the concept hierarchy; in
particular, it can be used to apply the so-called *geometric method* (see [GW99a],
69ff.).

M2.4 Determining a Context from an Ordered Collection of Ideas:
There are situations in which it is desirable to elaborate concepts from more
general ideas. This has caused a method for constructing a context from a collection of ideas ordered with respect to their generality. For such an idea collection
a *downward/upward refinement* is defined to be a subcollection of ideas which
contains with each idea all more general/special ideas of that idea and with ev-

ery two ideas an idea more special/general than those two ideas. A downward refinement is called *irreducible* if, with respect to a suitable upward refinement, it is maximal under all downward refinements having no idea in common with that upward refinement. Dually, an upward refinement is called *irreducible* if, with respect to a suitable downward refinement, it is maximal under all upward refinements having no idea in common with that downward refinement. Now, for the desired context, we take the irreducible downward refinements as objects, the irreducible upward refinements as attributes, and the pairs of a downward and an upward refinement having some ideas in common as the object-attribute-relationships (cf. [SW86]).

An interesting application of the method M2.4 is the conceptual analysis of Aristotle's conception of the time continuum (cf. [Wi04b], p.460ff.). There the basic ideas are the time durations, which do not consist of time points. Time points can be derived as concepts generated by irreducible downward refinements of durations in the respective context constructed as above. More elementary constructions of contexts are discussed in the next subsection.

2.3 Conceptual Scaling

Scaling is the development of formal patterns and their use for analyzing empirical data. In *Conceptual Scaling* these formal patterns consists of contexts and their concept hierarchies which have a clear structure and reflect some meaning. Such a context is said to be a *conceptual scale* and its objects and attributes are called *scale values* and *scale attributes*, respectively (cf. [GW89], p.142ff.).

M3.1 Conceptual Scaling of a Context: A context may be connected with a conceptual scale by a *scale measure* which assigns to each object of the context a scale value in such a way that the collection of all objects assigned to values in any fixed extension of the conceptual scale is an extension too, which is called the *preimage* of the fixed scale extension under the considered scale measure. A system of scale measures on a logical context with values in respective conceptual scales is said to be a *full conceptual scaling* if every extension of the context is the intersection of the preimages of some scale extensions under the respective scale measures (cf. [GW89]).

The context of a *repertory grid test of an anorectic patient* discussed in [Wi00b] on p.365 permits a full conceptual scaling into three one-dimensional ordinal scales having the values 0, 1, 2 (cf. M3.4). The three scale measures assign to the object MYSELF the scale values 0,0,2, to IDEAL 0,1,1, to FATHER 1,0,0, to MOTHER 2,0,0, to SISTER 1,0,1, and to BROTHER-IN-LAW 0,2,0. The described full conceptual scaling leads to a well-readable diagram of the corresponding concept hierarchy in a 3-dimensional grid.

M3.2 Conceptual Scaling of a Many-valued Context: In a (complete) many-valued context every many-valued attribute assigns to an object a unique attribute value. Therefore, for turning a many-valued context into a context to obtain a related concept hierarchy, it is natural to interpret the many-valued attributes as scale measures and the attribute values as scale values of suitable conceptual scales (cf. M3.1). This motivates *conceptual scaling* of a many-valued context by which a (meaningful) conceptual scale is assigned to each many-valued attribute so that the corresponding attribute values are objects of that scale. The *derived context* of such conceptual scaling has the same objects as the many-valued context and has as its attributes the attributes of all assigned conceptual scales. An object of the derived context has an attribute of a specific scale if, in that scale, the attribute applies to the scale value which the respective many-valued attribute assigns to the object in the many-valued context (cf. [GW99a], p.36ff.).

The specific scaling methods which are mostly used are listed below. Further scaling methods are described in Section 1.3 and 1.4 of [GW99a].

M3.3 Nominal Scaling of a Many-valued Context: A context is called a *nominal scale* if each of its objects has exactly one attribute and each of its attributes applies to exactly one object. A conceptual scaling of a many-valued context is said to be *nominal* if all conceptual scales of the scaling are nominal (cf. [GW99a], p.42).

A nominally scaled many-valued context, having the former presidents of the Federal Republic of Germany as objects, is discussed in [Wi00b]. Its many-valued attributes are the *age of entrance* with the values < 60 and > 60, the *terms of office* with the values 1 and 2, and the *party* with the values CDU, SPD, and FDP. Therefore, the derived context has the seven attributes *age of entrance: < 60, age of entrance: > 60, terms of office: 1, terms of office: 2, party: CDU, party: SPD,* and *party: FDP.* Each president has three attributes, namely the value of his age of entrance, of his terms of office, and of his party, respectively.

M3.4 Ordinal Scaling of a Many-valued Context: A context is called an *ordinal scale* if its objects and its attributes carry hierarchical order relations which are in one-to-one correspondence and if an object has an attribute exactly in case the object is in the order relation with the object corresponding to the attribute (or, equivalently, in case the attribute is in the opposite order relation with the attribute corresponding to the object). An ordinal scale is *one-dimensional* if the objects and attributes with their hierarchical order relations form corresponding increasing chains. A conceptual scaling of a many-valued context is said to be *(one-dimensional) ordinal* if all conceptual scales of the scaling are (one-dimensional) ordinal (cf. [GW99a], p.48 and p.42).

An ordinally scaled many-valued context, having 26 places along the Canadian Coast of Lake Ontario as objects, is discussed in [SW92]. Its many-valued attributes are five *tests concerning water pollution*, the attribute values of which are six *segments of potential measurement values*, respectively. Therefore, the derived context has 30 attributes: each of the 5 tests combined with one of its 6 segments. A place has a segment of a test as its attribute if the measurement value of the test at this place lies in the segment or is larger than all values of that segment. Clearly, each test represents a one-dimensional ordinal scale.

M3.5 Interordinal Scaling of a Many-valued Context: A context is called a (one-dimensional) *interordinal scale* if it is the juxtaposition of a (one-dimensional) ordinal scale and its opposite scale, i.e., an object has an attribute exactly in case the object is in the opposite order relation with the object corresponding to the attribute. A conceptual scaling of a many-valued context is said to be (*one-dimensional*) *interordinal* if all conceptual scales of the scaling are (one-dimensional) interordinal (cf. [GW99a], p.57 and p.42).

In the retrieval system developed for the library of the center of interdisciplinary technology research at Darmstadt University of Technology (see M6.3), a one-dimensional interordinal scale is used to represent time periods in which the books of the library have been published. The chosen objects of that scale are the time periods *before 1945, 1945-1959, 1960-1969, 1970-1979, 1980-1984, 1985-1989, 1990-1993, from 1994* and the scale attributes are the time periods *before 1945, before 1960, before 1970, before 1980, before 1985, before 1990, before 1994*, and *from 1945, from 1960, from 1970, from 1980, from 1985, from 1990, from 1994*. Naturally, a time period object is considered to have a time period attribute if the object period is contained in the attribute period. The concept hierarchy of the defined interordinal scale has a well-readable line diagram which is shown in [RW00] on p.250.

M3.6 Contraordinal Scaling of a Many-valued Context: A logical context is called a *contraordinal scale* if its objects and its attributes carry hierarchical order relations which are in one-to-one correspondence so that an object has an attribute exactly in case the object is not in the opposite order relation with the object corresponding to the attribute (or, equivalently, in case the attribute is not in the order relation with the attribute corresponding to the object). A conceptual scaling of a many-valued context is said to be *contraordinal* if all conceptual scales of the scaling are contraordinal (cf. [GW99a], p.49). The special case that the order relations are just the equality relations yields the so-called *contranominal scales* in which all subcollections of objects are extensions and all subcollections of attributes are intensions (cf. [GW99a], p.48).

The ordinally scaled many-valued context and its corresponding concept hierarchy presented in [GW99a] on p.44/45 reports ratings of sights on the Forum Romanum in Rome taken from the travel guides *Baedecker* (*B*), *Les Guides Bleus* (*GB*), *Michelin* (*M*), and *Polyglott* (*P*). The four-dimensional structure caused by the four guides could be made more transparent by a contraordinal scaling of the many-valued context as shown in [Wi87] on p.196. This scaling yields a derived context with the seven attributes [no star in B], [no star in GB], [no or one star in GB], [no star in M], [no or one star in M], [no or one or two stars in M], and [no star in P].

M3.7 Convex-Ordinal Scaling of a Many-valued Context: A logical context is called a *convex-ordinal scale* if it is the juxtaposition of a contraordinal scale and its opposite scale, i.e., a scale in which an object has an attribute exactly in case the object is in the opposite negated order relation with the object corresponding to the attribute. A conceptual scaling of a many-valued context is said to be *convex-ordinal* if all conceptual scales of the scaling are convex-ordinal (cf. [GW99a], p.52).

Convex-ordinal scales are often derived from hierarchically ordered structures. Such a structure is, for instance, presented in [SW93] in Figure 1 by a diagram representing 35 dyslexics ordered by their numerical scores obtained from three tests. The ordering locates a person below another one if her scores do not exceed the corresponding scores of the other person, but at least one score is even less the corresponding score of the other person. For dissecting the 35 dyslexics into widely uniform training groups it is desirable that each person located by the ordering between two persons of a group should also belong to that group. This rule has as consequence that the groups are extensions of the convex-ordinal scale canonically derivable from the described ordering (cf. [GW99a], p.52).

2.4   Conceptual Classification

Classifying objects is an important activity of human thinking which is basic for interpreting realities. There is a wide spectrum of methods to perform classifications and the interest is even to develop further methods. Especially, there is a strong demand for *mathematical methods of classification* which can be implemented on computers. This has stimulated a rich development of numerical classification methods which are of extensive use today. But those methods are also criticized because of a major limitation, in that the resulting classes may not be well characterized in some human-comprehensible language (cf. [SW93]). *Conceptual Classification*, which uses concept hierarchies of contexts, overcomes this limitation by incorporating a conceptual language based on attributes and attribute values.

M4.1 Concept Classification of Objects: The first step of *conceptually classifying objects* is to choose appropriate attributes according to the purpose of the approached classification. Then the logical context for the considered objects and attributes has to be established and after that its concept hierarchy. This hierarchy yields the desired classification, the object classes of which are just the non-empty extensions of the concepts forming the hierarchy.

An example of a concept classification is the logical support of the educational film *"Living Being and Water"* mentioned in [GW99a] on p.18 and p.24. This film was produced by the Institute of Educational Technology in Veszprém/Hungary. For developing the film, the first decision was to emphasize on the general objects leech, bream, frog, dog, spike-weed, reed, bean, and maize as well as on nine attributes from "needs water to live" to "suckle its offsprings". After determining the respective context, its concept hierarchy with its object classification was derived which supported not only the design and production of the film, but also the evaluation of its perception.

It has to be mentioned that, in our example, the object extensions do not form a tree as is often required for classifications (see e.g. [RS84]). In the German Standard DIN 2331 from 1976 about concept systems, classifications being tree-like are called *monohierarchical systems*, otherwise *polyhierarchical systems*; in this way the standard respects that classifications in practice are quite often not trees.

M4.2 Many-valued Classification of Objects: *Conceptually classifying objects* of many-valued contexts presupposes a conceptual scaling, the specific method of which is appropriately chosen according to the purpose of the approached classification, respectively. The *derived context* of such conceptual scaling has already been described in method M3.2. The special quality of a many-valued classification is that each many-valued attribute can function as a semantic criterion for which the attributes of the respective scale represent meanings specifying the criterion.

An example of a many-valued classification is the logical support of an *investigation in developmental psychology* (cf. [SW93]). The data of that investigation have been concentrated in a many-valued context, the objects of which are 62 children from the age of 5 to 13 and the attributes of which are 9 general criteria of concept development and the attribute age. The investigation was performed with the aim to reconstruct the developmental sequences of the concept *work*. The analysis of those sequences was based on the criteria *quality of motives*, *generalization*, and *structural differentiations* which give the most differentiated view of changes and advances in development. The many-valued context was convex-ordinally scaled by method M3.7 to a logical context so that the children could be classified in seven meaningful extensions representing levels of concept development. The most interesting result was that some children reached earlier a higher level of generalization than others who reached earlier a higher level of quality of motives. Such a kind of branching in concept development has not been proven before.

2.5   Analysis of Concept Hierarchies

The term "Analysis" means an investigation by dissecting a whole into suitable parts to obtain a better understanding of the whole. Thus, *analyzing a concept hierarchy* consists in partitioning its concepts into meaningful parts which together form a subdivided conceptual structure leading to an improved understanding of the concept hierarchy.

M5.1 Partitioning the Attributes of a Context (Nested Line Diagram): For studying larger concept hierarchies of contexts it has been proven useful to *partition the attributes* of the given context in classes and to identify the subcontexts formed by one of those classes and the objects of the whole context, respectively. Then, each concept of the whole context is represented by a *sequence of subcontext concepts*, just one from each identified subcontext; the intension of such a subcontext concept consists of all attributes of the represented concept which are also attributes of the subcontext concept. The resulting subdivided structure of the whole concept hierarchy can be visualized by a *nested line diagram* constructed as follows (cf. [GW99a], p.75ff.): First, line diagrams of the concept lattices of the subcontexts are prepared and ordered in a sequence of the same kind as the corresponding subcontexts. Then, the line diagram being second in the sequence is copied into each circle of the line diagram being first in the sequence; next, the line diagram being third in the sequence is copied into each circle of each copy of the line diagram being second in the sequence; and so on, until the line diagram being last in the sequence is copied into each circle of each copy of the line diagram being last but one in the sequence. Finally, each concept of the whole context and its sequence of subcontext concepts is indicated by a corresponding sequence of circles each of which contains the next one and represents the corresponding subcontext concept; such sequence of circles can be marked by only distinguishing the last circle of the sequence. This method is used, in particular, for applying the TOSCANA-aggregation (s. M6.3).

A well-readable nested line diagram of a concept hierarchy with 139 concepts concerning *old Chinese urns* is presented in [Wi84] on p.42. The underlying context has eight pairs of dichotomic attributes from which five pairs form a one-dimensional interordinal scale (cf. M3.5). For the attribute partition, those ten attributes were taken as the first attribute class, two further pairs as the second attribute class, and the last pair as the third attribute class. The concept hierarchies of the corresponding subcontexts consist of 22, 10, and 4 concepts, respectively. The subdivided structure diagram underlying the nested diagram has 880 very small circles, 220 small circles containing 4 very small circles, and 22 larger circles containing 10 small circles, respectively. Since the smallest concepts of the three hierarchies have an empty extension, one can erase all circles representing a concept with an empty extension (except the lowest very small circle) so that structure diagram consist of only 661 very small circles.

M5.2 Atlas-Decomposition of a Concept Hierarchy: Analyzing larger concept hierarchies may be stimulated by the atlas metaphor. Such approach can be based on the notion of a *block relation* which relates objects and attributes, in particular, if the object has the attribute in the underlying context; furthermore, the block relation derivation of each object/attribute is an extension/intension of the original context (cf. [GW99a], p.121ff.). This guarantees that each concept of the block relation context gives rise to an interval in the concept hierarchy of the original context consisting of all concepts the extension/intension of which is contained in the extension/intension of the block relation concept. Metaphorically, those intervals are the maps of the respective atlas. As in an atlas, for many applications it is desirable that neighbouring maps overlap.

A meaningful atlas-decomposition of a concept hierarchy concerning the *harmonic forms of the diatonic scale* is discussed in [Wi84] on p.45ff. For that hierarchy the used block relation yields the largest decomposition with overlapping neighbouring maps. Each map clarifies the relationships between harmonic forms differing just by one tone.

M5.3 Concept Patterns in a Concept Hierarchy: Concept hierarchies may be understood as source of well-interpretable *concept patterns*, for instance, as concept chains, ladders, trees, grids etc. Such patterns are considered as *conceptual measurement structures* as discussed in [GW99a] in section 7.3 and 7.4. A specific method of identifying concept patterns is based on the search of respective *subcontexts* constituted by suitable objects and attributes of the underlying context (cf. [GW99a], section 3.1). Such search is quite often successful if one tries to find long sequences of attributes, the extensions of which form a chain with respect to containment, and completes those attributes to a subcontext having enough objects to represent those chains.

A meaningful example about the support of designing working places for handicapped people is discussed in [Wi87], p.188ff. The established logical context indicates which part of the human body is affected by which demand of work. A well-drawn line diagram of the respective concept lattice shows a dominant two-dimensional grid pattern which is generated by two sequences of attributes concerned with body movements, one from climbing over waking and squating to foot moving and the other from climbing over reaching and holding to seizing. Another instructive example is the concept hierarchy in [Wi92] about the colour perception of a gold fish. This hierarchy which consists of 141 concepts could only be well-drawn and well-interpreted because of the discovery of two long attribute chains representing parts of the colour circle.

2.6   Aggregation of Contexts and Concept Hierarchies

Knowledge is often represented not in one, but in several contexts. Clearly, it is desirable to aggregate those contexts to a common context, so that the single contexts are derivable as direct as possibly from the common context. Furthermore, the construction of the concept hierarchy of the common context by the concept hierarchies of the single contexts should be known, as well as the projections from the common concept hierarchy onto the single concept hierarchies, respectively.

M6.1 Juxtaposition of Contexts with Common Object Collection: Forming the *juxtaposition of given contexts with common object collection* means to establish the context having as objects those of the common object collection and as attributes those which are attributes of one of the given contexts (attributes from different contexts are considered to be different too); in the juxtaposition, an object has an attribute if it has the attribute in the context containing that attribute (cf. [GW99a], p.40). The *concepts of the juxtaposition* are generated by the intersections of the extensions of concepts of the single contexts (cf. M2.1). Conversely, each *concept of a single context* is the projection of all concepts of the juxtaposition having the same extension as that concept.

An extensive project of Conceptual Knowledge Processing highly dependent on the juxtaposition aggregation was the development of an *information system about laws and regulations concerning building constructions* requested by the Department for Building and Housing of the State Nordrhein-Westfalen. The necessary knowledge for that project was represented in contexts and their concept hierarchies concerned with specific themes such as "fundamental construction of a family house" [EKSW00], "functional rooms in a hospital" [Wi05b], "operation and fire security" [KSVW94] etc. The concept hierarchy of the juxtaposition of the mentioned hospital and security context represented by a nested diagram can also be found in [KSVW94] (the common object collection was formed by all relevant information units about laws and regulations concerning building constructions).

M6.2 Aggregation Based on Object Families: A general framework for this method is the so-called "semiproduct of contexts" (cf. [GW99a], p.46). The *semiproduct* of a collection of contexts is a context the objects of which are all object families having exactly one object from each context of the collection and the attributes of which are just the attributes of the given contexts (attributes from different contexts are viewed to be different); in the semiproduct, an object family has an attribute if that attribute applies in its respective context to the unique object belonging to the respective context and to the considered object family. An *Aggregation Based on Object Families* is a subcontext of a semiproduct of contexts having meaningful object families as its objects, while its attributes are just all attributes of the semiproduct.

The method "Aggregation Based on Object Families" plays an important role in [BS97] (cf. also [GW99b]); in particular, applications of this method to switching network are suggested. Such applications and their theoretical background have been elaborated in [Kr99], where the method "Aggregation Based on Object Families" is, for instance, used to analyse a *lighting circuit with emergency light*. The analysis yields four contexts: one for the main switch with three states, another one for a switch with two states linking either to the main-light or to the emergency light, and two further contexts with two states indicating whether the main light or the emergency light is on or out, respectively. The semiproduct of the four contexts has as objects the $24(= 3 \cdot 2 \cdot 2 \cdot 2)$ quadruples of states and as attributes the $7(= 3 + 2 + 1 + 1)$ attributes of the four contexts. Only 6 of the 24 quadruples are meaningful (i.e. they satisfy the so-called network rule); hence a 6×7-subcontext of the semiproduct represents the logic of the analysed lighting circuit with emergency light.

M6.3 TOSCANA-Aggregation of Concept Hierarchies: The idea of a TOSCANA-aggregation is to view a related system of concept hierarchies metaphorically as a *conceptual landscape of knowledge* [Wi97b] which can be explored by a purpose-oriented combination and inspection of suitable selections of the given concept hierarchies. This is logically supported by line diagrams representing concept hierarchies of juxtapositions of contexts in the sense of M6.1 and suitable restrictions of those hierarchies (cf. [KSVW94], [EKSW00]). For performing the TOSCANA-aggregation method, software has been developed since 1990; the most advanced software is available by the programs of the TOSCANAJ *Suite* [BH05] which are developed as Open Source project on Sourceforge (http://sourceforge/projects/toscanaj).

The development of the TOSCANA-aggregation method was stimulated by a research project with political scientists in the late 1980th. The task was to analyse a data context with 18 objects, namely norm- and rule-guided international cooperations, so-called *regimes*, and 24 many-valued attributes representing factors of influence, typological properties, and regime impacts (cf. [KV00]). This many-valued context was conceptually scaled by the method M3.2, where the used scales arose as result of an interdisziplinary co-operation between the coworking mathematicians and political scientists. Then more then fifty subcontexts of the scaled many-valued context together with the corresponding concept hierarchies were produced by hand for answering special research questions. This made clear that a mathematically founded construction method and its implementation would be desirable which allows the aggregation of arbitrarily chosen conceptual scales. The TOSCANA method and software met this desire and gave rise to the development of many TOSCANA-systems in a wide spectrum of disciplines. In particular, the research on international regimes has benefited from this development which is witnessed by a new TOSCANA-system about 90 regime components the data of which were elaborated by a great number of international political scientists over more than four years (cf. [Ks05]).

2.7   Conceptual Identification

A method is considered to be a *conceptual identification* if it determines concepts which classifies given instances. A well-known example of a conceptual identification is to determine the position of an individual plant in a taxonomy of plants.

M7.1 Identifying a Concept: The elementary type of conceptual identification is the *classification of an instance* by a given system of concepts. Methodollogically, such identification can be well performed on the basis of a context, the objects of which are the classes of the given classification system; the attributes of the context are used for the identification process that increasingly determines those attributes which apply to the considered instances. It is advantageous to visualize this process in a line diagram of the concept hierarchy of the context by indicating the decreasing path from concept to concept generated by the determined attributes until no further attribute which apply to the considered instance leads to a new subconcept. Then the concept generated by all determined attributes is the identified concept for the considered instance (cf. [KW86]).

The described identification process can be effectively represented and supported by a computer. That has, for instance, been done for identifying the symmetry types of two-dimensional patterns. In this case, the computer screen shows the user a line diagram representing the concept hierarchy having as objects the considered symmetry classes. For a given two-dimensional symmetry pattern, the user tries to find enough attributes which apply to the given pattern. After each input of such an attribute, the screen highlights the concept which is generated by the already fed attributes. This process may, for instance, identify the symmetry type the concept of which is generated by the attributes "admitting two reflections with non-parallel axes", "admitting a rotation of 90°", and "admitting a rotation of 90° the center of which is not on a reflection axis" (see [Wi00b], p.362f.). A well-designed computer implementation for identifying the symmetry types of two-dimensional patterns has been successfully offered to the visitors of the large Symmetry Exhibition at the Institute Mathildenhöhe in Darmstadt 1986 (cf. [Wi87], p.183ff.).

M7.2 Identifying Concept Patterns: There are many families of contexts the concept hierarchies of which offer a specific type of *regular concept patterns* for interpreting conceptual structures; in particular, the so-called *standard scales* (discussed in [GW99a], Section 1.4) are such concept patterns. In a context, a concept pattern is *identified* by a collection of objects of the context if each concept of the pattern is generated by a subcollection of those objects; it is *strongly identified* if, in addition, each subcollection of the objects generates a concept of the pattern.

How the identification of concept patterns may support the interpretation of empirical data shall be briefly demonstrated by an example from linguistics. In the late 1980th, the dialectician H. Goebl from the University Salzburg became interested in the application of Formal Concept Analysis to his empirical data; in particular, he offered the Darmstadt research group data about *phonemes of French in Swizerland*. Those data were transformed in a context having 63 measurements points as objects and 40 phonetic characteristics as attributes. Since it is of special interest to study modifications of the phonemes along measurement points, one-dimensional interordinal scales (cf. M3.5) yield well-interpretable concept patterns. Indeed, the multifarious modifications of phonemes could be shown by the concept hierarchies of such scales strongly identified by 3, 4, and 5 consecutive measurement points, respectively (see [FW89]).

2.8   Conceptual Knowledge Inferences

Conceptual Knowledge Processing does not only rely on the representation of conceptual structures, but also on *conceptual inferences* which are inherent in knowledge structures. The importance of inferences for human thinking has been, in particular, underlined by R. Brandom in his influential book "Making it explicit. Reasoning, representing, and discursive commitment" [Br94]. According to Brandom, knowledge is founded on an *inferential semantics* which rests on material inferences based on a normative pragmatics. In Formal Concept Analysis, up to now, the research on inferences has been dominantly concentrated on implications and dependences (cf. [GW99a], Section 2.3 and 2.4).

M8.1 Determining the Attribute Implications of a Context: In a given context, the attributes $m_1, \ldots, m_k$ *imply* the attributes $n_1, \ldots, n_l$ if each object having the attributes $m_1, \ldots, m_k$ also has the attributes $n_1, \ldots, n_l$. Such implication can be determined within a line diagram of the concept hierarchy of the given context as follows: First one identifies the circle representing the largest common subconcept c of the attribute concepts generated by the attributes $m_1, \ldots, m_k$, respectively; then the attributes implied by $m_1, \ldots, m_k$ are recognizable as those attributes n which generate a superconcept of c, i.e. there is an ascending path from the circle representing c to the circle representing the attribute concept generated by n.

There is a number of *implemented algorithms* for determining bases of attribute implications, the mostly used algorithm of which has been developed by B. Ganter [Ga87] (see also [GW99a], Section 2.3). The Ganter algorithm determines the *stem basis* for all attribute implications of a given context (also called the Duquenne-Guigues-Basis [GD86]); all other bases can be easily derived from the stem basis. Applying the stem basis shall be briefly demonstrated by the analysis of properties of drive concepts for motorcars presented in [Wi87], p.174ff.: The context for this analysis has as objects the drive concepts "Conventional", "Front-wheel", "Rear-wheel", "Mid-engine", and "All-wheel" and as attributes 25 properties such as "good road holding", "under-steering", "high cost of construction" etc. The Ganter algorithm yields 31 attribute implications as, for instance, "good economy of space" implies "good road holding", "bad economy of space" implies "low cost of construction", "good drive efficiency unloaded" and "good maintainability" imply "low cost of construction". The few cited implications may already show how valuable the stem basis can be for the interpretation of the given data context.

M8.2 Determining Many-valued Attribute Dependencies: Dependencies between many-valued attributes are of great interest in many fields of empirical research. A basic type of such dependencies are the functional dependencies: In a (complete) many-valued context, many-valued attributes $n_1, \ldots, n_l$ are *functionally dependent* on many-valued attributes $m_1, \ldots, m_k$ if, for every two objects g and h of the many-valued context, the corresponding attribute values $m_i(g)$ and $m_i(h)$ $(i = 1, \ldots, k)$ are equal then the corresponding attribute values $n_i(g)$ and $n_i(h)$ $(i = 1, \ldots, l)$ are equal too. For determining functional dependencies, a method has been proven useful which is based on the following context derived from the given many-valued context: the *derived context* has as objects all pairs of two (different) objects of the many-valued context and as attributes all its many-valued attributes where a pair of objects g and h is related to a many-valued attribute m if the corresponding attribute values $m(g)$ and $m(h)$ are equal. Then it can be proved that many-valued attributes $n_1, \ldots, n_l$ are *functionally dependent* on many-valued attributes $m_1, \ldots, m_k$ in the many-valued context exactly if $m_1, \ldots, m_k$ implies $n_1, \ldots, n_l$ in the derived logical context. This equivalence allows us now to use method M8.1 to determine all functional dependences of the given many-valued context. If one replaces equality by the inequality $\leq$, respectively, one gets the analogous result for *ordinal dependency* (cf.[GW99a], p.91f.).

A prominent field for applying functional dependencies is Database Theory. Ordinal dependencies have been successfully applied in Measurement Theory. For instance, in [WW96], it is shown that enough ordinal dependencies in an ordinal many-valued context guarantees a linear representation of the context in a vector space over the field of real power series. This is demonstrated by a two-dimensional representation of a data context about the colour perception of a goldfish (cf. also [WW04]).

2.9   Conceptual Knowledge Acquisition

The *central idea of knowledge acquisition* in the frame of Formal Concept Analysis lies in the assumption that, in the field of exploration, the conceptual knowledge can be thought to be represented by a context with finitely many attributes and by its concept hierarchy; such a context is called a *universe*. The exploration of knowledge starts with some partial information about the considered universe and acquires more information by phrasing questions which are answered by experts. For this procedure it is a main concern not to ask questions which can already be answered by the acquired knowledge (cf. [Wi89b]).

M9.1 Attribute Exploration: For an attribute exploration, first a *universe* is specified as a context the attribute of which are explicitly given, but the objects of which are only known to belong to a certain type of objects. It is often helpful to choose at the beginning of the exploration some objects and to determine the subcontext of the universe based on those objects together with the attributes of the universe. Then an *implementation of the algorithm* described in [GW99a], p.85, should be used which leads to questions whether certain attribute implications are valid in the universe or not. If yes, then the actual attribute implication is added to the list of already recognized valid attribute implications of the universe. If not, then an object of the universe has to be made explicit which has all attributes of the premise of the actual implication, but has not at least one attribute of its conclusion; such a new object is used to extend the actual explicit subcontext to a new subcontext of the universe. Since the universe has only finitely many attributes, the exploration ends after finitely many steps. Then the resulting subcontext has the same concept intensions as the assumed universe.

In [GW99a], p.86ff., the *attribute exploration* is demonstrated by the universe which has as objects all binary relations between natural numbers and as attributes the properties "reflexive", "irreflexive", "symmetric", "asymmetric", "antisymmetric", "transitive", "negatively transitive", "connex", and "strictly connex". The concept hierarchy of the resulting subcontext consists of 50 concepts the structure of which clarifies completely the *implication logic* of the given nine properties of binary relations (cf. M8.1).

M9.2 Concept Exploration: For a concept exploration, first a *universe* is specified as a context the objects and attributes of which are only known to belong to a type of objects and a type of attributes, respectively; in addition, a finite number of concepts of the universe are specified by their names. Then the *aim of the concept exploration* is to identify all concepts of the universe which can be deduced from the specified concepts by iteratively forming the largest common subconcept and the smallest common superconcepts of already constructed concepts in the universe. This procedure is accompanied by the questions whether for two concepts one is a subconcept of the other or not. If yes, then this order relationship is added to the table of already recognized pairs of subconcept-superconcept of the universe. If not, then an object belonging to one concept and an attribute belonging to the other concept and not applying to the object have to be made explicit. The acquired knowledge, as it is accumulating, is represented in a context which has as objects the explicitly made objects together with the constructed concepts and as attributes the explicitly made attributes with the constructed concepts too; the context relation indicates which explicit object has which explicit attribute, which explicit object belongs to which constructed concept, which explicit attribute belongs to which constructed concept, and which constructed concept is the subconcept of another constructed concept (cf. [Wi89b], p.375ff.).

The concept exploration is demonstrated in [Wi89b] within the universe having as objects all countable relational structures with one binary relation R and as attributes all universal sentences of first order logic with the relational symbol 'R' and equality; a relational structure as object has a sentence as attribute if the structure satisfies the sentence. In particular, the attribute exploration is explicitly performed with the three specified concepts "orthogonality", "dominance", and "covering". The acquired knowledge is represented in a 14×14-context based on 4 objects, 4 attributes and 10 concepts. More examples and theoretical developments can be found in [St97].

M9.3 Discovering Association Rules: In a context, an *association rule* is an ordered pair $(X \to Y)$ of attribute collections X and Y for which the following relative frequencies are computed: the *support* of $(X \to Y)$ is the number of attributes which are in X or Y divided by the number of objects in the context, and the *confidence* of $(X \to Y)$ is the number of attributes which are in X or Y divided by the number of attributes in X. The task is to determine all ordered pairs $(X \to Y)$ for which the support of $(X \to Y)$ is above a given *support threshold* chosen from the interval $[0, 1]$ and the confidence of $(X \to Y)$ is above a chosen *confidence threshold* chosen from the interval $[0, 1]$. In solving the task, the crucial part is the determination of all key-attribute-collections, which are attribute collections being minimal in generating a concept (cf. M2.1). The method of doing that is based on the observation that each subcollection of a key-attribute-collection is also a key-attribute-collection. Thus, an effective procedure can be designed which tests first the one-element attribute collections of being key-attribute-collections, then the two-element attribute collections not properly containing a key-attribute-collection and so on (for more details see [LS05]).

Association rules are, for instance, used in warehouse basket analysis, which is carried out to learn which products are frequently bought together. A general overview about discovering and applying association rules can be found in [LS05], Section 6.

2.10   Conceptual Knowledge Retrieval

Since *Information and Knowledge Retrieval* deals with organizing, searching, and mining information and knowledge, methods of Conceptual Knowledge Processing may support retrieval activities. They can do this by effectively complementing the existing search systems, in particular, by visualizing retrieval results, improving individual search strategies, and hosting multiple integrated search strategies (cf. [CR04], [CR05]).

M10.1 Retrieval with Contexts and Concept Hierarchies: The retrieval of documents can be seen to take place in a *context* the objects of which are the available documents and the attributes of which are the constituents of queries. Then the intension of a concept of such context contains all *queries* having as retrieval result exactly all documents in the extension of the considered concept. The concept hierarchy of the context shows especially which queries yield neighbour concepts causing only minimal changes between the retrieved extensions (cf. [GSJ86]). In general, a concept hierarchy of a context based on a number of retrieval results may give a useful overview which faster leads to fulfill the purpose of the search (cf. [CR05]).

In [Ko05], the method M10.1 is applied and elaborated to develop an improved *front end to the standard Google search*. The basic idea of this development is to use Google's three-row result itemset consisting of title, short description, and the uniform resource locators (URL) to build a context and its concept hierarchy. The context has as objects the first n URLs and as attributes the meaningful feature terms extracted from Google's first n three-row results (the number n is eligible). The context is presented best by a cross table in which the names of the attributes, applying to the same objects, are heading the same column (cf. M1.2). It turns out that already this presentation of the retrieved results is often very useful because a larger manifold of information units can be viewed at once and selectively compared. This effect can even be increased if the corresponding concept hierarchy is visualized.

M10.2 Retrieval with a TOSCANA-System: Conceptual knowledge retrieval is often a process in which humans search for something which they only vaguely imagine. Therefore humans organize such processes not only by a sequence of queries in advance, they also learn step by step how to specify further what they are actually searching for. Such interactive retrieval and learning process can be successfully supported by a suitable *TOSCANA-system* established by the TOSCANA-aggregation method (M6.3). The TOSCANA-system is structured by a multitude of conceptual scales (cf. Section 2.3) which are applied as search structures to the objects under considerations. The line diagrams of the activated scales are shown to the user who learns by inspecting them how to act further (cf. [BH05]).

Retrieval with a TOSCANA-system has been, for instance, established by developing a *retrieval system for the library of the "Center of Interdisciplinary Technology Research"* (ZIT) at the TU Darmstadt using the method M6.3 (cf. [RW00]). For supporting the search of literature, a related system of 137 concept hierarchies was developed. The underlying contexts of those hierarchies have as objects all books of the library and as attributes well-chosen catch words which represent a specific theme, respectively. In [Wi05b], p.17, there is a report on a literature search in the ZIT-library concerning the theme "expert systems dealing with traffic". This search starts with the concept hierarchy "Informatics and Knowledge Processing" which has "Expert Systems" as one of its attributes; the corresponding line diagram shows that there are 60 books in the library having "Expert Systems" as assigned catchword. This suggests to the user to consider the concept hierarchy "Town and Traffic" restricted to those 60 books; then the resulting line diagram shows that 9 of the 60 books have also "Traffic" as an assigned catchword and, additionally, 4 resp. 1 of those 9 books "Means of Transportation" resp. "Town" as assigned catchword.

2.11   Conceptual Theory Building

Empirical theory building, in particular in the human and social sciences, may
be logically supported by methods of Conceptual Knowledge Processing. The
basic models used by those methods are contexts and their concept hierarchies
which allow the *representation of scientific theories* in a way that the theories
become structurally transparent and communicable (cf. [SWW01]).

M11.1 Theory Building with Concept Hierarchies: Conceptual theory
building starts from *data and information* which are mostly represented in data
tables, texts, images, and inferential connections. The goal is to generate a rich,
tightly woven, explanatory theory that closely approximates the reality it represents. Methodologically, this aims at a suitable representation of the considered data and information by a unifying concept hierarchy. That often leads to
question the data and information and to work further with their improvements.
Thus, *conceptual theory building* is an inductive process which stepwise improves
theories which are always represented by concept hierarchies (cf. [SC90]).

Interesting examples of conceptual theory building are the development of
everyday theories of logical relationships. There is a kind of surprising evidence
that a great deal of those theories are determined by attribute implications with
one-element premise and by incompatibilities between attributes (cf. [Wi04a]).
For instance, the theory about the core of the *lexical field of waters* can be characterized by the dichotomic pairs of attributes "natural - artifical", "running
- stagnant", "constant - temporary", and "inland - maritime". Applying those
attributes to words describing the types of waters like "plash", "channel", "sea"
etc. teaches that the types of waters satisfy the following attribute implications
with one-element premise:

"temporary" ⇒ "natural", "stagnant", "inland";
"running"   ⇒ "constant", "inland";
"artificial" ⇒ "constant", "inland";
"maritime" ⇒ "natural", "stagnant", "constant".

These implications together with the incompatibilities described by the four
dichotomic attribute pairs completely determine the theory of implicational relationships in the considered core of the lexical field of waters (this has been
shown in [Wi04a] by using the empirical results of [Kc79]). The concept hierarchy representing that theory is presented on the cover of [GW99a].

M11.2 Theory Building with TOSCANA: Conceptual theory building
can be based on the method "TOSCANA-Aggregation of Concept Hierarchies"
(M6.3) applied to an empirically derived collection of objects. For *building-up the
aimed theory*, this object collection is structured by justified conceptual scales.
Then interesting aggregations of those scales and their concept hierarchies are
tested with regard to their meaningfulness concerning the approached theory.
This testing might suggest improvements of the scales and their corresponding
concept hierarchies which are then tested again. The goal is to reach a well-
founded *TOSCANA-system* which adequately represents the aimed theory.

Theory building with TOSCANA has been substantially applied to support a dissertation about *"Simplicity. Reconstruction of a Conceptual Landscape in the Esthetics of Music of the 18th Century"* [Ma00]. The methodological foundation for this application was elaborated in [MW99]. The empirical collection of objects was given by 270 historical documents which were made accessible by a normed vocabulary of more then 400 text attributes. Those text attributes were used to form more general attributes for the conceptual scales of the approached TOSCANA-system. By repeatedly examining and improving aggregations of scales and their concept hierarchies, a well-founded TOSCANA-system was established which successfully supported the musicological research.

2.12   Contextual Logic

Contextual Logic has been introduced with the aim to support knowledge representation and knowledge processing. It is grounded on the traditional philosophical understanding of logic as the doctrine of the forms of human thinking. Therefore, Contextual Logic is developed by mathematizing the philosophical doctrines of concepts, judgments, and conclusions (cf. [Ka88], p.6). The mathematization of concepts follows the approach of *Formal Concept Analysis* [GW99a], and the mathematization of judgments uses, in addition, the *Theory of Conceptual Graphs* [So84]. The understanding of logic as the doctrine of the forms of human thinking has as consequence that main efforts are undertaken to investigate the mathematical and logical structures formed by (formal) concepts and concept(ual) graphs (cf. [Wi00a]).

M12.1 Conceptual Graphs Derived from Natural Language: A *conceptual graph* is a labelled graph that represents the literal meaning of a sentence or even a longer text. It shows the *concepts*, represented by boxes, and the *relations* between them, represented by ovals. The boxes contain always a name of a concept and, optionally, a name of an *object* belonging to that concept; no object name in the box means that there exist an object belonging to the concept named in the box. The ovals contain always a name of a relation which relates all the objects the names of which are contained in the boxes linked to the oval of that relation. In stead of repeating an object name in several boxes, it is allowed to write the name in only one box and to link this box to all those other boxes by broken lines (for more information see [So92]).

The representation by conceptual graphs has been practiced in many application projects concerning conceptual knowledge processing and has stimulated further useful theories (cf. contributions to the *International Conferences on Conceptual Structures* documented in the Springer Lecture Notes in Computer Science since 1993). One of such theories is the *Contextual Judgment Logic* the start of which was stimulated by a conceptual graph representation of a text about Seattle's central business district [Wi97a]. A quite special project was performed in a classroom of grade 6 with 32 boys and girls to clarify the question: Can already young pupils be trained in the ability of formal abstraction by transforming natural language into conceptual graphs? It turned out that most of the pupils learned very fast to turn simple sentences into a graphically presented conceptual graph. Already in the third lesson they were able to glue rectangular and oval pieces of paper on a cardboard in a way that they could inscribe and link those pieces to represent a little story by a conceptual graph (some of them built even little bridges of paper for the broken lines between equal object names to avoid misinterpretations) (cf. [SW99]).

M12.2 Derivation of Judgments from Power Context Families: It is worthwhile to understand the relations in a conceptual graph also as concepts of suitably chosen contexts. This understanding is basic for the derivation of judgments, represented by conceptual graphs, from so-called *power context families* which are composed by contexts $\mathbb{K}_0, \mathbb{K}_1, \mathbb{K}_2, \mathbb{K}_3, \ldots$ where $\mathbb{K}_0$ yields the concepts in the boxes and $\mathbb{K}_1, \mathbb{K}_2, \mathbb{K}_3, \ldots$ yield the concepts of relations of arity $k = 1, 2, 3, \ldots$ in the ovals, respectively (cf. M12.1); clearly, the objects of the relational context $\mathbb{K}_k$ ($k \geq 1$) are sequences of k objects belonging to the basic context $\mathbb{K}_0$, while the attributes of $\mathbb{K}_k$ have the function to give meaning to those object sequences (cf. [PW99]).

The sketched method can be effectively applied to develop information systems based on power context families representing the relevant knowledge. Such systems have been designed for flight information in Austria and Australia, respectively. The central idea of those information systems is to present to the user, who has inputed his constraints, a conceptual graph representing all flights which might be still relevant. In [PW99], Fig.6, a well-readable output graph is shown to a person who lives in Innsbruck and works in Vienna where he wants to arrive between 7 and 9 a.m. and to depart between 5 and 7 p.m. For more complex requests, the standard diagrams of conceptual graphs might become extremely complicated as shown in [EGSW00], Fig.7, for a customer who lives in Vienna and wants to visit partners in Salzburg, Innsbruck, and Graz at the weekend. But, using background knowledge which can be assumed for the customer, a much better readable diagram of the requested conceptual graph can be offered as shown in [EGSW00], Fig.8. Thus, conceptual graphs should be understood as logical structures which may have many different graphical representations useful for quite different purposes.

3   Supporting Human Thought, Judgment, and Action

As pointed out at the beginning of this paper, the main aim of *Conceptual Knowledge Processing* and its methods is to support rational thought, judgment and action of human beings and to promote the critical discourse. Since Conceptual Knowledge Processing treats knowledge based on actual realities, it relies on the *philosophical logic* as the science of thought in general, its general laws and kinds (cf. [Pe92], p.116). This understanding of philosophical logic has been developed since the 16th century, founded on the doctrines of concept, judgment, and conclusion. The assistance which Conceptual Knowledge Processing obtains from the philosophical logic becomes substantially intensified by the mathematical methods of *Contextual Logic*, which is based on a mathematization of the philosophical doctrines of concept, judgment, and conclusion (cf. [Wi00a]).

Thus, for applying and elaborating the discussed methods of Conceptual Knowledge Processing, it is worth-while not only to work on the level of actual realities in the frame of philosophical logic, but also on the level of potential realities activating mathematical methods. This is, in particular, necessary for the development of new software and theoretical extensions. Nevertheless, the *logical level* should have the primacy over the *mathematical level* because applying methods of Conceptual Knowledge Processing should primarily support human thought, judgment, and action.

Methods of knowledge processing always presuppose, consciously or unconsciously, some understanding of what knowledge is. Different from *ambitious knowledge*, specified for Conceptual Knowledge Processing in Section 1, a quite dominant understanding views knowledge as a collection of facts, rules, and procedures justifiable by objectively founded reasoning. K.-O. Apel criticizes this cognitive-instrumental understanding and advocates for a *pragmatic understanding of knowledge*:

> "In view of this problematic situation [of rational argumentation] it is more obvious not to give up reasoning entirely, but rather to break with the concept of reasoning which is orientated by the pattern of logic-mathematical proofs. In accordance with a new foundation of critical rationalism, Kant's question of transcendental reasoning has to be taken up again as the question about the normative conditions of the possibility of discursive communication and understanding (and therewith discursive criticism too). Reasoning then appears primarily not as deduction of propositions out of propositions within an objectivizable system of propositions in which one has already abstracted from the actual pragmatic dimension of argumentation, but as answering of why-questions of all sorts within the scope of argumentative discourse." (cf. [Ap89], p.19)

In [Wi96], a restructuring of mathematical logic is proposed which locates reasoning within the intersubjective community of communication and argumentation. Only the process of discourse and understanding in the intersubjective community leads to comprehensive states of rationality. Such process does not exclude logic-mathematical proofs, but they can be only part of a broader argumentative discourse (cf. [Wi97b]).

Methods of Conceptual Knowledge Processing can only be successfully applied if *discourses* can be made possible which allow the users and the persons concerned to understand and even to criticize the methods, their performances, and their effects. This does not mean an understanding of all technical details, but the gained competence to judge about the effects which the involved persons and institutions have to expect. A method of conceptual knowledge processing should be transparent in such a manner that persons affected could even successfully fight against the use of that method. An important precondition for critical discourses is that the methods can be communicated in a language which can be understood by the persons concerned; but establishing such languages needs *transdisciplinary efforts* (cf. [Wi02]).

References

[Ap76]   K.-O. Apel: Das Apriori der Kommunikationsgemeinschaft und die Grund-
          lagen der Ethik. In: *Transformation der Philosophie*. Band 2: Das Apriori
          der Kommunikationsgemeinschaft. Suhrkamp Taschenbuch Wissenschaft
          165, Frankfurt 1976.

[Ap89]   K.-O. Apel: Begründung. In: H. Seiffert, G. Radninzky (Hrsg.): *Handlexi-
          kon der Wissenschaftstheorie*. Ehrenwirth, München 1989, 14–19.

[BS97]   J. Barwise, J. Seligman: *Information flow: the logic of distributive systems*.
          Cambridge University Press, Cambridge 1997.

[BH05]   P. Becker, J. Hereth Correia: The ToscanaJ Suite for implementing con-
          ceptual information systems. In: [GSW05], 324–348.

[Br94]   R. B. Brandom: *Making it explicit. Reasoning, representing, and discursive
          commitment*. Havard University Press, Cambridge 1994.

[Bu00]   P. Burmeister: ConImp - Ein Programm zur Formalen Begriffsanalyse. In:
          [SW00], 25–56.

[CR04]   C. Carpineto, G. Romano: *Concept data analysis: theory and applications*.
          Wiley, London 2004.

[CR05]   C. Carpineto, G. Romano: Using concept lattices for text retrieval and
          mining. In: [GSW05], 161–179.

[EGSW00] P. W. Eklund, B. Groh, G. Stumme, R. Wille: A contextual-logic extension
          of TOSCANA. In: B. Ganter, G. Mineau (eds.): *Conceptual structures: log-
          ical, linguistic and computational issues*. LNAI **1867**. Springer, Heidelberg
          2000, 453-467.

[EKSW00] D. Eschenfelder, W. Kollewe, M. Skorsky, R. Wille: Ein Erkundungssystem
          zum Baurecht: Methoden der Entwicklung eines TOSCANA-Systems. In:
          [SW00], 254–272.

[FW89]   S. Felix, R. Wille: Phonemes of French in Switzerland: an application of
          Formal Concept Analysis. Unpublished material. TH Darmstadt 1989.

[Ga87]   B. Ganter: Algorithmen zur Formalen Begriffsanalyse. In: B. Ganter,
          R. Wille, K. E. Wolff (Hrsg.): *Beiträge zur Begriffsanalyse*. B.I.-Wissen-
          schaftsverlag, Mannheim 1986, 241–254.

[GSW05]   B. Ganter, G. Stumme, R. Wille (eds.): *Formal Concept Analysis: foun-
          dations and applications*. State-of-the-Art Survey. LNAI **3626**. Springer,
          Heidelberg 2005.

[GW89]   B. Ganter, R. Wille: Conceptual scaling. In: F. Roberts (ed.): *Applications
          of combinatorics and graph theory in the biological and social sciences*.
          Springer-Verlag, New York 1989, 139–167.

[GW99a]   B. Ganter, R. Wille: *Formal Concept Analysis: mathematical foundations*.
          Springer, Heidelberg 1999.

[GW99b]   B. Ganter, R. Wille: Contextual Attribute Logic. In: W. Tepfenhart,
          W. Cyre (eds.): Conceptual structures: standards and practices. LNAI
          1640. Springer, Heidelberg 1999 , 377–388.

[GSJ86]   R. Godin, E. Saunders, J. Jecsei: Lattice model of browsable data spaces.
          Journal of Information Sciences **40** (1986), 89–116.

[GD86]   J.-L. Guigues, V. Duquenne: Familles minimales d'implications informa-
          tive resultant d'un tableau de données binaires. Math. Sci. Humaines 95
          (1986), 5–18.

[Ha81]   J. Habermas: *Theorie des kommunikativen Handelns*. 2 Bände. Suhrkamp,
          Frankfurt 1981.

[Ks05]   T. Kaiser: A TOSCANA-System for the International Regimes Database (IRD). In: H. Breitmeier, O. R. Young, M. Zürn (eds.): *Analyzing international environmental regimes: from case study to database* (to appear)

[Ka88]   I. Kant: *Logic*. Dover, New York 1988.

[Kc79]   G. L. Karcher: *Konstrastive Untersuchungen von Wortfeldern im Deutschen und Englischen*. Peter Lang, Frankfurt 1979.

[Kr99]   M. Karl: *Eine Logik verteilter Systeme und deren Anwendung auf Schaltnetzwerke*. Diplomarbeit. FB Mathematik, TU Darmstadt 1999.

[KW86]   U. Kipke, R. Wille: Begriffsverbände als Ablaufschemata zur Gegenstandsbestimmung. In: P. O. Degens, H.-J. Hermes, O. Opitz (Hrsg.): *Die Klassifikation und ihr Umfeld*. Indeks Verlag, Frankfurt 1986, 164–170.

[KW87]   U. Kipke, R. Wille: Formale Begriffsanalyse erläutert an einem Wortfeld. *LDV-Forum* **5** (1987), 31–36.

[Ko05]   B. Koester: FooCA: Enhacing Google information research by means of Formal Concept Analysis. Preprint. TU Darmstadt 2005.

[KV00]   B. Kohler-Koch, F. Vogt: Normen- und regelgeleitete internationale Kooperationen - Formale Begriffsanalyse in der Politikwissenschaft. In: [SW00], 325–340.

[KSVW94] W. Kollewe, M. Skorsky, F. Vogt, R. Wille: TOSCANA - ein Werkzeug zur begrifflichen Analyse und Erkundung von Daten. In: R. Wille, M. Zickwolff (Hrsg.): *Begriffliche Wissensverarbeitung - Grundfragen und Aufgaben*. B.I.-Wissenschaftsverlag, Mannheim 1994, 267–288.

[LS05]   L. Lakhal, G. Stumme: Efficient mining of association rules based on Formal Concept Analysis. In: [GSW05], 180–195.

[Lo84]   K. Lorenz: Methode. In: J. Mittelstraß (Hrsg.): *Enzyklopädie Philosophie und Wissenschaftstheorie*. Bd.2. B.I.-Wissenschaftsverlag, Mannheim 1984, 876–879.

[Ma00]   K. Mackensen: *Simplizität - Genese und Wandel einer musikästhetischen Kategorie des 18. Jahrhunderts*. Bärenreiter, Kassel 2000.

[MW99]   K. Mackensen, U. Wille: Qualitative text analysis supported by conceptual data systems. *Quality & Quantity* **33** (1999), 135–156.

[Na96]   NaviCon GmbH: ANACONDA für Windows. Frankfurt 1996.

[Pe35]   Ch. S. Peirce: *Collected papers*. Harvard Univ. Press, Cambridge 1931–35.

[Pe92]   Ch. S. Peirce: *Reasoning and the logic of things*. Edited by K. L. Ketner; with an introduction by K. L. Ketner and H. Putnam. Havard University Press, Cambridge 1992.

[PW99]   S. Prediger, R. Wille: The lattice of concept graphs of a relationally scaled context. In: W. Tepfenhart, W. Cyre (eds.): *Conceptual structures: standards and practices*. LNAI **1640**. Springer, Heidelberg 1999, 401-414.

[RW00]   T. Rock, R. Wille: Ein TOSCANA-Erkundungssystem zur Literatursuche. In: [SW00], 239–253.

[RS84]   S. Rückl, G. Schmoll (Hrsg.): *Lexikon der Information und Dokumentation*. VEB Bibliographisches Institut, Leipzig 1984.

[SW99]   F. Siebel, R. Wille: Unpublished material. TU Darmstadt 1999.

[So84]   J. F. Sowa: *Conceptual structures: information processing in mind and machine*. Adison-Wesley, Reading 1984.

[So92]   J. F. Sowa: Conceptual Graphs summary. In: T. E. Nagle, J. A. Nagle, L. L. Gerholz, P. W. Eklund (eds.): *Conceptual structures: current research and practice*. Ellis Horwood, 1992, 3–51.

[SW86]   J. Stahl, R. Wille: Preconcepts and set representations of contexts. In: W. Gaul, M. Schader (eds.): *Classification as a tool of research.* North-Holland, Amsterdam 1986, 431–438.

[SW92]   S. Strahringer, R. Wille: Towards a structure theory of ordinal data. In: M. Schader (ed.): *Analyzing and modeling data and knowledge.* Spinger, Heidelberg 1992, 129–139.

[SW93]   S. Strahringer, R. Wille: Conceptual clustering via convex-ordinal structures. In: O. Opitz, B. Lausen, R. KLar (eds.): *Information and classification. Concepts, methods and applications.* Springer, Heidelberg 1993, 85–98.

[SWW01]   S. Strahringer, R. Wille, U. Wille: Mathematical support for empirical theory building. In: H. S. Delugach, G. Stumme (eds.): *Conceptual structures: broadening the base.* LNAI 2120. Springer, Heidelberg 2001, 169–186.

[SC90]   A. Strauss. J. Corbin: Basics of qualitative research: grounded theory procedures and techniques. Sage Publ., Newbury Park 1990.

[St97]   G. Stumme: *Concept exploration: knowledge acquisition in conceptual knowledge systems.* Dissertation, TU Darmstadt. Shaker Verlag, Aachen 1997.

[SW00]   G. Stumme, R. Wille (Hrsg.): *Begriffliche Wissensverarbeitung: Methoden und Anwendungen.* Springer, Heidelberg 2000.

[VW95]   F. Vogt, R. Wille: TOSCANA – A graphical tool for analyzing and exploring data. In: R. Tamassia, I. G. Tollis (eds.): *Graph drawing '94.* LNCS **894**. Springer, Heidelberg 1995, 226–233.

[Wi84]   R. Wille: Liniendiagramme hierarchischer Begriffssysteme. In: H. H. Bock (Hrsg.): Anwendungen der Klassifikation: Datenanalyse und numerische Klassifikation. Indeks-Verlag, Frankfurt 1984, 32–51; English translation: Line diagrams of hierachical concept systems. *International Classiffication* **11** (1984), 77–86.

[Wi87]   R. Wille: Bedeutungen von Begriffsverbänden. In: B. Ganter, R. Wille, K. E. Wolff (Hrsg.): *Beiträge zur Begriffsanalyse.* B.I.-Wissenschaftsverlag, Mannheim 1986, 161–211.

[Wi89a]   R. Wille: Lattices in data analysis: how to draw them with a computer. In: I. Rival (Ed.): *Algorithms and order.* Kluwer, Dordrecht 1989, 33–58.

[Wi89b]   R. Wille: Knowledge acquisition by methods of Formal Concept Analysis. In: E. Diday (ed.): *Data analysis and learning symbolic and numeric knowledge.* Nova Science Publisher, New York–Budapest 1989, 365–380.

[Wi92]   R. Wille: Concept lattices and conceptual knowledge systems. *Computers & mathematics with applications* **23** (1992), 493–515.

[Wi94]   R. Wille: Plädoyer für eine philosophische Grundlegung der Begrifflichen Wissensverarbeitung. In: R. Wille, M. Zickwolff (Hrsg.): *Begriffliche Wissensverarbeitung — Grundfragen und Aufgaben.* B.I.-Wissenschaftsverlag, Mannheim 1994, 11–25.

[Wi96]   R. Wille: Restructuring mathematical logic: an approach based on Peirce's pragmatism. In: A. Ursini, P. Agliano (eds.): *Logic and algebra.* Marcel Dekker, New York 1996, 267–281.

[Wi97a]   R. Wille: Conceptual graphs and formal concept analysis. In: D. Lukose, H. Delugach, M. Keeler, L. Searle, J. F. Sowa (eds.): *Conceptual structures: fulfilling Peirce's dream.* LNAI **1257**. Springer, Heidelberg 1997, 290–303.

[Wi97b]   R. Wille: Conceptual landscapes of knowledge: a pragmatic paradigm for knowledge processing. In: G. Mineau, A. Fall (eds.): *Proceedings of the International Symposium on Knowledge Representation, Use, and Storage Efficiency.* Simon Fraser University, Vancouver 1997, 2–13; also in: W. Gaul, H. Locarek-Junge (Eds.): Classification in the Information Age. Springer, Berlin-Heidelberg 1999, 344–356.

[Wi00a]   R. Wille: Contextual Logic summary. In: G. Stumme (ed.): *Working with conceptual structures. Contributions to ICCS 2000.* Shaker-Verlag, Aachen 2000, 265–276.

[Wi00b]   R. Wille: Begriffliche Wissensverarbeitung: Theorie und Praxis. *Informatik Spektrum* **23** (2000), 357–369; revised version in: B. Schmitz (Hrsg.): *Thema Forschung: Information, Wissen, Kompetenz.* Heft 2/2000, TU Darmstadt, 128–140.

[Wi01]   R. Wille: Mensch und Mathematik: Logisches und mathematisches Denken. In: K. Lengnink, S. Prediger, F. Siebel (Hrsg.): *Mathematik und Mensch: Sichtweisen der Allgemeinen Mathematik.* Verlag Allgemeine Wissenschaft, Mühltal 2001, 141–160.

[Wi02]   R. Wille: Transdisziplinarität und Allgemeine Wissenschaft. In: H. Krebs, U. Gehrlein, J. Pfeifer, J. C. Schmidt (Hrsg.): *Perspektiven Interdisziplinärer Technikforschung: Konzepte, Analysen, Erfahrungen.* Agenda-Verlag, Münster 2002, 73–84.

[Wi04a]   R. Wille: Truncated distributive lattices: conceptual structures of simple-implicational theories. Order **20** (2004), 229–238.

[Wi04b]   R. Wille: Dyadic mathematics - abstractions of logical thought. In: K. Denecke, M. Erné, S. L. Wismath (eds.): *Galois connections and applications.* Kluwer, Dordrecht 2004, 453–498.

[Wi05a]   R. Wille: Mathematik präsentieren, reflektieren, beurteilen. In: K. Lengnink, F. Siebel (Hrsg.): *Mathematik präsentieren, reflektieren, beurteilen.* Verlag Allgemeine Wissenschaft, Mühltal 2005, 3–19.

[Wi05b]   R. Wille: Formal Concept Analysis as mathematical theory of concepts and concept hierarchies. In: [GSW05], 1–33.

[WW96]   R. Wille, U. Wille: Coordinatization of ordinal structures. *Order* **13**, (1996), 281–294.

[WW04]   R. Wille, U. Wille: Restructuring General Geometry: measurement and visualization of spatial structures. *Contributions to General Algebra* **14**. Johannes Heyn Verlag, Klagenfurt 2004, 189–203.

[Ye00]   S. Yevtushenko: System of data analysis "Concept Explorer". In: *Proceedings of the sevens national conference on Artificial Intelligence KII-2000,* Russia 2000, 127–134. (in Russian)

An Enumeration Problem in Ordered Sets Leads to Possible Benchmarks for Run-Time Prediction Algorithms[*]

Tushar S. Kulkarni[1] and Bernd S.W. Schröder[2]

[1] Program of Computer Science, Louisiana Tech University,
Ruston, LA 71272
tushar_kul@hotmail.com
[2] Program of Mathematics & Statistics, Louisiana Tech University,
Ruston, LA 71272
schroder@coes.LaTech.edu

Abstract. Motivated by the desire to estimate the number of order-preserving self maps of an ordered set, we compare three algorithms (Simple Sampling [4], Partial Backtracking [10] and Heuristic Sampling [1]) which predict how many nodes of a search tree are visited. The comparison is for the original algorithms that apply to backtracking and for modifications that apply to forward checking. We identify generic tree types and concrete, natural problems on which the algorithms predict incorrectly. We show that incorrect predictions not only occur because of large statistical variations but also because of (perceived) systemic biases of the prediction. Moreover, the quality of the prediction depends on the order of the variables. Our observations give new benchmarks for estimation and seem to make heuristic sampling the estimation algorithm of choice.

Keywords: constraint satisfaction, search, enumeration.

1 Authors' Motivation: The Automorphism Problem

An **ordered set** is a set P equipped with a reflexive, antisymmetric and transitive binary relation $\leq$, the order relation. Order-preserving maps are the natural morphisms for these structures.

Definition 1.1. *Let P and Q be finite ordered sets. A map $f : P \to Q$ is called a **homomorphism** (or an **order-preserving map**) iff f preserves order, that is, $x \leq y$ implies $f(x) \leq f(y)$. An **endomorphism** is a homomorphism from P to P. An **automorphism** is a bijective endomorphism whose inverse also is an endomorphism.*

The set of endomorphisms of P will be denoted $\mathrm{End}(P)$ and the set of automorphisms will be denoted $\mathrm{Aut}(P)$. The set of homomorphisms from P to Q will be denoted $\mathrm{Hom}(P, Q)$.

[*] This work was sponsored by Louisiana Board of Regents RCS grant LEQSF (1999-02)-RD-A-27. Part of this work is part of Tushar Kulkarni's Master's Thesis.

R. Missaoui and J. Schmid (Eds.): ICFCA 2006, LNAI 3874, pp. 30–44, 2006.

This work is originally motivated by the following question first asked in [11].

Open Question 1.2. The Automorphism Problem. *What is*

$$\lim_{|P|\to\infty} \frac{|\mathrm{Aut}(P)|}{|\mathrm{End}(P)|} := \lim_{n\to\infty} \max_{|P|=n} \frac{|\mathrm{Aut}(P)|}{|\mathrm{End}(P)|}?$$

The common conjecture is that the limit is zero.

Conjecture 1.3. The Automorphism Conjecture [11].

$$\lim_{|P|\to\infty} \frac{|\mathrm{Aut}(P)|}{|\mathrm{End}(P)|} = 0.$$

The automorphism conjecture was proved for the class of finite lattices in [7]. There are also various exercises in [12], p.288, 289 that prove the automorphism conjecture for some easy classes of ordered sets, such as interval ordered sets and sets of width ≤ 3.

To get some experimental insight into the problem, it is natural to use an enumeration algorithm to compute the numbers of endomorphisms and automorphisms for specific ordered sets and investigate these quotients. One of the authors (Schröder) had such algorithms coded from earlier work on the fixed point property inspired by [16]. The enumeration of endomorphisms and automorphisms for small sets led to useful experimental data and brought about the results in [13].

For larger ordered sets, the enumeration takes too long, but it is equally natural to try to obtain reliable estimates. This idea led to the investigation presented here. The authors implemented the three well-established tree-size estimation techniques and tested them on ordered sets. Similar to [9], it was observed that for tall, narrow ordered sets the estimates were not necessarily reliable. To date, estimation has not been used in connection with the automorphism problem. Instead, we started an investigation into possible weaknesses of the estimation algorithms and as we will see, the estimation of the tree searched to enumerate all endomorphisms of a chain already is challenging.

It should be noted that the typical tree search is a search for the first solution, not an enumeration of all solutions. This paper exclusively considers the enumeration of all solutions. The weaknesses observed should however also exist in searches for the first solution. Trivial examples would be to add another level to the searches presented here and hiding a small number of solutions at that level. More reasonably, we can say that estimates for problems with deep hierarchies that contain many "partial solutions" (consistent instantiations, see below) should be susceptible to the effects presented here.

2   Tree Search

Tree searching is a well-established solution technique for decision and optimization problems. Because the search time for a particular problem can be unacceptably long, various algorithms (cf. [1, 4, 10]) to estimate the size of the search tree have been proposed. Such estimates could be used to determine which (if any) algorithm is more feasible or if some variable order is preferable over another. For a more exhaustive list of possible applications cf. [14]. These algorithms have been successful in all test cases reported in [1, 4, 10], with the main concern for accuracy being the variance of the reported results. In [9], section 7.4, Simple Sampling was used in a variable ordering heuristic. In their setting Simple Sampling gave unreliable estimates and in some cases gross underestimates consistently. On the other hand, in [14] the use of a sampling technique based on Partial Backtracking in a variable ordering scheme produced satisfactory results. Indeed, variable ordering heuristics are considered one of the prime applications of estimation algorithms. To shed some light on the contradictory observations above, we investigate three estimation algorithms (Simple Sampling, Partial Backtracking and Heuristic Sampling) in the framework of binary constraint satisfaction problems. After giving the necessary background on binary constraint satisfaction problems and the estimation algorithms, we provide a class of natural examples for which the estimation algorithms fail. The examples show similar behavior for forward checking as well as for backtracking, thus covering two major search paradigms. Our examples could thus become benchmarks on which to measure the quality of future prediction algorithms.

2.1   (Binary) Constraint Networks

A **binary constraint network** or **binary constraint satisfaction problem (CSP)** consists of the following.

A. A set of variables, which we denote $1, \ldots, n$,
B. A set of domains $D_1, \ldots, D_n$, one for each variable; we will assume each domain is the set $\{1, \ldots, m\}$,
C. A set $\mathcal{C}$ of binary constraints. Each binary constraint consists of a set of two variables $\{i, j\}$ and a binary relation $C_{ij} \subseteq D_i \times D_j$. For each set of variables we have at most one constraint.

For a given CSP, let $Y \subseteq \{1, \ldots, n\}$. Any element $x = \{(j, x_j) : j \in Y\} \in \Pi_{j \in Y} D_j$ is an **instantiation** of the variables in Y. (If the set Y is equal to $\{1, \ldots, k\}$ we will also denote our tuples as $(x_1, \ldots, x_k)$.) An instantiation is called **consistent** if and only if for all $i, j \in Y$ we have that $(x_i, x_j) \in C_{ij}$. A **solution** to a CSP is a consistent instantiation of all variables. The typical questions about CSPs are if there is a solution and how many solutions there are. For a good introduction to CSPs consider [2, 8, 15]. We will use the terms CSP and constraint network interchangeably. All CSPs in this paper are binary, so that we will also often drop this specification.

2.2   Backtracking

Backtracking (BT) is a search algorithm to find a solution of a CSP. The backtracking algorithm maintains a consistent instantiation of the first k variables

at all times. (At the start this is the empty set.) Given a consistent instantiation CI of the first k variables, backtracking instantiates $k+1$ to the first
value $x_{k+1,1}$ of D_{k+1}. If $CI \cup \{(k+1, x_{k+1,1})\}$ is consistent, then CI is replaced
with $CI \cup \{(k+1, x_{k+1,1})\}$ and backtracking tries to instantiate $k+2$. If not,
the next value in D_{k+1} is tried. If backtracking does not find any instantiation
of $k+1$ that extends the current instantiation, then k is uninstantiated, that
is, CI is replaced with $CI \setminus \{(k, x_{k,\text{current}})\}$. The search then resumes as above
by instantiating k to the next element of the kth domain, $x_{k,\text{current}+1}$. Every
instantiation that is checked by BT will be called a **node** in the BT search
tree.

2.3 Forward Checking

Forward Checking (FC) can be viewed as refinement of BT. Call a consistent instantiation $(x_1, \ldots, x_k)$ **forward consistent** if and only if for all variables $j \in \{k+1, \ldots, n\}$ there is an instantiation x_j of j so that the instantiation $(x_1, \ldots, x_k)$ with $\{(j, x_j)\}$ added is consistent. As FC reaches a forward consistent instantiation $(x_1, \ldots, x_k)$, it has stored for all variables $j \in$
$\{k+1, \ldots, n\}$ the values x_j in D_j for which the instantiation $(x_1, \ldots, x_k)$ with
$\{(j, x_j)\}$ added is consistent. FC instantiates $k+1$ to the first $x_{k+1} \in D_{k+1}$ so
that $(x_1, \ldots, x_k, x_{k+1})$ is consistent. It is then checked if $(x_1, \ldots, x_k, x_{k+1})$ is
forward consistent also. The result of this checking procedure is stored as the
list of values that lead to consistent extensions of $(x_1, \ldots, x_k, x_{k+1})$. If there
is a $j \in \{k+2, \ldots, n\}$ for which there is no x_j so that $(x_1, \ldots, x_k, x_{k+1})$ with
$\{(j, x_j)\}$ added is consistent, one speaks of a **domain annihilation**. In this
case FC instantiates $k+1$ to the next element of D_{k+1} that consistently extends $(x_1, \ldots, x_k)$. If $(x_1, \ldots, x_k, x_{k+1})$ is forward consistent, FC instantiates
$k+2$. If there are no further values to be tried, FC uninstantiates k and $k+1$
and instantiates k to the next consistent extension of the (forward consistent)
$(x_1, \ldots, x_{k-1})$. Any instantiation of all variables that is visited by FC automatically is a solution, since FC only visits consistent instantiations. If FC finds
a solution the search either stops (if only one solution was to be found) or it
continues (if all solutions were to be found).

  For an excellent description of various search algorithms and their relative
comparisons, cf. [6].

3 Prediction Algorithms

The focus of this paper is the comparison of prediction algorithms for the number
of nodes that FC or BT visits in a search for all solutions. These algorithms search
a subtree and compute an estimate based on the data gathered in the subtree.
In the following, the ordering of the variables will be considered fixed unless a
reordering is explicitly indicated.

  For each algorithm we will first describe the original version (for BT) and
then how to adapt it to estimate FC.

3.1   Simple Sampling

In [4] **"Simple Sampling" (SS)** (terminology from [1]) is introduced to predict
how many nodes BT visits. Instantiate the first variable with a random element
of D_1 and set $e_0 := 1$, $e_1 := |D_1|$, and $e_2 := \cdots := e_n := 0$. After variables
$1, \ldots, k$ have been instantiated to $x_1, \ldots, x_k$, compute the number $c_{x_1,\ldots,x_k}$ of
instantiations x_{k+1} for $k + 1$ so that $(x_1, \ldots, x_k, x_{k+1})$ is consistent. There are
two possibilities.

A. $c_{x_1,\ldots,x_k} \neq 0$. In this case instantiate variable $k + 1$ to a randomly chosen
   value x_{k+1} so that $(x_1, \ldots, x_k, x_{k+1})$ is consistent. Set $e_{k+1} := e_k \cdot c_{x_1,\ldots,x_k}$
   and continue.
B. $c_{x_1,\ldots,x_k} = 0$. In this case we stop.

At the end of this algorithm the number e_k is an estimate for the number of
consistent nodes at depth k in the search space. Since (cf. [6], Theorems 8 and
9) BT visits all nodes whose parent is consistent, the number

$$v := \sum_{k=0}^{n-1} e_k |D_{k+1}|$$

is an estimate for the number of nodes that BT visits.

If, for FC, the word "consistent" above is replaced with "forward consistent",
then the algorithm gives an estimate of the number of forward consistent nodes.
Unfortunately, there is no simple formula to compute the number of nodes FC
visits from the number of forward consistent nodes. Thus for SS (and the following algorithms) the adaptation to FC is slightly more subtle. The initial step
for SS remains the same, except that we set $e_1 := 0$. Let $f_{x_1,\ldots,x_k}$ be the number
of forward consistent children of $(x_1, \ldots, x_k)$. Run the algorithm until k_0 with
$f_{x_1,\ldots,x_{k_0}} = 0$ is reached. For $k = 1, \ldots, k_0$ set $e_k := f_{x_1} \cdots f_{x_1,\ldots,x_{k-1}} c_{x_1,\ldots,x_k}$.
The sum of the e_k is an estimate how many nodes FC visits. (This notation is
motivated by the fact that FC visits a subset of the set of consistent nodes.)

3.2   Partial Backtracking

Partial Backtracking $(\mathbf{PB}^b)$, cf. [10], for $b > 1$ is a refinement of SS. Instead
of randomly selecting *one* consistent child of every visited node, for a fixed $b \geq 1$,
one randomly selects b consistent children (all if there are fewer than b) of every
consistent node that is visited. In the tree that PB^b visits, every node $(x_1, \ldots, x_k)$
is consistent and has exactly $\min\{b, c_{x_1,\ldots,x_k}\}$ children.

To implement PB^b, one modifies BT as follows. At each depth k, maintain a set
S_k of possible instantiations for the kth variable. Whenever one goes deeper into
the tree, say from depth k to $k + 1$, one computes first the set of all consistent
extensions of the current instantiation $(x_1, \ldots, x_k)$. Then one picks b random
elements out of this set (or takes all if the set has $\leq b$ elements). These elements
form the set S_{k+1}. Whenever k is reinstantiated, one instantiates k to the next
unused element of S_k (instead of the next domain element).

All e_k are initially set to zero. For every visited node $(x_1, \ldots, x_k)$ we add to e_k the product $\Pi_{l=1}^{k-2} \max\left(\frac{c_{x_1,\ldots,x_l}}{b}, 1\right)$. Again e_k is an estimate of the number of consistent nodes at depth k. The estimate for the total number of nodes visited by BT is computed as for SS. Note that PB^b is actually a collection of algorithms, since for every value of b we have a different algorithm. The smaller the b, the faster the estimate (for $b = 1$ PB^1 is SS). The larger the b, the more accurate the estimate. (Indeed, if $b \geq |D_k|$ for all k, then PB^b performs regular BT.)

To have some consistency in language we shall call the FC modification of this algorithm **Partial Forward Checking (PF^b)**. The modification is that we use only forward consistent nodes when the sets S_k are formed. For every visited node (consistent nodes that are not forward consistent are visited, too; they are just not used to form the sets S_k) we add the product $\Pi_{l=1}^{k-2} \max\left(\frac{f_{x_1,\ldots,x_l}}{b}, 1\right)$ to e_k. Then the sum of the e_k is an estimate of the number of nodes visited by FC.

3.3 Heuristic Sampling

Heuristic Sampling (HS), cf. [1], is another generalization of SS. Sampled nodes are still chosen randomly, but some information about the nodes is used in the choices. When a set of nodes contains only nodes that are considered similar, then only the children of *one* representative of this set of nodes will be investigated.

The mechanism that identifies similar nodes is a function called the **stratifier**. Any function that is defined on all nodes and strictly decreases along each edge of the search tree can be used as a stratifier. Nodes for which the stratifier gives the same value are considered similar. This idea makes HS adaptable to particular situations, since a problem specific stratifier can be used. Yet we are interested in an algorithm that is applicable to arbitrary CSPs. The needed domain-independent stratifiers are not too hard to generate. Yet they seem to not have played a great role in the literature so far. The stratifiers that we will be particularly interested in are

A. **Depth.** Nodes at the same depth are considered similar. HS reduces to SS.
B. **(Depth, number of consistent children).** Nodes at the same depth and with the same number of consistent children are considered similar. This stratifier is mentioned in [1], p. 297 as something to "*perhaps* [our emph.] give us better stratification".
C. **(Depth, number of consistent children, number of consistent grandchildren).** Generalization of B. Nodes at the same depth, with the same number of consistent children and with the same number of consistent grandchildren are similar.

In the algorithm, the stratifier function needs to be numerical. This can always be achieved by hashing multiple values into one number. For example, one can hash the stratifier in B as $s_B := -(m\langle\text{depth}\rangle + \langle\text{nr. consistent children}\rangle)$ and the stratifier in C as $s_C := -(m^3\langle\text{depth}\rangle + m^2\langle\text{nr. consistent children}\rangle + \langle\text{nr. consistent grandchildren}\rangle)$.

The algorithm itself maintains a queue of triples (n, s, w) consisting of a node n, its stratifier value s and its weight w. HS starts with a queue that contains only the triple consisting of the root node, the stratifier value zero and the weight 1. HS then executes until the queue is empty. In every iteration the element (n_p, s_p, w_p) with the largest stratifier value is removed from the queue and all its consistent children are examined. For each consistent child c the stratifier value $s(c)$ is computed. If there is no element in the queue with stratifier value $s(c)$, add $(c, s(c), w_p)$ to the queue. If there is an element $(n, s(c), w)$ in the queue, replace w with $w + w_p$ and then with probability $\frac{w_p}{w}$ replace n with c. For each consistent child at depth k, add w_p to e_k. Again e_k will be an estimate of the number of consistent nodes at depth k.

Just like PB^b, HS is a set of algorithms, one for each stratifier. We will refer to the algorithms with stratifiers as above as HS^c (depth, consistent children) and HS^g (depth, consistent children, consistent grandchildren).

HS was conceived as an algorithm for general trees, so it is not surprising that it adapts most easily to FC. For every element in the queue first check if it is forward consistent. If it is not, discard the element. If it is, process it as above. The sum of the e_k will be an estimate of the number of nodes visited by FC.

3.4   Failure Modes

All sampling algorithms described above will ideally visit only a small fraction of the nodes in the search tree in each iteration (**individual sample**). While the individual samples for each algorithm are potentially very inaccurate there is theoretical as well as experimental evidence (cf. [1, 4, 10]) that the average of a sufficient number of estimates is close to the actual value. Indeed (cf. [1], Theorem 4; [4], Theorem 1) for all algorithms the individual samples are unbiased estimators for the number of nodes visited. That is, the expected value of the individual sample estimates is the number of nodes visited. Normally several individual samples are averaged to arrive at an estimate of the number of nodes (**sample average**). In practice, under the *assumption that the sampled nodes are representative*, the sample averages are expected to be close to the actual number of nodes. Lack of uniformity in the properties of the nodes can lead to two possible modes of failure.

Large statistical fluctuations. The sample averages spread symmetrically about the actual value. However the number of samples needed to obtain a sample average that is with high probability close to the actual value is so large that it may be faster to search the whole tree. Essentially the distribution of the individual samples has an unacceptably large standard deviation.

(Perceived) systemic bias. In this mode, even sample averages of many samples are consistently bounded away from the actual number of nodes. The distribution of the sample values *appears* to have an expected value that differs from the actual number of nodes. This failure mode is more dangerous than the large fluctuations. The algorithm appears to give a statistically valid, stable estimate that should be close to the actual tree size, while it actually delivers an underestimate.

The possibility of (perceived) systemic bias is mentioned in [4], section 4, using lottery type distributions. (If an algorithm returns a large value with small probability and small values with high probability, then even averages of larger samples can be consistently lower than the actual mean.) An example involving trees on p. 481 in [10] shows the main idea for causing (perceived) systemic bias. If the children of most nodes have vastly different numbers of consistent descendants, then SS is likely to exhibit (perceived) systemic bias. In the example noted in [10] each node has two children with differing numbers of descendants. Essentially one "wrong" turn will cause SS to miss most of the nodes of the tree. This type of example inspired the development of PB^b, which showed better performance than SS in "tall, skinny trees".

It is noted in [4] that no natural example of (perceived) systemic bias was ever found. Indeed it is mentioned ([4], section 6) that the right order of magnitude in all tests was found in 10-1,000 trials. Neither [10] nor [1] give any indication of natural examples in which (perceived) systemic bias was observed for any algorithm. The only concrete recorded statement of a (perceived) systemic bias is in section 7.4 of [9] and details are not given.

3.5   Order-Preserving Maps (again)

(Perceived) systemic bias can occur, when for most nodes the next instantiation partitions the descendants of the node into subsets of vastly different sizes. This type of behavior is typical for hierarchical structures. Ordered sets are the natural tool to describe hierachies.

A **chain** is a partially ordered set for which any two elements are comparable. We shall denote a generic chain with n elements by C_n and assume it is of the form $1 < 2 < 3 < \cdots < n$. That is, it consists of the first n natural numbers with their usual order.

The search tree for all order-preserving self maps of a chain is a natural occurrance of an imbalanced search tree. Instantiate the variables in their natural order. For this problem and variable order an instantiation of the first k variables is consistent exactly when it is forward consistent. (Any missing elements can always be mapped to the top element n.) This, in turn, is the case exactly when the instantiation corresponds to an order-preserving map from $\{1, \ldots, k\} = C_k$ to $\{1, \ldots, n\} = C_n$.

Let $C(k, n)$ be the number of order-preserving maps from C_k to C_n. Note that an order-preserving map from C_k to C_n can map the smallest element of C_k to the smallest element of C_n or strictly above that element. This provides a natural partition of the order-preserving maps from C_k to C_n into two subsets. For the first subset there is only one image for the smallest element, so it has $C(k-1, n)$ elements. For the second subset there are at most $n-1$ images for each element, so it has $C(k, n-1)$ elements. This means $C(k, n) = C(k-1, n) + C(k, n-1)$ and the $C(k, n)$ are computable. This was helpful for larger experiments in which the search algorithm would run too long. Equally important, the above idea can be used to show that the instantiation of the last variable determines the number of (forward) consistent descendants of the node. If the first k variables have been instantiated and k is instantiated to $f(k)$, then the node will have exactly $C(n - k, n - f(k) + 1)$ consistent descendants. While $n - k$ is determined by the depth, the number $n - f(k) + 1$ can typically span a wider range of possibilities. Focusing our attention back on children, we see that the numbers $C(n - (k + 1), n - a + 1)$ with $a \geq f(k)$ span a wide range of sizes. Thus all consistent nodes with at least two children have children with vastly different numbers of consistent descendants.

The above appears to be the first recorded example of a structured family of problems (rather than isolated examples as in [9]) in which (perceived) systemic bias can be observed. This structure allows for a more organized investigation of the phenomenon. While chains have a very special structure, they abound in ordered sets. Thus any ordered set that contains long chains appears to be a set on which sampling algorithms could produce erroneous estimates. Our experimental observations will confirm that sampling search is problematic in this setting.

4   Experimental Results

Results for estimation of FC and BT were similar, so we will only report results for FC. For each experiment we ran each estimation algorithm 100 times. To allow fair performance comparison, for each algorithm we bounded the number of nodes that are checked for (forward) consistency by $10,000nm$. Once more than the maximum number of nodes is checked for consistency, the current individual sample is finished and then the average is computed. In this fashion, each algorithm gathers roughly the same amount of information. With the specific bound, SS will collect at least 10,000 individual samples. Run times increase for the same number of nodes checked depending on the overhead per node, with SS being fastest, then PF^2, PF^3, HS^c and HS^g.

In recorded results we present a histogram of the quotients

$$r = \frac{\text{estimated nodes}}{\text{actual nodes}}.$$

This appears preferable to presenting the relative error. Consider that a -50% error means the estimate is off by a factor 2, while a +50% error means the estimate is off by a factor 1.5, giving a skewed picture of the quality of the estimate.

4.1   (Perceived) Systemic Bias

We first detected a (perceived) systemic bias in SS for the BT search for all order-preserving self-maps for the 14-chain. In Table 1 we show the data for the FC search on the 15-chain.

We see SS tends to underestimate. This bias is worse when only $1,000nm$ nodes are checked. The systemic bias remains even when allowing $100,000nm$

Table 1. Estimating the FC search tree for order-preserving maps of the 15-chain

estimated nodes / actual nodes	SS	PF²	PF³	HSᶜ	HSᵍ
$< \frac{1}{4}$	16	0	0	0	0
$[\frac{1}{4}, \frac{1}{3})$	17	0	0	0	0
$[\frac{1}{3}, \frac{1}{2})$	21	2	10	0	0
$[\frac{1}{2}, \frac{2}{3})$	15	23	16	0	0
$[\frac{2}{3}, \frac{4}{5})$	10	18	14	0	3
$[\frac{4}{5}, 1)$	7	25	23	100	97
$[1, \frac{5}{4})$	3	15	14	0	0
$[\frac{5}{4}, \frac{3}{2})$	6	6	13	0	0
$[\frac{3}{2}, 2)$	2	4	4	0	0
$[2, 3)$	1	5	5	0	0
$[3, 4)$	1	1	1	0	0
≥ 4	1	1	0	0	0

nodes to be checked for SS. The effect is not yet visible for PF² and PF³ (though it was visible in the measurement for BT and PB² and PB³). Also note that PF² is performing better than PF³, which can be explained by the smaller number of repetitions PF³ can afford with the allowed number of checks. The HS algorithms gave underestimates that were less than 5% off for HSᵍ and less than 1% off for HSᶜ, so they were extremely accurate.

How does the (perceived) systemic bias manifest itself as the size of the chain increases? Consider Table 2 below, which shows the results for SS with $10,000nm$ nodes allowed to be checked for consistency and increasing chain size. As the size of the chain increases we observe a transition in the results of the estimation algorithms. First the sample variance increases and then more and more sample averages become underestimates. This is despite the fact that the increase in number of nodes allowed to be checked actually allows slightly more repetitions of the algorithm for the larger chains.

Did we simply not perform enough repetitions? The central limit theorem guarantees that for *sufficiently large* sample sizes, sample averages are approximately normally distributed around the actual mean. Hence, the bias in SS must eventually go away as the sample size increases. This is not practical for these problems. We established experimentally that for the 20-chain one must use sample sizes that force the experimenter to sample more nodes than there are in the full search tree (!) before the systemic bias in SS (for FC and BT) is overcome. Again, the (perceived) systemic bias worsens as the length of the chain increases, so eventually arbitrarily bad underestimates are obtained with arbitrarily large sample sizes.

The same type of transition indicated above for SS is also observable for PBᵇ and PFᵇ. In Table 3 we list the size n of the chain C_n for which we first observe a systemic bias when estimating the size of the FC tree for finding all order-preserving self maps of C_n. Since the bias is increasing with the size of the chain we had to establish a starting point. We considered the experiment to show

Table 2. Using SS to estimate the FC search tree for order-preserving maps of the n-chain, allowed $10,000nm$ nodes to be checked

$\frac{\text{estimated nodes}}{\text{actual nodes}}$	C_{11}	C_{12}	C_{13}	C_{14}	C_{15}	C_{16}
$< \frac{1}{4}$	0	0	0	2	16	41
$\left[\frac{1}{4}, \frac{1}{3}\right)$	0	0	1	4	17	19
$\left[\frac{1}{3}, \frac{1}{2}\right)$	0	1	18	31	21	10
$\left[\frac{1}{2}, \frac{2}{3}\right)$	3	15	22	13	15	5
$\left[\frac{2}{3}, \frac{4}{5}\right)$	11	15	18	18	10	5
$\left[\frac{4}{5}, 1\right)$	26	25	11	6	7	6
$\left[1, \frac{5}{4}\right)$	36	18	9	10	3	2
$\left[\frac{5}{4}, \frac{3}{2}\right)$	12	9	7	2	6	3
$\left[\frac{3}{2}, 2\right)$	11	7	3	4	2	3
$[2, 3)$	1	8	7	7	1	0
$[3, 4)$	0	1	1	2	1	0
≥ 4	0	1	3	1	1	6

Table 3. Estimation of the FC and BT search trees for order-preserving maps of the n-chain shows systemic bias for $n > \langle\text{cutoff}\rangle$

algorithm	SS	PF/B^2	PF/B^3	HS^c	HS^g
FC $\langle\text{cutoff}\rangle$	16	19	20	> 30	> 30
BT $\langle\text{cutoff}\rangle$	15	17	18	> 30	> 30

(perceived) systemic bias when at least 60% of the estimates were less than half the actual number.

HS^c and HS^g consistently gave estimates (for the total number of nodes as well as for the number of solutions!) that were less than 1% off for all chains up to the 30-chain! However we also found examples of more complex ordered sets (a "stack of five six-crowns" as in Figure 1 to be precise) for which HS^c showed systemic bias worse than PF^2 and PF^3. (Thus PF^2 and PF^3 can outperform HS^c.) We suspect that HS^g also is susceptible to this phenomenon, but the experiments became too large. For the stack of five six-crowns, the ratio of estimated to actual nodes in 100 runs of HS^g was 11 times in $[0.8, 1)$, 85 times in $[1, 1.25)$, and 4 times in $[1.25, 1.5)$.

Generally, as long as the hierarchy did not get too wide, (perceived) systemic bias could be observed for sufficiently large ordered sets.

These examples show that the specific structure of a long chain embedded within a problem can result in (perceived) systemic biases for this problem. Systemic bias can turn into a practical problem if estimation is used to choose between FC and BT. An estimate that predicts BT to visit fewer nodes than FC can still be identified as nonsensical. Yet the possible use of estimation algorithms to predict the number of consistency checks performed (which is often proportional to run time and which can be lower for BT than for FC) could lead to a situation where an underestimate for BT causes BT to be chosen over the actually more efficient FC.

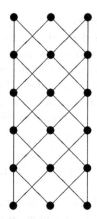

Fig. 1. When estimating the search tree for enumerating the order-preserving self maps of a stack of five six-crowns, HS^c is outperformed by PF^2 and PF^3

4.2 The Effect of Reordering the Variables

To demonstrate how the quality of the estimation also depends on the variable order, we reordered the domain variables in the search for the order-preserving maps of the 15-chain as follows. From left to right we go from the bottom up, numbers indicate the number in the variable ordering. The original order in 4.1 was the natural order $1, 2, 3, \ldots, 15$. The reorder for which we report results in Table 4 was 1,9,2,10,3,11,4,12,5,13,6,14,7,15,8. Both the original FC tree as well as the FC tree on the reordered chain have the same number of nodes (155,117,519). This is a specific property of this type of problem, which does not occur widely.

All algorithms performed better in the reordered problem than in the original. SS, HS^c and HS^g gave near equivalent and satisfactory results. PF^2 and PF^3 exhibit larger variance than the other algorithms. This is interesting, as we are not aware of any recorded examples in which SS outperforms PF^b/PB^b. Results for BT were similar, indeed, more pronounced.

Table 4. Estimating the FC search tree for order-preserving maps of the renumbered 15-chain

$\dfrac{\text{estimated nodes}}{\text{actual nodes}}$	SS	PF²	PF³	HSᶜ	HSᵍ
$< \frac{1}{2}$	0	0	0	0	0
$[\frac{1}{2}, \frac{2}{3})$	0	0	10	0	0
$[\frac{2}{3}, \frac{4}{5})$	0	6	13	0	0
$[\frac{4}{5}, 1)$	19	47	22	0	0
$[1, \frac{5}{4})$	81	42	36	100	100
$[\frac{5}{4}, \frac{3}{2})$	0	5	13	0	0
$[\frac{3}{2}, 2)$	0	0	6	0	0
≥ 2	0	4	0	0	0

The data is indicative of another potentially serious problem for applications. If the quality of the estimate depends on the variable ordering, then a bad quality underestimate of the search time for one variable ordering could cause this variable ordering to be chosen over another, which may have been more efficient, but was estimated more accurately. The equal sizes of the search trees above show that there is no proportionality, say, in such a way that smaller trees have worse underestimates than larger trees. Observations of this nature led to the discarding of SS as a variable order heuristic in [9].

4.3   Variances of the Different Algorithms

Fortunately, many problems do not exhibit the (perceived) systemic bias reported above. Tables 1 and 4 already give an indication how variances behave (ignore the SS column in Table 1). Another set of measurements that indicates how the variances of the different sample averages behave when there is no perceived bias is given in Table 5 for a search for Hamiltonian cycles. The estimates are for the number of nodes FC visits when enumerating the number of knight's tours on a 6×6 chessboard. Since predictions were very good for this problem we only allowed $100nm$ nodes to be checked.

It is interesting to note here that SS outperforms all other algorithms. The PF algorithms are marked with a * because both needed more than the allowed number of nodes even for one repetition. For PF^2 this was not too critical (it used about 4 times the allowed number of nodes). For PF^3 it was (it used more than 350 times the allowed number of nodes).

Data sets in which SS was outperformed by PF, which in turn was outperformed by HS were found for various searches for relation preserving maps in structures in which chains are not too long, randomly seeded CSPs, and the n-queens problem ($n \leq 17$). Supplementary measurements with fewer or more nodes allowed to be checked show (as reported in the literature) that the vari-

Table 5. Estimating the FC search tree for knight's tours on a 6×6 chessboard

$\frac{\text{estimated nodes}}{\text{actual nodes}}$	SS	*PF^2	*PF^3	HS^c	HS^g
$< \frac{1}{4}$	0	0	0	0	0
$[\frac{1}{4}, \frac{1}{3})$	0	0	0	0	0
$[\frac{1}{3}, \frac{1}{2})$	0	5	0	0	0
$[\frac{1}{2}, \frac{2}{3})$	0	11	1	1	0
$[\frac{2}{3}, \frac{4}{5})$	0	12	13	13	10
$[\frac{4}{5}, 1)$	51	24	30	45	44
$[1, \frac{5}{4})$	49	30	47	24	36
$[\frac{5}{4}, \frac{3}{2})$	0	17	9	14	7
$[\frac{3}{2}, 2)$	0	1	0	2	3
$[2, 3)$	0	0	0	1	0
$[3, 4)$	0	0	0	0	0
≥ 4	0	0	0	0	0

ance decreases as the number of nodes checked increases. For a report on all experiments, consider [5].

5 Conclusion and Outlook

We considered basic versions of SS, PB and HS for BT and FC. This appears to be the first recorded application of these algorithms to FC.

The most important experiments were the enumeration of order-preserving self maps of chains and other "narrow" ordered sets. In these experiments SS, PB^b, PF^b and HS^c exhibit (perceived) systemic bias for sufficiently large sets. This effect occurs for BT as well as for FC, though it occurs a little later for FC. This suggests that estimations of the BT/FC tree size for any problem that contains deep hierarchical structures could be susceptible to (perceived) systemic bias. Order-preserving self maps of narrow ordered sets can be considered a future benchmark for CSP estimation algorithms.

In regards to ranking the algorithms, we can say that a clear ranking may not be possible. It was surprising to find examples in which SS outperforms PB/PF, examples in which SS outperforms HS^c and HS^g and examples in which PF^2 and PF^3 outperform HS^c. Overall, HS^g consistently gave estimates that were near the best among the algorithms, though not always the best. Thus, among the algorithms presented, HS^g would be our algorithm of choice.

Variable ordering affects the performance of estimation algorithms in a fashion that is not proportional to the effect on the run time. Thus the choice of a variable ordering based on an estimation algorithm (cf. [14], p. 39) has to be treated with care.

Future experiments should be paralleled by theoretical investigations into what general types of problems will be hard to estimate. The examples given here show some natural structures that cause failure. Problems similar to the examples and problems that contain our examples as substructures should be equally affected. Is there a general description of "deep hierarchies", which would allow the detection of possible problems with estimation before an estimation algorithm is run? Is there is a graphical description of a CSP similar to how formal concept analysis describes relations that could warn of the problems described here or suggest a good variable ordering? One of the authors' (B. Schröder) original starting point was [16], which is about formal concept analysis. Insights in this direction could clarify the scope of estimation algorithms for tasks such as algorithm selection or variable reordering. Any translation of constraint satisfaction problems into the formal concept domain would likely lead to very large concept lattices, which could only be "drawn" by solving the problem in question. Yet maybe some of a specific problem's properties translate to properties of the concept lattice that can guide search algorithms.

Finally, with a refined approach to search, called *maintaining arc consistency*, showing great potential, it may be interesting to translate the estimation ideas to this setting and then re-consider the experiments reported here.

References

1. P. C. Chen. Heuristic Sampling: A method for predicting the performance of tree searching programs. *SIAM J. Comput.*, 21:295-315, 1992.
2. R. Dechter. Constraint networks. In *Encyclopedia of Artificial Intelligence*, pages 276-284, Wiley, New York, 1992.
3. D. Duffus, V. Rödl, B. Sands, R. Woodrow. Enumeration of order-preserving maps. *Order*, 9:15-29, 1992.
4. D. E. Knuth. Estimating the efficiency of backtrack programs. *Math. Comp.*, 29:121-136, 1975.
5. T. Kulkarni. Experimental evaluation of selected algorithms for estimating the cost of solving a constraint satisfaction problem. MS. Thesis, Louisiana Tech University, 2001.
6. G. Kondrak and P. van Beek. A theoretical evaluation of selected backtracking algorithms. *Artificial Intelligence*, 89:365-387, 1997.
7. W.-P. Liu and H. Wan. Automorphisms and Isotone Self-Maps of Ordered Sets with Top and Bottom. *Order*, 10:105-110, 1993.
8. A.K. Mackworth. Constraint Satisfaction. In *Encyclopedia of Artificial Intelligence*, pages 284-293 Wiley, New York, 1992.
9. H. A. Priestley and M. P. Ward. A multipurpose backtracking algorithm. *Journal of Symbolic Computation*, 18:1-40, 1994.
10. P. W. Purdom. Tree size by partial backtracking. *SIAM J. Comput.*, 7:481-491, 1978.
11. I. Rival and A. Rutkowski (1991). Does almost every isotone self-map have a fixed point? In: Bolyai Math. Soc.. *Extremal Problems for Finite Sets*. Bolyai Soc. Math. Studies 3, Viségrad, Hungary. p. 413-422
12. B. Schröder (2003). Ordered Sets – An Introduction. Birkhäuser Verlag. Boston, Basel, Berlin
13. B.Schröder (2005). The Automorphism Conjecture for Small Sets and Series Parallel Sets. To appear in ORDER.
14. J. Sillito. Arc consistency for general constraint satisfaction problems and estimating the cost of solving constraint satisfaction problems. M. Sc. thesis, University of Alberta, 2000.
15. E. Tsang. Foundations of Constraint Satisfaction. Academic Press, New York, New York, 1993.
16. W. Xia. Fixed point property and formal concept analysis. *Order*, 9:255–264, 1992.

Attribute Implications in a Fuzzy Setting[*]

Radim Bělohlávek and Vilém Vychodil

Department of Computer Science, Palacky University, Olomouc,
Tomkova 40, CZ-779 00 Olomouc, Czech Republic
{radim.belohlavek, vilem.vychodil}@upol.cz

Abstract. The paper is an overview of recent developments concerning attribute implications in a fuzzy setting. Attribute implications are formulas of the form $A \Rightarrow B$, where A and B are collections of attributes, which describe dependencies between attributes. Attribute implications are studied in several areas of computer science and mathematics. We focus on two of them, namely, formal concept analysis and databases.

Keywords: attribute implication, fuzzy logic, functional dependency, concept lattice.

1 Introduction

Formulas of the form $A \Rightarrow B$ where A and B are collections of attributes have been studied for a long time in computer science and mathematics. In formal concept analysis (FCA), formulas $A \Rightarrow B$ are called attribute implications. Attribute implications are interpreted in formal contexts, i.e. in data tables with binary attributes, and have the following meaning: Each object having all attributes from A has also all attributes from B, see e.g. [22, 25]. In databases, formulas $A \Rightarrow B$ are called functional dependencies. Functional dependencies are interpreted in relations on relation schemes, i.e. in data tables with arbitrarily-valued attributes and have the following meaning: Any two objects which have the same values of attributes from A have also the same values of attributes from B, see e.g. [2, 29].

In what follows, we present an overview of some recent results on attribute implications and functional dependencies developed from the point of view of fuzzy logic. Section 2 provides an overview to some notions of fuzzy logic which will be needed. Section 3 deals with attribute implications in a fuzzy setting. Section 4 deals with functional dependencies in a fuzzy setting. Section 5 discusses Armstrong-like rules. Section 6 contains concluding remarks.

2 Preliminaries in Fuzzy Logic and Fuzzy Sets

Contrary to classical logic, fuzzy logic uses a scale L of truth degrees, the most favorite choice being $L = [0, 1]$ (real unit interval) or some subchain of $[0, 1]$.

[*] Supported by grant No. 1ET101370417 of GA AV ČR, by grant No. 201/05/0079 of the Czech Science Foundation, and by institutional support, research plan MSM 6198959214.

R. Missaoui and J. Schmid (Eds.): ICFCA 2006, LNAI 3874, pp. 45–60, 2006.

This enables to consider intermediate truth degrees of propositions, e.g. "object x has attribute y" has a truth degree 0.8 indicating that the proposition is almost true. In addition to L, one has to pick an appropriate collection of logical connectives (implication, conjunction, ...). A general choice of a set of truth degrees plus logical connectives is represented by so-called complete residuated lattices (equipped possibly with additional operations). The rest of this section presents an introduction to fuzzy logic notions we will need. Details can be found e.g. in [4, 24, 26], a good introduction to fuzzy logic and fuzzy sets is presented in [28].

A complete residuated lattice is an algebra $\mathbf{L} = \langle L, \wedge, \vee, \otimes, \rightarrow, 0, 1 \rangle$ such that $\langle L, \wedge, \vee, 0, 1 \rangle$ is a complete lattice with 0 and 1 being the least and greatest element of L, respectively; $\langle L, \otimes, 1 \rangle$ is a commutative monoid (i.e. $\otimes$ is commutative, associative, and $a \otimes 1 = 1 \otimes a = a$ for each $a \in L$); $\otimes$ and $\rightarrow$ satisfy so-called adjointness property:

$$a \otimes b \leq c \quad \text{iff} \quad a \leq b \rightarrow c$$

for each $a, b, c \in L$. Elements a of L are called truth degrees. Fuzzy logic is truth-functional and $\otimes$ and $\rightarrow$ are truth functions of ("fuzzy") conjunction and ("fuzzy") implication. That is, if $||\varphi||$ and $||\psi||$ are truth degrees of formulas φ and ψ then $||\varphi|| \otimes ||\psi||$ is a truth degree of formula $\varphi \& \psi$ (& is a symbol of conjunction connective); analogously for implication.

A useful connective is that of a truth-stressing hedge (shortly, a hedge) [26, 27]. A hedge is a unary function $^* : L \rightarrow L$ satisfying $1^* = 1$, $a^* \leq a$, $(a \rightarrow b)^* \leq a^* \rightarrow b^*$, $a^{**} = a^*$, for each $a, b \in L$. Hedge * is a truth function of logical connective "very true", see [26, 27]. The properties of hedges have natural interpretations, see [27].

A common choice of $\mathbf{L}$ is a structure with $L = [0, 1]$ (unit interval) or L being a finite chain. We refer to [4, 26] for details.

Two boundary cases of (truth-stressing) hedges are (i) identity, i.e. $a^* = a$ $(a \in L)$; (ii) globalization [34]:

$$a^* = \begin{cases} 1 & \text{if } a = 1, \\ 0 & \text{otherwise.} \end{cases}$$

A special case of a complete residuated lattice with hedge is the two-element Boolean algebra $\langle \{0, 1\}, \wedge, \vee, \otimes, \rightarrow, ^*, 0, 1 \rangle$, denoted by $\mathbf{2}$, which is the structure of truth degrees of the classical logic. That is, the operations $\wedge, \vee, \otimes, \rightarrow$ of $\mathbf{2}$ are the truth functions (interpretations) of the corresponding logical connectives of the classical logic and $0^* = 0$, $1^* = 1$. Note that if we prove an assertion for general $\mathbf{L}$, then, as a particular example, we get a "crisp version" of this assertion for $\mathbf{L}$ being $\mathbf{2}$.

Having $\mathbf{L}$, we define usual notions: an $\mathbf{L}$-set (fuzzy set) A in universe U is a mapping $A : U \rightarrow L$, $A(u)$ being interpreted as "the degree to which u belongs to A". We say "fuzzy set" instead of "$\mathbf{L}$-set" if $\mathbf{L}$ is obvious. If $U = \{u_1, \ldots, u_n\}$ then A can be denoted by $A = \{^{a_1}/u_1, \ldots, ^{a_n}/u_n\}$ meaning that $A(u_i)$ equals a_i for each $i = 1, \ldots, n$. For brevity, we introduce the following convention:

we write $\{\ldots, u, \ldots\}$ instead of $\{\ldots, {}^1\!/u, \ldots\}$, and we also omit elements of U whose membership degree is zero. For example, we write $\{u, {}^{0.5}\!/v\}$ instead of $\{{}^1\!/u, {}^{0.5}\!/v, {}^0\!/w\}$, etc. Let $\mathbf{L}^U$ denote the collection of all $\mathbf{L}$-sets in U. The operations with $\mathbf{L}$-sets are defined componentwise. For instance, the intersection of $\mathbf{L}$-sets $A, B \in \mathbf{L}^U$ is an $\mathbf{L}$-set $A \cap B$ in U such that $(A \cap B)(u) = A(u) \wedge B(u)$ for each $u \in U$, etc. Binary $\mathbf{L}$-relations (binary fuzzy relations) between X and Y can be thought of as $\mathbf{L}$-sets in the universe $X \times Y$. That is, a binary $\mathbf{L}$-relation $I \in \mathbf{L}^{X \times Y}$ between a set X and a set Y is a mapping assigning to each $x \in X$ and each $y \in Y$ a truth degree $I(x, y) \in L$ (a degree to which x and y are related by I). An $\mathbf{L}$-set $A \in \mathbf{L}^X$ is called crisp if $A(x) \in \{0, 1\}$ for each $x \in X$. Crisp $\mathbf{L}$-sets can be identified with ordinary sets. For a crisp A, we also write $x \in A$ for $A(x) = 1$ and $x \notin A$ for $A(x) = 0$. An $\mathbf{L}$-set $A \in \mathbf{L}^X$ is called empty (denoted by $\emptyset$) if $A(x) = 0$ for each $x \in X$. For $a \in L$ and $A \in \mathbf{L}^X$, $a \otimes A \in \mathbf{L}^X$ is defined by $(a \otimes A)(x) = a \otimes A(x)$.

Given $A, B \in \mathbf{L}^U$, we define a subsethood degree

$$S(A, B) = \bigwedge_{u \in U} \big(A(u) \to B(u) \big), \tag{1}$$

which generalizes the classical subsethood relation $\subseteq$. $S(A, B)$ represents a degree to which A is a subset of B. In particular, we write $A \subseteq B$ iff $S(A, B) = 1$ (A is fully contained in B). As a consequence, $A \subseteq B$ iff $A(u) \leq B(u)$ for each $u \in U$.

A binary $\mathbf{L}$-relation $\approx$ in U (i.e., between U and U) is called reflexive if for each $u \in U$ we have $u \approx u = 1$; symmetric if for each $u, v \in U$ we have $u \approx v = v \approx u$; transitive if for each $u, v, w \in U$ we have $(u \approx v) \otimes (v \approx w) \leq (u \approx w)$; $\mathbf{L}$-equivalence if it is reflexive, symmetric, and transitive; $\mathbf{L}$-equality if it is an $\mathbf{L}$-equivalence for which $u \approx v = 1$ iff $u = v$.

In the following we use well-known properties of residuated lattices and fuzzy structures which can be found in monographs [4, 26]. Throughout the rest of the paper, $\mathbf{L}$ denotes an arbitrary complete residuated lattice and $*$ (possibly with indices) denotes a hedge.

3 Attribute Implications

3.1 Attribute Implications, Validity, Theories and Models

We first introduce attribute implications. Suppose Y is a finite set (of attributes).

Definition 1. *A (fuzzy) attribute implication (over Y) is an expression $A \Rightarrow B$, where $A, B \in \mathbf{L}^Y$ (A and B are fuzzy sets of attributes).*

Fuzzy attribute implications are our basic formulas. The intended meaning of $A \Rightarrow B$ is: "if it is true that an object has all attributes from A, then it has also all attributes from B".

Remark 1. For an fuzzy attribute implication $A \Rightarrow B$, both A and B are fuzzy sets of attributes. Particularly, A and B can both be ordinary sets (i.e. $A(y) \in \{0, 1\}$ and $B(y) \in \{0, 1\}$ for each $y \in Y$), i.e. ordinary attribute implications are a special case of fuzzy attribute implications.

A fuzzy attribute implication does not have any kind of "validity" on its own—it is a syntactic notion. In order to consider validity, we introduce an interpretation of fuzzy attribute implications. Fuzzy attribute implications are meant to be interpreted in data tables with fuzzy attributes. A *data table with fuzzy attributes* (called also a *formal fuzzy context*) can be seen as a triplet $\langle X, Y, I \rangle$ where X is a set of objects, Y is a finite set of attributes (the same as above in the definition of a fuzzy attribute implication), and $I \in \mathbf{L}^{X \times Y}$ is a binary $\mathbf{L}$-relation between X and Y assigning to each object $x \in X$ and each attribute $y \in Y$ a degree $I(x, y)$ to which x has y. $\langle X, Y, I \rangle$ can be thought of as a table with rows and columns corresponding to objects $x \in X$ and attributes $y \in Y$, respectively, and table entries containing degrees $I(x, y)$. A row of a table $\langle X, Y, I \rangle$ corresponding to an object $x \in X$ can be seen as a fuzzy set I_x of attributes to which an attribute $y \in Y$ belongs to a degree $I_x(y) = I(x, y)$.

The basic step in the definition of a validity of a fuzzy attribute implication $A \Rightarrow B$ is its validity in a fuzzy set M of attributes.

Definition 2. *For a fuzzy attribute implication $A \Rightarrow B$ over Y and a fuzzy set $M \in \mathbf{L}^Y$ of attributes, we define a degree $||A \Rightarrow B||_M \in L$ to which $A \Rightarrow B$ is valid in M by*

$$||A \Rightarrow B||_M = S(A, M)^* \rightarrow S(B, M). \tag{2}$$

Remark 2. (1) $S(A, M)$ and $S(B, M)$ are the degrees to which A and B are contained in M, as defined by (1); * is a truth-stressing hedge; $\rightarrow$ is a truth function of implication. Therefore, it is easily seen that if M is a fuzzy set of attributes of some object x then $||A \Rightarrow B||_M$ is a truth degree of a proposition "if it is (very) true that x has all attributes from A then x has all attributes from B".

(2) A hedge * plays a role of a parameter controlling the semantics. Consider the particular forms of (2) for the boundary choices of *. First, if * is identity, (2) becomes

$$||A \Rightarrow B||_M = S(A, M) \rightarrow S(B, M).$$

In this case, $||A \Rightarrow B||_M$ is a truth degree of "if A is contained in M then B is contained in M". Second, if * is globalization, (2) becomes

$$||A \Rightarrow B||_M = \begin{cases} S(B, M) & \text{if } A \subseteq M, \\ 1 & \text{otherwise.} \end{cases}$$

In this case, $||A \Rightarrow B||_M$ is a truth degree of "B is contained in M" provided A is fully contained in M (i.e. $A(y) \leq M(y)$ for each $y \in Y$), and $||A \Rightarrow B||_M$ is 1 otherwise. Therefore, compared to the former case (* being identity), partial truth degrees of "A is contained in M" are disregarded for * being globalization.

(3) Consider now the case $\mathbf{L} = \mathbf{2}$ (i.e., the structure of truth degrees is a two-element Boolean algebra of classical logic). In this case, $||A \Rightarrow B||_M = 1$ iff we have that if $A \subseteq M$ then $B \subseteq M$. Hence, for $\mathbf{L} = \mathbf{2}$, Definition 2 yields the well-known definition of validity of an attribute implication in a set of attributes, cf. [22].

We now extend the definition of a validity of attribute implications to validity in systems of fuzzy sets of attributes and to validity in data tables with fuzzy attributes.

Definition 3. *For a system* $\mathcal{M}$ *of* **L**-*sets in* Y, *define a degree* $||A \Rightarrow B||_{\mathcal{M}}$ *to which* $A \Rightarrow B$ *is true in (each* M *from)* $\mathcal{M}$ *by*

$$||A \Rightarrow B||_{\mathcal{M}} = \bigwedge_{M \in \mathcal{M}} ||A \Rightarrow B||_{M}. \tag{3}$$

Given a data table $\langle X, Y, I \rangle$ *with fuzzy attributes, define a degree* $||A \Rightarrow B||_{\langle X,Y,I \rangle}$ *to which* $A \Rightarrow B$ *is true in* $\langle X, Y, I \rangle$ *by*

$$||A \Rightarrow B||_{\langle X,Y,I \rangle} = ||A \Rightarrow B||_{\{I_x \mid x \in X\}}. \tag{4}$$

Remark 3. Since I_x represents a row of table $\langle X, Y, I \rangle$ corresponding to x (recall that $I_x(y) = I(x, y)$ for each $y \in Y$), $||A \Rightarrow B||_{\langle X,Y,I \rangle}$ is, in fact, a degree to which $A \Rightarrow B$ is true in a system $\mathcal{M} = \{I_x \mid x \in X\}$ of all rows of table $\langle X, Y, I \rangle$.

Remark 4. For a fuzzy attribute implication $A \Rightarrow B$, degrees $A(y) \in L$ and $B(y) \in L$ can be seen as thresholds. This is best seen when $*$ is globalization, i.e. $1^* = 1$ and $a^* = 0$ for $a < 1$. Since for $a, b \in L$ we have $a \leq b$ iff $a \rightarrow b = 1$, we have

$$(a \rightarrow b)^* = \begin{cases} 1 & \text{iff } a \leq b, \\ 0 & \text{iff } a \not\leq b. \end{cases}$$

Therefore, $||A \Rightarrow B||_{\langle X,Y,I \rangle} = 1$ means that a proposition "for each object $x \in X$: if for each attribute $y \in Y$, x has y in degree greater than or equal to (a threshold) $A(y)$, then for each $y \in Y$, x has y in degree at least $B(y)$" is true. In general, $||A \Rightarrow B||_{\langle X,Y,I \rangle}$ is a truth degree of the latter proposition. As a particular example, if $A(y) = a$ for $y \in Y_A \subseteq Y$ (and $A(y) = 0$ for $y \notin Y_A$) $B(y) = b$ for $y \in Y_B \subseteq Y$ (and $B(y) = 0$ for $y \notin Y_B$), the proposition says "for each object $x \in X$: if x has all attributes from Y_A in degree at least a, then x has all attributes from Y_B in degree at least b", etc. That is, having A and B fuzzy sets allows for a rich expressibility of relationships between attributes which is why we want A and B to be fuzzy sets in general.

Example 1. For illustration, consider Tab. 1, where table entries are taken from **L** defined on the real unit interval $L = [0, 1]$ with $*$ being globalization. Consider now the following fuzzy attribute implications.

(1) $\{^{0.3}/y_3, {}^{0.7}/y_4\} \Rightarrow \{y_1, {}^{0.3}/y_2, {}^{0.8}/y_4, {}^{0.4}/y_6\}$ is true in degree 1 in data table from Tab. 1. On the other hand, implication $\{y_1, {}^{0.3}/y_3\} \Rightarrow \{^{0.1}/y_2, {}^{0.7}/y_5, {}^{0.4}/y_6\}$ is not true in degree 1 in Tab. 1—object x_2 can be taken as a counterexample: x_2 does not have attribute y_5 in degree greater than or equal to 0.7.

(2) $\{y_1, y_2\} \Rightarrow \{y_4, y_5\}$ is a crisp attribute implication which is true in degree 1 in the table. On the contrary, $\{y_5\} \Rightarrow \{y_4\}$ is also crisp but it is not true in degree 1 (object x_3 is a counterexample).

Table 1. Data table with fuzzy attributes

I	y_1	y_2	y_3	y_4	y_5	y_6
x_1	1.0	1.0	0.0	1.0	1.0	0.2
x_2	1.0	0.4	0.3	0.8	0.5	1.0
x_3	0.2	0.9	0.7	0.5	1.0	0.6
x_4	1.0	1.0	0.8	1.0	1.0	0.5

$X = \{x_1, \ldots, x_4\}$

$Y = \{y_1, \ldots, y_6\}$

(3) Implication $\{^{0.5}/y_5, ^{0.5}/y_6\} \Rightarrow \{^{0.3}/y_2, ^{0.3}/y_3\}$ is in the above-mentioned form for $Y_A = \{y_5, y_6\}$, $Y_B = \{y_2, y_3\}$, $a = 0.5$, and $b = 0.3$. The implication is true in data table in degree 1. $\{^{0.5}/y_5, ^{0.5}/y_6\} \Rightarrow \{^{0.3}/y_1, ^{0.3}/y_2\}$ is also in this form (for $Y_B = \{y_1, y_2\}$) but it is not true in the data table in degree 1 (again, take x_3 as a counterexample).

We now come to the notions of a *theory* and a *model*. In logic, a theory is considered as a collection of formulas. The formulas are considered as valid formulas we can use when making inferences. In fuzzy logic, a theory T can be considered as a fuzzy set of formulas, see [30] and also [23, 26]. Then, for a formula φ, a degree $T(\varphi)$ to which φ belongs to T can be seen as a degree to which we assume φ valid (think of φ as expressing "Mary likes John", "temperature is high", etc.). This will be also our approach. In general, we will deal with fuzzy sets T of attribute implications. Sometimes, we use only sets T of attribute implications (particularly when interested only in fully true implications). The following definition introduces the notion of a model.

Definition 4. *For a fuzzy set T of fuzzy attribute implications, the set* $\mathrm{Mod}(T)$ *of all* models *of T is defined by*

$$\mathrm{Mod}(T) = \{M \in \mathbf{L}^Y \mid \text{for each } A, B \in \mathbf{L}^Y : T(A \Rightarrow B) \leq \|A \Rightarrow B\|_M\}.$$

That is, $M \in \mathrm{Mod}(T)$ means that for each attribute implication $A \Rightarrow B$, a degree to which $A \Rightarrow B$ holds in M is higher than or at least equal to a degree $T(A \Rightarrow B)$ prescribed by T. Particularly, for a crisp T, $\mathrm{Mod}(T) = \{M \in \mathbf{L}^Y \mid \text{for each } A \Rightarrow B \in T : \|A \Rightarrow B\|_M = 1\}$.

3.2 Relationship to Fuzzy Concept Lattices

Analogously as in the ordinary case, there is a close relationship between attribute implications and concept lattices. A useful structure derived from $\langle X, Y, I \rangle$ which is related to attribute implications is a so-called fuzzy concept lattice with hedges [10]. Let $*_X$ and $*_Y$ be hedges (their meaning will become apparent later). For **L**-sets $A \in \mathbf{L}^X$ (**L**-set of objects), $B \in \mathbf{L}^Y$ (**L**-set of attributes) we define **L**-sets $A^\uparrow \in \mathbf{L}^Y$ (**L**-set of attributes), $B^\downarrow \in \mathbf{L}^X$ (**L**-set of objects) by

$$A^\uparrow(y) = \bigwedge_{x \in X} \left(A(x)^{*_X} \rightarrow I(x, y) \right) \quad \text{and} \quad B^\downarrow(x) = \bigwedge_{y \in Y} \left(B(y)^{*_Y} \rightarrow I(x, y) \right).$$

We put $\mathcal{B}(X^{*_X}, Y^{*_Y}, I) = \{\langle A, B \rangle \in \mathbf{L}^X \times \mathbf{L}^Y \mid A^\uparrow = B, B^\downarrow = A\}$. For $\langle A_1, B_1 \rangle$, $\langle A_2, B_2 \rangle \in \mathcal{B}(X^{*_X}, Y^{*_Y}, I)$, put $\langle A_1, B_1 \rangle \leq \langle A_2, B_2 \rangle$ iff $A_1 \subseteq A_2$ (or, iff $B_2 \subseteq$

B_1; both ways are equivalent). Operators $\downarrow, \uparrow$ form a Galois connection with hedges [10]. $\langle \mathcal{B}(X^{*X}, Y^{*Y}, I), \leq \rangle$ is called a *fuzzy concept lattice (with hedges)* induced by $\langle X, Y, I \rangle$; $\langle A, B \rangle \in \mathcal{B}(X^{*X}, Y^{*Y}, I)$ are called formal concepts. For $*_Y = \mathrm{id}_L$ (identity), we write only $\mathcal{B}(X^{*X}, Y, I)$.

Remark 5. (1) Fuzzy concept lattices with hedges generalize some of the approaches to concept lattices from the point of view of a fuzzy approach, see [14] for details.

(2) Hedges can be seen as parameters which control the size of a fuzzy concept lattice (the stronger the hedges, the smaller $\mathcal{B}(X^{*X}, Y^{*Y}, I)$). See [10] for details.

(3) For $\mathbf{L} = \mathbf{2}$, a fuzzy concept lattice with hedges coincides with the ordinary concept lattice.

For each $\langle X, Y, I \rangle$ we consider a set $\mathrm{Int}(X^{*X}, Y^{*Y}, I) \subseteq \mathbf{L}^Y$ of all intents of concepts of $\mathcal{B}(X^{*X}, Y^{*Y}, I)$, i.e.

$$\mathrm{Int}(X^{*X}, Y^{*Y}, I) = \{B \in \mathbf{L}^Y \mid \langle A, B \rangle \in \mathcal{B}(X^{*X}, Y^{*Y}, I) \text{ for some } A \in \mathbf{L}^X\}.$$

For $*^X = *$ (the hedge used in (2)) and $*^Y = \mathrm{id}_L$ (identity on L), $\mathcal{B}(X^*, Y, I)$ and $\mathrm{Int}(X^*, Y, I)$ play analogous roles for fuzzy attribute implications to the roles of ordinary concept lattices and systems of intents for ordinary attribute implications.

We close this section by a theorem showing some formulas expressing a degree $\|A \Rightarrow B\|_M$ in terms of fuzzy concept lattices with hedges and the operators $\uparrow$ and $\downarrow$. For hedges $\bullet, * : L \to L$ put $\bullet \leq *$ iff $a^\bullet \leq a^*$ for each $a \in L$.

Theorem 1 ([13]). *For a data table $\langle X, Y, I \rangle$ with fuzzy attributes, hedges $\bullet$ and $*$ with $\bullet \leq *$, and an attribute implication $A \Rightarrow B$, the following values are equal:*

$$\|A \Rightarrow B\|_{\langle X, Y, I \rangle}, \quad \|A \Rightarrow B\|_{\mathrm{Int}(X^*, Y, I)}, \quad S(B, A^{\downarrow\uparrow}),$$
$$\bigwedge_{x \in X, a \in L} S(a^* \otimes A, \{{}^{1}\!/x\}^\uparrow)^\bullet \to S(a^* \otimes B, \{{}^{1}\!/x\}^\uparrow),$$
$$\bigwedge_{x \in X, a \in L} S(A, \{{}^{a}\!/x\}^\uparrow)^\bullet \to S(B, \{{}^{a}\!/x\}^\uparrow),$$
$$\bigwedge_{a \in L} \|a^* \otimes A \Rightarrow a^* \otimes B\|_{\langle X, Y, I \rangle},$$
$$\bigwedge_{M \in \mathrm{Int}(X^*, Y, I)} S(A, M)^\bullet \to S(B, M).$$

3.3   Complete Sets and Guigues-Duquenne Bases

We now turn our attention to the notions of semantic entailment, completeness in data tables, non-redundant basis, etc.

Definition 5. *A degree $\|A \Rightarrow B\|_T \in L$ to which $A \Rightarrow B$ semantically follows from a fuzzy set T of attribute implications is defined by*

$$\|A \Rightarrow B\|_T = \|A \Rightarrow B\|_{\mathrm{Mod}(T)}. \tag{5}$$

That is, $\|A \Rightarrow B\|_T$ can be seen as a degree to which $A \Rightarrow B$ is true in each model of T. From now on in this section, we will assume that T is an ordinary set of fuzzy attribute implications.

Definition 6. *A set T of attribute implications is called* complete (in $\langle X, Y, I \rangle$) *if $||A \Rightarrow B||_T = ||A \Rightarrow B||_{\langle X,Y,I \rangle}$ for each $A \Rightarrow B$. If T is complete and no proper subset of T is complete, then T is called a* non-redundant basis (of $\langle X, Y, I \rangle$).

Note that both the notions of a complete set and a non-redundant basis refer to a given data table with fuzzy attributes.

Since we are primarily interested in implications which are fully true in $\langle X, Y, I \rangle$, the following notion seems to be of interest. Call T *1-complete in* $\langle X, Y, I \rangle$ if we have that $||A \Rightarrow B||_T = 1$ iff $||A \Rightarrow B||_{\langle X,Y,I \rangle} = 1$ for each $A \Rightarrow B$. Clearly, if T is complete then it is also 1-complete. Surprisingly, we have also

Theorem 2 ([12]). *T is 1-complete in $\langle X, Y, I \rangle$ iff T is complete in $\langle X, Y, I \rangle$.*

The following assertion shows that the models of a complete set of fuzzy attribute implications are exactly the intents of the corresponding fuzzy concept lattice.

Theorem 3 ([7]). *T is complete iff $\mathrm{Mod}(T) = \mathrm{Int}(X^*, Y, I)$.*

In the following, we focus on so-called Guigues-Duquenne basis, i.e. a non-redundant basis based on the notion of a pseudointent, see [21, 22, 25]. As we will see, the situation is somewhat different from what we know from the ordinary case. We start by the notion of a system of pseudointents.

Definition 7. *Given $\langle X, Y, I \rangle$, $\mathcal{P} \subseteq \mathbf{L}^Y$ (a system of fuzzy sets of attributes) is called a* system of pseudo-intents *of $\langle X, Y, I \rangle$ if for each $P \in \mathbf{L}^Y$ we have:*

$$P \in \mathcal{P} \quad \textit{iff} \quad P \neq P^{\downarrow\uparrow} \textit{ and } ||Q \Rightarrow Q^{\downarrow\uparrow}||_P = 1 \textit{ for each } Q \in \mathcal{P} \textit{ with } Q \neq P.$$

It is easily seen that if $\mathbf{L}$ is a complete residuated lattice with globalization then $\mathcal{P}$ is a system of pseudo-intents of $\langle X, Y, I \rangle$ if for each $P \in \mathbf{L}^Y$ we have:

$$P \in \mathcal{P} \quad \textit{iff} \quad P \neq P^{\downarrow\uparrow} \textit{ and } Q^{\downarrow\uparrow} \subseteq P \textit{ for each } Q \in \mathcal{P} \textit{ with } Q \subset P.$$

In addition to that, in case of finite $\mathbf{L}$, for each data table with finite set of attributes there is exactly one system of pseudo-intents which can be described recursively the same way as in the classical case [22, 25]:

Theorem 4 ([11]). *Let $\mathbf{L}$ be a finite residuated lattice with globalization. Then for each $\langle X, Y, I \rangle$ with finite Y there is a unique system of pseudo-intents $\mathcal{P}$ of $\langle X, Y, I \rangle$ and*

$$\mathcal{P} = \{P \in \mathbf{L}^Y \mid P \neq P^{\downarrow\uparrow} \textit{ and } Q^{\downarrow\uparrow} \subseteq P \textit{ holds for each } Q \in \mathcal{P} \textit{ such that } Q \subset P\}.$$

Remark 6. (1) Neither the uniqueness of $\mathcal{P}$ nor the existence of $\mathcal{P}$ is assured in general, see [11].

(2) For $\mathbf{L} = \mathbf{2}$, the system of pseudointents described by Theorem 4 coincides with the ordinary one.

The following theorem shows that each system of pseudointents induces a non-redundant basis.

Theorem 5 ([11]). *Let $\mathcal{P}$ be a system of pseudointents of $\langle X, Y, I \rangle$. Then $T = \{P \Rightarrow P^{\downarrow\uparrow} \mid P \in \mathcal{P}\}$ is a non-redundant basis of $\langle X, Y, I \rangle$ (so-called Guigues-Duquenne basis).*

Non-redundancy of T does not ensure that T is minimal in terms of its size. The following theorem shows a generalization of a well-known result saying that Guigues-Duquenne basis is minimal in terms of its size.

Theorem 6 ([11]). *Let $\mathbf{L}$ be a finite residuated lattice with * being the globalization, Y be finite. Let T be a Guigues-Duquenne basis of $\langle X, Y, I \rangle$, i.e. $T = \{P \Rightarrow P^{\downarrow\uparrow} \mid P \in \mathcal{P}\}$ where $\mathcal{P}$ is the system of pseudointents of $\langle X, Y, I \rangle$. If T' is complete in $\langle X, Y, I \rangle$ then $|T| \leq |T'|$.*

Remark 7. For hedges other than globalization we can have several systems of pseudointents. The systems of pseudointents may have different numbers of elements, see [11].

3.4   Algorithms for Generating Systems of Pseudointents

CASE 1: Finite $\mathbf{L}$ and * being globalization. If $\mathbf{L}$ is finite and * is globalization, there is a unique system $\mathcal{P}$ of pseudointents for $\langle X, Y, I \rangle$, see Theorem 4. In what follows we describe an algorithm for computing this $\mathcal{P}$. The algorithm is based on the ideas of Ganter's algorithm for computing ordinary pseudointents, see [21, 22]. Details can be found in [7].

For simplicity, let us assume that $\mathbf{L}$ is, moreover, linearly ordered. For $Z \in \mathbf{L}^Y$ put

$$Z^{T^*} = Z \cup \bigcup \{B \otimes S(A, Z)^* \mid A \Rightarrow B \in T \text{ and } A \neq Z\},$$
$$Z^{T_0^*} = Z,$$
$$Z^{T_n^*} = (Z^{T_{n-1}^*})^{T^*}, \quad \text{for } n \geq 1,$$

and define an operator cl_{T^*} on $\mathbf{L}$-sets in Y by

$$cl_{T^*}(Z) = \bigcup_{n=0}^{\infty} Z^{T_n^*}.$$

Theorem 7 ([7]). *cl_{T^*} is a fuzzy closure operator, and*

$$\{cl_{T^*}(Z) \mid Z \in \mathbf{L}^Y\} = \mathcal{P} \cup \text{Int}(X^*, Y, I).$$

Using Theorem 7, we can get all intents and all pseudo-intents (of a given data table with fuzzy attributes) by computing the fixed points of cl_{T^*}. This can be done with polynomial time delay using a "fuzzy extension" of Ganter's algorithm for computing all fixed points of a closure operator, see [6]. We refer to [7] for details.

CASE 2: Finite $\mathbf{L}$ and arbitrary *. If $\mathbf{L}$ is finite and * is an arbitrary hedge (not necessarily globalization), the systems of pseudointents for $\langle X, Y, I \rangle$ can be computed using algorithms for generating maximal independent sets in graphs.

Namely, as we show in the following, systems of pseudointents in this case can be identified with particular maximal independent sets. (details can be found in [15]): For $\langle X, Y, I \rangle$ define a set V of fuzzy sets of attributes by

$$V = \{P \in \mathbf{L}^Y \mid P \neq P^{\downarrow\uparrow}\}. \tag{6}$$

If $V \neq \emptyset$, define a binary relation E on V by

$$E = \{\langle P, Q \rangle \in V \mid P \neq Q \text{ and } \|Q \Rightarrow Q^{\downarrow\uparrow}\|_P \neq 1\}. \tag{7}$$

In this case, $\mathbf{G} = \langle V, E \cup E^{-1} \rangle$ is a graph. For any $Q \in V$ and $\mathcal{P} \subseteq V$ define the following subsets of V: $\mathrm{Pred}(Q) = \{P \in V \mid \langle P, Q \rangle \in E\}$, and $\mathrm{Pred}(\mathcal{P}) = \bigcup_{Q \in \mathcal{P}} \mathrm{Pred}(Q)$.

Theorem 8 ([15]). *Let $\mathbf{L}$ be finite, * be any hedge, $\langle X, Y, I \rangle$ be a data table with fuzzy attributes, $\mathcal{P} \subseteq \mathbf{L}^Y$, V and E be defined by (6) and (7), respectively. Then the following statements are equivalent.*

(i) $\mathcal{P}$ *is a system of pseudo-intents;*
(ii) $V - \mathcal{P} = \mathrm{Pred}(\mathcal{P})$*;*
(iii) $\mathcal{P}$ *is a maximal independent set in $\mathbf{G}$ such that $V - \mathcal{P} = \mathrm{Pred}(\mathcal{P})$.*

The Theorem gives a way to compute systems of pseudo-intents. It suffices to find all maximal independent sets in $\mathbf{G}$ and check which of them satisfy additional condition $V - \mathcal{P} = \mathrm{Pred}(\mathcal{P})$.

4   Functional Dependencies over Domains with Similarity Relations

As we mentioned in Section 1, ordinary attribute implications have been used in databases under the name functional dependencies. Functional dependencies are interpreted in data tables with arbitrarily-valued attributes. A table entry corresponding to an object (row) x and an attribute (column) y contains an arbitrary value from a so-called domain D_y (set of all possible values for y). Then, $A \Rightarrow B$ is considered true in such a table if any two objects (rows) which agree in their values of attributes from A agree also in their values of attributes from B. In this section we consider functional dependencies from the point of view of a fuzzy approach. We show several relationships to fuzzy attribute implications. Most importantly, we argue that in a fuzzy setting, the concept of a functional dependence is an interesting one for the theory of databases.

Definition 8. *A (fuzzy) functional dependence (over attributes Y) is an expression $A \Rightarrow B$, where $A, B \in \mathbf{L}^Y$ (A and B are fuzzy sets of attributes).*

Therefore, the notion of a fuzzy functional dependence coincides with the notion of a fuzzy attribute implication. We prefer using both of the terms, depending on the context of usage. Fuzzy functional dependencies will be interpreted in data tables over domains with similarities.

Definition 9. *A data table over domains with similarity relations is a tuple* $\mathcal{D} = \langle X, Y, \{\langle D_y, \approx_y \rangle \mid y \in Y\}, T \rangle$ *where*

- *X is a non-empty set (of objects, table items),*
- *Y is a non-empty finite set (of attributes),*
- *for each* $y \in Y$, D_y *is a non-empty set (of values of attribute y) and* $\approx_y$ *is a binary fuzzy relation which is reflexive and symmetric,*
- *T is a mapping assigning to each* $x \in X$ *and* $y \in Y$ *a value* $T(x, y) \in D_y$ *(value of attribute y on object x).*

$\mathcal{D}$ will always denote some data table over domains with similarity relations with its components denoted as above.

Remark 8. (1) Consider $L = \{0, 1\}$ (case of classical logic). If each $\approx_y$ is an equality (i.e. $a \approx_y b = 1$ iff $a = b$), then $\mathcal{D}$ can be identified with what is called a relation on relation scheme Y with domains D_y ($y \in Y$) [29].

(2) For $x \in X$ and $Z \subseteq Y$, $x[Z]$ denotes a tuple of values $T(x, y)$ for $y \in Z$. We may assume that attribute from Y are numbered, i.e. $Y = \{y_1, \ldots, y_n\}$, and thus linearly ordered by this numbering, and assume that attributes in $x[Z]$ are ordered in this way. Particularly, $x[y]$ is $x[\{y\}]$ which can be identified with $T(x, y)$.

(3) $\mathcal{D}$ can be seen as a table with rows and columns corresponding to $x \in X$ and $y \in Y$, respectively, and with table entries containing values $T(x, y) \in D_y$. Moreover, each domain D_y is equipped with an additional information about similarity of elements from D_y.

Given a data table $\mathcal{D} = \langle X, Y, \{\langle D_y, \approx_y \rangle \mid y \in Y\}, T \rangle$, we want to introduce a condition for a functional dependence $A \Rightarrow B$ to be true in $\mathcal{D}$ which says basically the following: "for any two objects $x_1, x_2 \in X$: if x_1 and x_2 have similar values on attributes from A then x_1 and x_2 have similar values on attributes from A". Define first for a given $\mathcal{D}$, objects $x_1, x_2 \in X$, and a fuzzy set $C \in \mathbf{L}^Y$ of attributes a *degree* $x_1(C) \approx x_2(C)$ *to which* x_1 *and* x_2 *have similar values on attributes from* C (agree on attributes from C) by

$$x_1(C) \approx x_2(C) = \bigwedge_{y \in Y} (C(y) \rightarrow (x_1[y] \approx_y x_2[y])). \tag{8}$$

That is, $x_1(C) \approx x_2(C)$ is truth degree of proposition "for each attribute $y \in Y$: if y belongs to C then the value $x_1[y]$ of x_1 on y is similar to the value $x_2[y]$ of x_2 on y", which can be seen as a degree to which x_1 and x_2 have similar values on attributes from C. Then, the above idea of validity of a functional dependence is then captured by the following definition.

Definition 10. *A degree* $\|A \Rightarrow B\|_{\mathcal{D}}$ *to which* $A \Rightarrow B$ *is true in* $\mathcal{D}$ *is defined by*

$$\|A \Rightarrow B\|_{\mathcal{D}} = \bigwedge_{x_1, x_2 \in X} ((x_1(A) \approx x_2(A))^* \rightarrow (x_1(B) \approx x_2(B))). \tag{9}$$

Remark 9. (1) If A and B are crisp sets then A and B may be considered as ordinary sets and $A \Rightarrow B$ may be seen as an ordinary functional dependence. Then, if $\approx_y$ is an ordinary equality for each $y \in Y$, we have that $||A \Rightarrow B||_{\mathcal{D}} = 1$ iff $A \Rightarrow B$ is true in $\mathcal{D}$ in the usual sense of validity of ordinary functional dependencies.

(2) For a functional dependence $A \Rightarrow B$, degrees $A(y) \in L$ and $B(y) \in L$ can be seen as thresholds. Namely, if $*$ is globalization, $||A \Rightarrow B||_{\mathcal{D}} = 1$ means that a proposition "for any objects $x_1, x_2 \in X$: if for each attribute $y \in Y$, $A(y) \leq (x_1[y] \approx_y x_2[y])$, then for each attribute $y' \in Y$, $B(y') \leq (x_1[y'] \approx_y x_2[y'])$" is true. That is, having A and B fuzzy sets allows for a rich expressibility, cf. Remark 4.

We now have two ways to interpret a fuzzy attribute implication (fuzzy functional dependence) $A \Rightarrow B$. First, given a data table $\mathcal{T} = \langle X, Y, I \rangle$ with fuzzy attributes, we can consider a degree $||A \Rightarrow B||_{\mathcal{T}}$ to which $A \Rightarrow B$ is true in $\mathcal{T}$, see (4). Second, given a data table $\mathcal{D}$ over domains with similarities, we can consider a degree $||A \Rightarrow B||_{\mathcal{D}}$ to which $A \Rightarrow B$ is true in $\mathcal{D}$, see (9). In the rest of this section, we focus on presenting the following relationship between the two kinds of semantics for our formulas $A \Rightarrow B$: The notion of semantic entailment based on data tables with fuzzy attributes coincides with the notion of semantic entailment based on data tables over domains with similarity relations.

As in case of fuzzy attribute implications, we introduce the notions of a model and semantic entailment for functional dependencies. For a fuzzy set T of fuzzy functional dependencies, the set $\mathrm{Mod}^{\mathrm{FD}}(T)$ of all *models* of T is defined by

$$\mathrm{Mod}^{\mathrm{FD}}(T) = \{ \mathcal{D} \,|\, \text{for each } A, B \in \mathbf{L}^Y : T(A \Rightarrow B) \leq ||A \Rightarrow B||_{\mathcal{D}} \},$$

where $\mathcal{D}$ stands for an arbitrary data table over domains with similarities. A degree $||A \Rightarrow B||_T^{\mathrm{FD}} \in L$ to which $A \Rightarrow B$ *semantically follows* from a fuzzy set T of functional dependencies is defined by

$$||A \Rightarrow B||_T^{\mathrm{FD}} = \bigwedge_{\mathcal{D} \in \mathrm{Mod}^{\mathrm{FD}}(T)} ||A \Rightarrow B||_{\mathcal{D}}.$$

Denoting now $||A \Rightarrow B||_T$, see (5), by $||A \Rightarrow B||_T^{\mathrm{AI}}$, one can prove the following theorem.

Theorem 9 ([17]). *For any fuzzy set T of fuzzy attribute implications and any fuzzy attribute implication $A \Rightarrow B$ we have*

$$||A \Rightarrow B||_T^{\mathrm{FD}} = ||A \Rightarrow B||_T^{\mathrm{AI}}. \tag{10}$$

5 Armstrong Rules and Provability

In this section we present a system of Armstrong-like rules for reasoning with fuzzy attribute implications. Throughout this section we assume that L is finite. We show that the system is complete w.r.t. the semantics of fuzzy attribute implications based on data tables with fuzzy attributes. Due to Theorem 9, this is equivalent to completeness w.r.t. the semantics based on data tables over domains with similarities. In fact, we show two kinds of completeness. The first one is a usual one and concerns provability and entailment of $A \Rightarrow B$ from a set T of attribute implications. Provability and entailment remain bivalent: $A \Rightarrow B$ is provable from T iff $A \Rightarrow B$ semantically follows from T in degree 1. The second one (called also graded completeness or Pavelka-style completeness) concerns provability and entailment of $A \Rightarrow B$ from a fuzzy set T of attribute implications. Provability and entailment themselves become graded: A degree to which $A \Rightarrow B$ is provable from T equals a degree to which $A \Rightarrow B$ semantically follows from T. Details can be found in [12, 16].

Our axiomatic system consists of the following *deduction rules*.

(Ax) infer $A \cup B \Rightarrow A$,

(Cut) from $A \Rightarrow B$ and $B \cup C \Rightarrow D$ infer $A \cup C \Rightarrow D$,

(Mul) from $A \Rightarrow B$ infer $c^* \otimes A \Rightarrow c^* \otimes B$

for each $A, B, C, D \in \mathbf{L}^Y$, and $c \in L$. Rules (Ax)–(Mul) are to be understood as follows: having functional dependencies which are of the form of functional dependencies in the input part (the part preceding "infer") of a rule, a rule allows us to infer (in one step) the corresponding functional dependence in the output part (the part following "infer") of a rule.

Completeness. A fuzzy attribute implication $A \Rightarrow B$ is called *provable* from a set T of fuzzy attribute implications using (Ax)–(Mul), written $T \vdash A \Rightarrow B$, if there is a sequence $\varphi_1, \ldots, \varphi_n$ of fuzzy attribute implications such that φ_n is $A \Rightarrow B$ and for each φ_i we either have $\varphi_i \in T$ or φ_i is inferred (in one step) from some of the preceding formulas (i.e., $\varphi_1, \ldots, \varphi_{i-1}$) using some of deduction rules (Ax)–(Mul). To comply to the notation $T \vdash A \Rightarrow B$, we write $T \models A \Rightarrow B$ to denote that $\|A \Rightarrow B\|_T = 1$ ($A \Rightarrow B$ semantically follows from T in degree 1). Then we have the first kind of completeness:

Theorem 10 ([16]). *For any set T of fuzzy attribute implications and any fuzzy attribute implication $A \Rightarrow B$ we have*

$$T \vdash A \Rightarrow B \quad \text{iff} \quad T \models A \Rightarrow B.$$

Graded completeness. Now, we are going to define a notion of a degree $|A \Rightarrow B|_T$ of provability of a functional dependence of a fuzzy set T of functional dependencies. Then, we show that $|A \Rightarrow B|_T = \|A \Rightarrow B\|_T^{FD}$ which can be understood as a graded completeness (completeness in degrees). Note that graded completeness was introduced by Pavelka [30], see also [23, 26] for further information.

For a fuzzy set T of fuzzy attribute implications and for $A \Rightarrow B$ we define a *degree* $|A \Rightarrow B|_T \in L$ to which $A \Rightarrow B$ is provable from T by

$$|A \Rightarrow B|_T = \bigvee \{c \in L \mid c(T) \vdash A \Rightarrow c \otimes B\}, \tag{11}$$

where $c(T)$ is an ordinary set of fuzzy attribute implications defined by

$$c(T) = \{A \Rightarrow T(A \Rightarrow B) \otimes B \mid A, B \in \mathbf{L}^Y \text{ and } T(A \Rightarrow B) \otimes B \neq \emptyset\}. \tag{12}$$

Then we have the second kind of completeness:

Theorem 11 ([16]). *For any fuzzy set T of fuzzy attribute implications and any fuzzy attribute implication $A \Rightarrow B$ we have*

$$|A \Rightarrow B|_T = ||A \Rightarrow B||_T.$$

6   Concluding Remarks

6.1   Bibliographic Remarks

The first study on fuzzy attribute implications is S. Pollandt's [31]. Pollandt uses the same notion of a fuzzy attribute implication, i.e. $A \Rightarrow B$ where A, B are fuzzy sets, and obtains several results. Pollandt's notion of validity is a special case of ours, namely the one for $*$ being identity on L. On the other hand, her notion of a pseudointent corresponds to $*$ being globalization. That is why Pollandt did not get a proper generalization of results leading to Guigues-Duquenne basis. Pollandt's [31] contains some other results (proper premises, implications in fuzzy-valued contexts) which we did not discuss here. We will comment more on Pollandt's results elsewhere.

[19, 33, 35] are papers dealing with fuzzy functional dependencies. Our approach presented in this paper is more general. Namely, [33, 35] consider formulas $A \Rightarrow B$ with A and B being ordinary sets, i.e. A and B are not suitable for expressing thresholds. In [19], thresholds in A and B are present but are the same in A and the same in B. Furthermore, the degrees are restricted to values from $[0, 1]$ in [19, 33, 35].

Our paper is based on [4]–[17].

6.2   Further Issues

Due to a limited scope of this paper, we did not cover several interesting topics, some of which are still under investigation. For instance, it is shown in [9, 13] that a data table $\mathcal{T}$ with fuzzy attributes can be transformed to a data table $\mathcal{T}'$ with binary attributes in such a way that fuzzy attribute implications true in degree 1 in $\mathcal{T}$ correspond in a certain way to ordinary attribute implications which are true in $\mathcal{T}'$. The transformation of data tables and attribute implications makes it possible to obtain an ordinary non-redundant basis T' for $\mathcal{T}'$ and to obtain a corresponding set T of fuzzy attribute implications from T'. However, while T is always complete for $\mathcal{T}$, it may be redundant. Note that some results on transformations of data tables with fuzzy attributes to tables with binary attributes which are related to attribute implications are also present in [31].

Interesting open problems include: further study of relationships between attribute implications in a fuzzy setting and ordinary attribute implications (from both ordinary formal contexts and many-valued contexts); study of further problems of attribute implications in a fuzzy setting; further study of functional dependencies and other kinds of dependencies in databases in a fuzzy setting; development of agenda for databases where domains are equipped with similarity relations.

References

1. Abiteboui S. *et al.*: The Lowell database research self-assessment. *Communications of ACM* **48**(5)(2005), 111–118.
2. Armstrong W. W.: Dependency structures in data base relationships. *IFIP Congress*, Geneva, Switzerland, 1974, pp. 580–583.
3. Bělohlávek R.: Similarity relations in concept lattices. *J. Logic Comput.* 10(6):823–845, 2000.
4. Bělohlávek R.: *Fuzzy Relational Systems: Foundations and Principles.* Kluwer, Academic/Plenum Publishers, New York, 2002.
5. Bělohlávek R.: Concept lattices and order in fuzzy logic. *Ann. Pure Appl. Logic* **128**(2004), 277–298.
6. Bělohlávek R.: Algorithms for fuzzy concept lattices. *Proc. Fourth Int. Conf. on Recent Advances in Soft Computing.* Nottingham, United Kingdom, 12–13 December, 2002, pp. 200–205.
7. Bělohlávek R., Chlupová M., Vychodil V.: Implications from data with fuzzy attributes. AISTA 2004 in Cooperation with the IEEE Computer Society Proceedings, 2004, 5 pages, ISBN 2–9599776–8–8.
8. Bělohlávek R., Funioková T., Vychodil V.: Fuzzy closure operators with truth stressers. *Logic Journal of IGPL* (to appear).
9. Bělohlávek R., Vychodil V.: Implications from data with fuzzy attributes vs. scaled binary attributes. In: FUZZ-IEEE 2005, The IEEE International Conference on Fuzzy Systems, May 22–25, 2005, Reno (Nevada, USA), pp. 1050–1055 (proceedings on CD), abstract in printed proceedings, p. 53, ISBN 0–7803–9158–6.
10. Bělohlávek R., Vychodil V.: Reducing the size of fuzzy concept lattices by hedges. In: FUZZ-IEEE 2005, The IEEE International Conference on Fuzzy Systems, May 22–25, 2005, Reno (Nevada, USA), pp. 663–668 (proceedings on CD), abstract in printed proceedings, p. 44, ISBN 0–7803–9158–6.
11. Bělohlávek R., Vychodil V.: Fuzzy attribute logic: attribute implications, their validity, entailment, and non-redundant basis. In: Liu Y., Chen G., Ying M. (Eds.): *Fuzzy Logic, Soft Computing & Computational Intelligence: Eleventh International Fuzzy Systems Association World Congress* (Vol. I), 2005, pp. 622–627. Tsinghua University Press and Springer, ISBN 7–302–11377–7.
12. Bělohlávek R., Vychodil V.: Fuzzy attribute logic: syntactic entailment and completeness. In: JCIS 2005, 8th Joint Conference on Information Sciences, July 21–26, Salt Lake City, Utah, USA, pp. 78–81.
13. Bělohlávek R., Vychodil V.: Reducing attribute implications from data tables with fuzzy attributes to tables with binary attributes. In: JCIS 2005, 8th Joint Conference on Information Sciences, July 21–26, Salt Lake City, Utah, USA, pp. 82–85.
14. Bělohlávek R., Vychodil V.: What is a fuzzy concept lattice? Proc. CLA 2005, September 7–9, 2005, Olomouc, Czech Republic, pp. 34–45.
15. Bělohlávek R., Vychodil V.: Fuzzy attribute implications: computing non-redundant bases using maximal independent sets. In: S. Zhang and R. Jarvis (Eds.): AI 2005, *LNAI* **3809**, pp. 1126–1129, Springer-Verlag, Berlin/Heidelberg, 2005
16. Bělohlávek R., Vychodil V.: Axiomatizations of fuzzy attribute logic. IICAI 2005, Pune, India (to appear).
17. Bělohlávek R., Vychodil V.: Functional dependencies of data tables over domains with similarity relations. IICAI 2005, Pune, India (to appear).
18. Buckles B. P., Petry F. E.: Fuzzy databases in the new era. Proceedings of the 1995 ACM symposium on Applied computing, pp. 497–502, Nashville, Tennessee, ISBN 0-89791-658-1, 1995.

19. Cubero J. C., Vila M. A.: A new definition of fuzzy functional dependency in fuzzy relational datatabses. *Int. J. Intelligent Systems* **9**(5)(1994), 441–448.
20. Ganter B.: *Begriffe und Implikationen*, manuscript, 1998.
21. Ganter B.: Algorithmen zur formalen Begriffsanalyse. In: Ganter B., Wille R., Wolff K. E. (Hrsg.): *Beiträge zur Begriffsanalyse*. B. I. Wissenschaftsverlag, Mannheim, 1987, 241–254.
22. Ganter B., Wille R.:   *Formal Concept Analysis. Mathematical Foundations*. Springer, Berlin, 1999.
23. Gerla G.: *Fuzzy Logic. Mathematical Tools for Approximate Reasoning*. Kluwer, Dordrecht, 2001.
24. Goguen J. A.: The logic of inexact concepts. *Synthese* **18**(1968-9), 325–373.
25. Guigues J.-L., Duquenne V.: Familles minimales d'implications informatives resultant d'un tableau de données binaires. *Math. Sci. Humaines* **95**(1986), 5–18.
26. Hájek P.: *Metamathematics of Fuzzy Logic*. Kluwer, Dordrecht, 1998.
27. Hájek P.: On very true. *Fuzzy Sets and Systems* **124**(2001), 329–333.
28. Klir G. J., Yuan B.: *Fuzzy Sets and Fuzzy Logic. Theory and Applications*. Prentice Hall, 1995.
29. Maier D.: *The Theory of Relational Databases*. Computer Science Press, Rockville, 1983.
30. Pavelka J.: On fuzzy logic I, II, III. *Z. Math. Logik Grundlagen Math.* **25**(1979), 45–52, 119–134, 447–464.
31. Pollandt S.: *Fuzzy Begriffe*. Springer-Verlag, Berlin/Heidelberg, 1997.
32. Prade H., Testemale C.: Generalizing database relational algebra for the treatment of incomplete or uncertain information and vague queries. *Information Sciences* **34**(1984), 115–143.
33. Raju K. V. S. V. N., Majumdar A. K.: Fuzzy functional dependencies and lossless join decomposition of fuzzy relational database systems. *ACM Trans. Database Systems* Vol. 13, No. 2, 1988, pp. 129–166.
34. Takeuti G., Titani S.: Globalization of intuitionistic set theory. *Annals of Pure and Applied Logic* **33**(1987), 195–211.
35. Tyagi B. K., Sharfuddin A., Dutta R. N., Tayal D. K.: A complete axiomatization of fuzzy functional dependencies using fuzzy function. *Fuzzy Sets and Systems* **151**(2)(2005), 363–379.

The Assessment of Knowledge,
in Theory and in Practice*

Jean-Claude Falmagne[1,**], Eric Cosyn[2],
Jean-Paul Doignon[3], and Nicolas Thiéry[2]

[1] Dept. of Cognitive Sciences, University of California, Irvine, CA 92697
jcf@aris.ss.uci.edu
[2] ALEKS Corporation
{ecosyn, nthiery}@aleks.com
[3] Free University of Brussels
doignon@ulb.ac.be

Abstract. This paper is adapted from a book and many scholarly arti-
cles. It reviews the main ideas of a theory for the assessment of a student's
knowledge in a topic and gives details on a practical implementation in the
form of a software. The basic concept of the theory is the 'knowledge state,'
which is the complete set of problems that an individual is capable of solving in a particular topic, such as Arithmetic or Elementary Algebra. The
task of the assessor—which is always a computer—consists in uncovering
the particular state of the student being assessed, among all the feasible
states. Even though the number of knowledge states for a topic may exceed several hundred thousand, these large numbers are well within the
capacity of current home or school computers. The result of an assessment
consists in two short lists of problems which may be labelled: 'WHAT THE
STUDENT CAN DO' and 'WHAT THE STUDENT IS READY TO LEARN.' In the
most important applications of the theory, these two lists specify the exact
knowledge state of the individual being assessed. Moreover, the family of
feasible states is specified by two combinatorial axioms which are pedagog-
ically sound from the standpoint of learning. The resulting mathematical
structure is related to closure spaces and thus also to concept lattices. This
work is presented against the contrasting background of common methods
of assessing human competence through standardized tests providing numerical scores. The philosophy of these methods, and their scientific origin
in nineteenth century physics, are briefly examined.

The assessment of human competence, as it is still performed today by many
specialists in the schools and in the workplace, is almost systematically based
on the numerical evaluation of some 'aptitude.' Its philosophy owes much to
nineteenth century physics, whose methods were regarded as exemplary. The
success of classical physics was certainly grounded in its use of a number of
fundamental numerical scales, such as mass, time, or length, to describe basic

---

* We wish to thank Chris Doble, Dina Falmagne, and Lin Nutile for their reactions to
earlier drafts of this article.
** Corresponding author. Phone: (949) 824 4880; Fax: (949) 824 1670; e-mail:
jcf@uci.edu.

R. Missaoui and J. Schmid (Eds.): ICFCA 2006, LNAI 3874, pp. 61–79, 2006.
© Springer-Verlag Berlin Heidelberg 2006

aspects of objects or phenomena. In time, 'measurement' came to represent the *sine qua non* for precision and the essence of the scientific method, and physics the model for other sciences to imitate. In other words, for an academic endeavor to be called a 'science,' it had to resemble physics in critical ways. In particular, its basic observations had to be quantified in terms of measurement scales in the exact sense of classical physics.

Prominent advocates of this view were Francis Galton, Karl Pearson and William Thomson Kelvin. Because that position is still influential today, with a detrimental effect on fields such as 'psychological measurement,' which is relevant to our subject, it is worth quoting some opinions in detail. In Pearson's biography of Galton ([Pearson, 1924, Vol. II, p. 345]), we find the following definition:

> "**Anthropometry**, *or the art of measuring the physical and mental faculties of human beings, enables a shorthand description of any individual by measuring a small sample of his dimensions and qualities. This will sufficiently define his bodily proportions, his massiveness, strength, agility, keenness of senses, energy, health, intellectual capacity and mental character, and will constitute concise and exact* **numerical**[1] *values for verbose and disputable estimates*[2]*."*

For scientists of that era, it was hard to imagine a non-numerical approach to precise study of an empirical phenomenon. Karl Pearson himself, for instance— commenting on a piece critical of Galton's methods by the editor of the *Spectator*[3]—, wrote

> "*There might be difficulty in ranking Gladstone and Disraeli for 'candour,' but few would question John Morley's position relative to both of them in this quality. It would require an intellect their equal to rank truly in scholarship Henry Bradshaw, Robertson Smith and Lord Acton, but most judges would place all three above Sir John Seeley, as they would place Seeley above Oscar Browning. After all, there are such things as brackets, which only makes the statistical theory of ranking slightly less simple in the handling.*" ([Pearson, 1924, Vol. II, p. 345].)

In other words, measuring a psychical attribute such as 'candor' only requires fudging a little around the edges of the order relation of the real numbers[4].

---

[1] Our emphasis.

[2] This excerpt is from an address "Anthropometry at Schools" given in 1905 by Galton at the London Congress of the Royal Institute for Preventive Medicine. The text was published in the *Journal for Preventive Medicine*, Vol. XIV, p. 93-98, London, 1906.

[3] The *Spectator*, May 23, 1874. The editor was taking Galton to task for his method of ranking applied to psychical character. He used 'candour' and 'power of repartee' as examples.

[4] Making such a relation a 'weak order' or perhaps a 'semiorder' (in the current terminology of combinatorics). A binary relation $\precsim$ on a finite or countable set S is a *weak order* if there is a real valued function f defined on S such that $x \precsim y \Leftrightarrow f(x) \leq f(y)$ for all objects x and y in the set S. The relation $\precsim$ is a *semiorder* if the representation has the form: $x \precsim y \Leftrightarrow f(x) + 1 \leq f(y)$. For these concepts, see e.g. Roberts [1979] or Trotter [1992].

The point here is that real numbers are still used to represent 'quantity of attribute.'

As for Kelvin, his position on the subject is well known, and often represented in the form: "*If you cannot measure it, then it is not science.*" The full quotation is:

> "*When you can measure what you are speaking about, and express it in numbers, you know something about it; but when you cannot measure it, when you cannot express it in numbers, your knowledge is of a meager and unsatisfactory kind: it may be the beginning of knowledge, but you are scarcely, in your thoughts, advanced to the stage of* **science**, *whatever the matter may be.*" (Kelvin [1889].)

Such a position, which equates precision with the use of numbers, was not on the whole beneficial to the development of mature sciences outside of physics. It certainly had a costly impact on the assessment of mental traits. For instance, for the sake of scientific precision, the assessment of mathematical knowledge was superseded in the U.S. by the measurement of mathematical aptitude using instruments directly inspired from Galton via Alfred Binet in France. They are still used today in such forms as the S.A.T.[5], the G.R.E. (*Graduate Record Examination*), and other similar tests. The ubiquitous I.Q. test is of course part of the list. In the minds of those nineteenth century scientists and their followers, the numerical measurement of mental traits was to be a prelude to the establishment of sound, predictive scientific theories in the spirit of those used so successfully in classical physics. The planned constructions, however, never went much beyond the measurement stage[6].

The limitations of a purely numerical description of some phenomena can be illustrated by an analogy with sports. It is true that the success of an athlete in a particular sport is often described by a set of impressive numbers. So, imagine that some committee of experts has carefully designed an 'Athletic Quotient' or 'A.Q.' test, intended to measure athletic prowess. Suppose that three exceptional athletes have taken the test, say Michael Jordan, Tiger Woods and Pete Sampras. Conceivably, all three of them would get outstanding A.Q.'s. But these high scores equating them would completely misrepresent how essentially different from each other they are. One may be tempted to salvage the numerical representation and argue that the asssessment, in this case, should be multidimensional. However, adding a few numerical dimensions capable of differentiating Jordan, Woods and Sampras would only be the first step in a sequence. Including Greg Louganis or Pele to the evaluated lot would require more dimensions, and there is no satisfactory end in sight. Besides, assuming that one would

---

[5] Note that the meaning of the acronym S.A.T. has recently been changed by Education Testing Service from '*Scholastic Aptitude Test*' to '*Scholastic Assessment Test*,' suggesting that a different philosophy on the part of the test makers may be under development.

[6] Sophisticated theories can certainly be found in some areas of the behavioral sciences, for example, but they do not usually rely on measurement scales intrinsic to these sciences. One prominent exception in economics is the *money* scale.

settle for a representation in n dimensions, for some small n equal 3, 4 or 5 say, the numerical vectors representing these athletes would be poor, misleading expressions of the exquisite combination of skills making each of them a champion in his own specialty. Evidently, the same shortcomings of a numerical description also apply in mathematics education. Numerical test results may be appropriate to decide who is winning a race. As an evaluative prelude to college, intended to assess the students' readiness for further learning, they are very imprecise indeed. The conclusion should be that a different descriptive language is needed.

More generally, in many scientific areas, from chemistry to biology and especially the behavioral sciences, theories must often be built on a very different footing than that of classical physics. Evidently, the standard physical scales such as length, time, mass or energy, must be used in measuring aspects of phenomena. But the substrate proper to these other sciences may very well be, in most cases, of a fundamentally different nature.

Of course, we are enjoying the benefits of hindsight. In all fairness, there were important mitigating circumstances affecting those who upheld the cause of numerical measurement as a prerequisite to science. For one thing, the appropriate mathematical tools were not yet available to support different conceptions. Combinatorics, for example, was yet to be born as a mathematical topic. More importantly, the 'Analytical Engine' of Charles Babbage was still a dream, and close to another century had to pass before the appearance of computing machines capable of handling the symbolic manipulations that would be required for another approach.

The theory reviewed here[7] represents a sharp departure from other approaches to the assessment of knowledge. Its mathematics is in the spirit of current research in combinatorics. No attempt is made to obtain a numerical representation. We start from the concept of a possibly large but essentially discrete set of 'units of knowledge.' In the case of Elementary Algebra, for instance, one such unit might be a particular type of algebra problem. The full domain for High School Algebra may contain a few hundred such problems. Our two key concepts are the 'knowledge state,' a particular set of problems that some individual is capable of solving correctly, and the 'knowledge structure,' which is a distinguished collection of knowledge states. For High School Algebra, we shall see that a useful knowledge structure may contain several hundred thousand feasible knowledge states. Thus, precision is achieved by the intricacy of the representing structure.

Knowledge Structures: Main Concepts

The precedence relation. A natural starting point for an assessment theory stems from the observation that some pieces of knowledge normally precede, in time, other pieces of knowledge. In our context, some algebra problem may be solvable by a student only if some other problems have already been mastered by that student. This may be because some prerequisites are required to master a

---

[7] To lighten the presentation, we gather the references in the last paragraph of this paper.

problem, but may also be due to historical or other circumstances. For example, in a given environment, some concepts are always taught in a particular order, even though there may be no logical or pedagogical reason to do so. Whatever its genesis may be, this precedence relation may be used to design an efficient assessment mechanism.

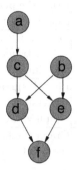

Fig. 1. Precedence diagram for the six types of algebra problems illustrated in Table 1

A simple example of a precedence relation between problems is illustrated by Figure1, which displays a plausible *precedence diagram* pertaining to the six types of algebra problems illustrated in Table 1. Note in passing that we distinguish between a **type** of problem and an **instance** of that type. Thus, a type of problem is an abstract formulation subsuming a possibly large class of instances. For the rest of this article, 'problem' is almost always intended to mean 'problem type.' The exceptions will be apparent from the context.

The precedence relation between problems is symbolized by the downward arrows. For example, Problem (e) is preceded by Problems (b), (c) and (a). In other words, the mastery of Problem (e) implies that of (b), (c) and (a). In the case of these six problems, the precedence relation proposed by the diagram of Figure 1 is a credible one. For example, observing a correct response to an instance of Problem (f), makes it highly plausible that the student has also mastered the other five problems. This precedence relation is part of a much bigger one, representing a substantial coverage of Beginning Algebra, starting with the solution of simple linear equations and ending with problem types such as (f) in Table 1. An example of such a larger precedence relation is represented by the diagram of Figure 2, which displays 88 vertices, for the 88 problems used, in principle, for the assessment. (For the sake of clarity of our diagrams, we are limiting the size of our examples. The full Beginning Algebra curriculum is larger, containing around 150 problems.) This larger precedence diagram is itself part of a still larger one, displayed in Figure 3, and comprising Arithmetic, Beginning Algebra, Intermediate Algebra, and Pre-Calculus.

For concreteness, we consider a particular situation in which the assessment is computer driven and the problems are presented on a monitor, via the Internet. All the virtual tools needed for providing the answers to the test—pencil, ruler,

Table 1. Six types of problems in Elementary Algebra

Name of problem type	Example of instance
(a) *Word problem on proportions (Type 1)*	A car travels on the freeway at an average speed of 52 miles per hour. How many miles does it travel in 5 hours and 30 minutes?
(b) *Plotting a point in the coordinate plane*	Using the pencil, mark the point at the coordinates $(1, 3)$.
(c) *Multiplication of monomials*	Perform the following multiplication: $$4x^4y^4 \cdot 2x \cdot 5y^2$$ and simplify your answer as much as possible.
(d) *Greatest common factor of two monomials*	Find the greatest common factor of the expressions $14t^6y$ and $4tu^5y^8$. Simplify your answer as much as possible.
(e) *Graphing the line through a given point with a given slope*	Graph the line with slope -7 passing through the point $(-3, -2)$.
(f) *Writing the equation of the line through a given point and perpendicular to a given line*	Write an equation for the line that passes through the point $(-5, 3)$ and is perpendicular to the line $8x + 5y = 11$.

graphical displays, calculators of various kinds when deemed necessary—, are part of the interface. In the course of a tutorial, the testees have been familiarized with these tools. In Problems (b) and (e), a coordinate plane is displayed on the computer monitor as part of the question, and the pencil and, for Problem (e), also the ruler, are provided. In this problem, the student must graph the line using the virtual pencil and ruler. We also suppose that all the problems have open responses (i.e. no multiple choice), and that 'lucky guesses' are unlikely. (Careless errors are always possible, of course, and a clever assessment procedure has to guard against them.)

We postpone for the moment the discussion of how to construct a valid precedence diagram for a realistically large problem set. (For example, how were the precedence diagrams of Figs. 2 or 3 obtained?) This question and other critical ones are considered later on in this article. For the time being, we focus on the miniature example of Table 1 which we use to introduce and illustrate the basic ideas.

The knowledge states. The precedence diagram of Figure 1 completely specifies the feasible knowledge states. The respondent can certainly have mastered just Problem **a**: having mastered **a** does not imply knowing anything else. But if he or she knows **e**, for example, then **a**, **b** and **c** must also have been mastered, forming a knowledge state which we represent as the set of problems $\{a, b, c, e\}$ or more compactly **abce**. Analyzing carefully the precedence diagram of Figure 1, we see that there are exactly 10 knowledge states consistent with it, forming the set

$$\mathcal{K} = \{\varnothing, a, b, ab, ac, abc, abcd, abce, abcde, abcdef\},$$

where ∅ symbolizes the empty state: the respondent is unable to solve any of the 6 problems. The set $\mathcal{K}$ is our basic concept, and is called the *knowledge struc-ture*. Note that a useful knowledge structure is not necessarily representable by a precedence diagram such as those of Figs. 1, 2 or 3 and may simply be specified by the collection of knowledge states.

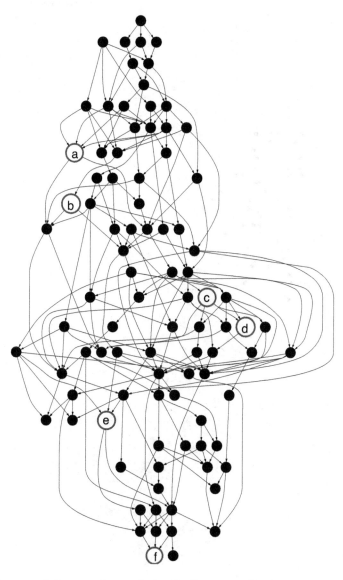

Fig. 2. Diagram of the precedence relation for Beginning Algebra. The vertices marked a-f refer to Problems (a)-(f) of Figure 1, whose diagram may be inferred from the one above.

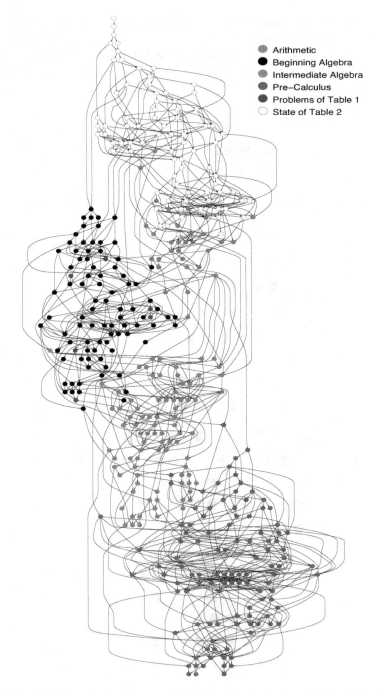

Fig. 3. Combined precedence diagram for Arithmetic, Beginning Algebra, Intermediate Algebra, and Pre-Calculus. Each of the 397 points represents a problem type. We recall that these 397 problem-types capture only a representative part of the standard curriculum. The full curriculum in these topics would contain around 650 problem-types.

The learning paths. This knowledge structure allows several *learning paths.* Starting from the naive state ∅, the full mastery of state **abcdef** can be achieved by mastering first **a**, and then successively the other problems in the order **b** ↦ **c** ↦ **d** ↦ **e** ↦ **f**. But there are other possible ways to learn. All in all, there are 6 possible learning paths consistent with the knowledge structure $\mathcal{K}$, which are displayed in Figure 4.

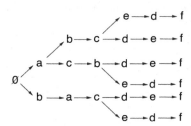

Fig. 4. The 6 possible learning paths consistent with the precedence diagram of Figure 1

In realistic knowledge structures such as those for Arithmetic or Elementary Algebra, the numbers of feasible knowledge states and of learning paths become very large. In the case of that part of Beginning Algebra whose precedence diagram was given in Figure 2, there are around 60,000 knowledge states and literally billions of feasible learning paths. These numbers may be puzzling. Where is the diversity coming from? After all, these mathematical subjects are highly structured and typically taught in roughly the same sequence. However, even though the school curriculum may be more or less standard in a particular region of the world, learning the material, and also forgetting it, follow their own haphazard course. Besides, 60,000 states form but a minute fraction of the 2^{88} possible subsets of the set of 88 problems. In any event, even in a highly structured mathematical topic, an accurate assessment of knowledge involves sorting out a considerable array of possibilities.

The outer and inner fringes of a knowlegde state. As suggested by the precedence diagrams and by the learning paths of Figure 4, the knowledge structures considered here have the property that learning can take place step by step, one problem type at a time. More precisely, each knowledge state (except the top one) has at least one *immediate successor* state, that is, a state containing all the same problems, plus exactly one. The knowledge state **abc** of $\mathcal{K}$, for instance, has the two states **abcd** and **abce** as immediate successors. Problems **d** and **e** form the 'outer fringe' of state **abc**. In general, the *outer fringe* of some knowledge state K is the set of all problems **p** such that adding **p** to K forms another knowledge state. The concept of outer fringe is critical because this is where progress is taking place: learning proceeds by mastering a new problem in the outer fringe, creating a new state, with its own outer fringe.

Conversely, each knowledge state (except the empty state) has at least one *predecessor state,* that is a state containing exactly the same problems, except one. The knowledge state **abc** that we just considered has two predecessor states, namely **ab** and **ac**. Problems **b** and **c** together form the *inner fringe* of state **abc**: removing either **b** or **c** from state **abc** creates other states in the structure, that is **ab** and **ac**. If for some reason a student experiences difficulties in mastering the problems in the outer fringe, reviewing previous material should normally take place in the inner fringe of a student's state. Figure 5 illustrates these concepts of fringes and others introduced so far. A state K is pictured with three problems in its outer fringe. Another state K' has two problems in its inner fringe.

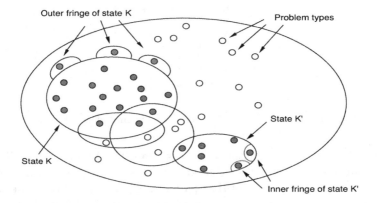

Fig. 5. The outer fringe of a state K and the inner fringe of another state K'

Thus, we can use the two fringes as the main building blocks of the 'navigation tool' of the system, with the outer fringes directing the progress, and the inner fringes monitoring temporary retreats, and making them profitable.

Interestingly, the fringes also play a less obvious, but equally important role in summarizing the results of an assessment. A knowledge state is essentially a list of all the problems mastered by a student at the time of an assessment. Such a list will often be unwieldy and contain several dozen problem names, not a very convenient description. It can be shown mathematically, however, that for the 'learning spaces' (a type of knowledge structures specified in our next section), the two fringes suffice to specify the knowledge state completely[8]. In other words, the result of an assessment can be given in the form of two short lists, one for the inner fringe (WHAT THE STUDENT CAN DO, which is understood here as the most advanced problems in the student's state), and one for the outer fringe (WHAT THE STUDENT IS READY TO LEARN). Experience with realistic knowledge structures in school mathematics has shown that these two lists together will contain on average 10-15 problems, enabling a very compact and faithful presentation of the result of an assessment.

Table 2 contains a typical example of the two fringes of a knowledge state, which is that of an actual student currently using the system in a middle school.

---

[8] The exact statement of this result is contained in [Doignon and Falmagne, 1999, Theorem 2.8, p. 48].

Table 2. A knowledge state in Arithmetic specified by its two fringes

Inner fringe: WHAT THE STUDENT CAN DO	**Outer fringe:** WHAT THE STUDENT IS READY TO LEARN
Double negation: $-(-12) - 7 =$	*Decimal division:* $5.2\,\overline{)7.54}$
Arithmetic with absolute value: $\left\| \,\|9 - 12\| - \|5\|\, \right\|$	*Word problem on percentage (Problem type 2):* A sofa is on sale for \$630 after a 28% discount. What was the price before discount?
Word problem with clocks: A clock runs too fast and gains 6 minutes every 5 days. How many minutes and seconds will it have gained at the end of 9 days?	*Word problem with inverse proportion:* If 4 machines are needed to complete a task in 21 days, how long will it take 7 machines to complete the same task?
Word problem on percentage (Problem type 1): A pair of sneakers usually sells for \$45. Find the sale price after a 20% discount.	*Average of two numbers:* What is the average value of 114 and 69?
Mixed number multiplication: $3\frac{3}{4} \times 2\frac{4}{9} =$ (Write your answer as a mixed number in lowest terms.)	

Taken together, the two fringes amount to 9 problems, which suffice to specify the 80 problems of that student's state which is represented in the top region of Figure 3. The economy is startling.

The information provided by such a table is a more meaningful result of an assessment than a couple of numerical scores from a standardized test. It is also considerably more precise. An assessment involving all of high school mathematics, based on the knowledge states consistent with the precedence diagram of Figure 3, would classify the students in hundreds of thousands of categories, each with its own unique table of inner and outer fringes. By contrast, a quantitative S.A.T. classifies the test taker into one of roughly 40 categories (from 400 to 800, in steps of 10).

Axioms for Learning Spaces

The precedence diagram illustrated by Figures 1, 2 and 3 offers an economical way of summarizing knowledge structures that may sometimes contain millions of states. However, such a summary represents a structure exactly only if that structure satisfies tight conditions, one of which is highly questionable, namely, that the collection of states of the structure is closed under intersection, to wit: if K and K' are states, then $K \cap K'$ must also be a state. (This point is closely related to a classical result of Birkhoff [1937].)

In fact, all the concepts and techniques discussed in this paper are consistent with the weaker conditions captured by the axioms [FC] and [LS] stated below, both of which appear to be eminently reasonable from a pedagogical standpoint. These two axioms may be regarded as forming the core of the theory. This section is slightly more formal than the rest of the paper.

We recall that knowledge structure on a set Q is a collection $\mathcal{K}$ of subsets of Q which are called the *states* of the structure $\mathcal{K}$. It is assumed that both the full set Q and the empty set $\varnothing$ are states.

Axioms

[FC] A knowledge structure $\mathcal{K}$ is *outer fringe coherent* if for any two states $K \subset L$, any problem type in the outer fringe of K is either in L or in the outer fringe of L.

Intuitively: *If a problem type* **p** *is learnable from some state K which is a subset of some state L, then either* **p** *is also in L or is learnable from L.*

[LS] A knowledge structure is *learning smooth* if whenever $L \subset K$ for two states L and K, then there is a sequence of states

$$K_0 = K \subset K_1 \subset \ldots \subset K_n = L$$

such that for some problem types $\mathbf{p}_1, \ldots, \mathbf{p}_n$, we have

$$K_0 \cup \{\mathbf{p}_1\} = K_1, K_1 \cup \{\mathbf{p}_2\} = K_2, \ldots, K_{n-1} \cup \{\mathbf{p}_n\} = K_n = L.$$

In words: *If some state K is a subset of some state L, then there is a way of learning successively the problem types in L which are missing from K.*[9]

Definition. A knowledge structure which is outer fringe coherent and learning smooth is called a *learning space*.

Thus, the assessment techniques discussed here are fully applicable to learning spaces. Some important consequences of Axioms [FC] and [LS] have been derived by Cosyn and Uzun [2005] (see also Doignon and Falmagne [1999]). Note that it is possible to represent a learning space by a precedence diagram. However, the cost of such a representation is that it involves fictitious states, namely, all those states that must be added to the learning space to ensure that the knowledge structure is closed under intersection (see above). The addition of such fictitious states may of course lengthen the assessment and may have other drawbacks.

Building a Knowledge Structure

We now turn to what is certainly the most demanding task in a realistic application of these ideas. It certainly makes sense to enroll experts, such as seasoned teachers or textbook writers, to find the knowledge states. This must be done for

---

[9] Knowledge structures which are learning smooth are called 1-*learnable* by Doignon and Falmagne [1999].

the first draft of a knowledge structure, which can then be refined by a painstaking analysis of student data. However, we cannot simply sit an expert in front of a computer terminal with the instruction: "provide a complete list of all the knowledge states in a given topic." Fortunately, an indirect approach is possible. An expert can reliably respond to questions such as these:

Q1. SUPPOSE THAT A STUDENT IS NOT CAPABLE OF SOLVING PROBLEM **p**. COULD THIS STUDENT NEVERTHELESS SOLVE PROBLEM **p'**?

It can be proven that a knowledge structure represented by a precedence diagram such as the one of Figure 2 can be inferred exactly from the responses to a complete collection of questions of the type **Q1**. (For a very large precedence diagram, such as the one of Figure 3, several diagrams are first constructed by querying experts on each of the fields of knowledge, like Arithmetic, Beginning Algebra, *etc.* Those diagrams are then 'glued' together, relying again on experts' judgment.)

In the case of the precedence diagram of Figure 1, the mastery of problem **e**, for instance, implies that of a single minimum set of precedent problems, namely **a**, **b** and **c**. In other words, all learning paths in Figure 4 progress through these three problems before reaching **e**. There are important cases, however, in which the mastery of a problem may be achieved *via* anyone of several distinct minimum sets of precedent problems. Such structures, which generalize those that can be represented by precedence diagrams, are called *knowledge spaces*. They are derived from the responses to the collection of more difficult questions of the following type:

Q1. SUPPOSE THAT A STUDENT HAS NOT MASTERED PROBLEMS $p_1, p_2, \ldots, p_n$. COULD THIS STUDENT NEVERTHELESS SOLVE PROBLEM **p'**?

In practice, not all questions of type **Q1** or **Q2** must be asked because, in many cases, responses to some questions can be inferred from responses to other questions. For typical knowledge structures encountered in education, an expert may be required to respond to a few thousand questions to get a complete description of all the knowledge states. Note that the knowledge structures resulting from responses to the more elaborate questions of type **Q2** are consistent with Axioms [FC] and [LS].

By interviewing several experts and combining their answers, one can build a knowledge structure which reflects their consensual view of the field. This alone does not guarantee the validity of the knowledge structure, that is, the agreement between the states in the structure and the actual states in the student population. Actual student data are essential to complete the picture. With an Internet based, largely distributed assessment system such as the one discussed here, data from tens of thousands users can be collected in the span of a year, providing a bounty of information. Such data can be used to refine a knowledge structure obtained from experts' judgments via the questions of type **Q1** or **Q2**. To begin with, states occurring rarely or not at all in the empirical applications can be deleted from the knowledge structure. More importantly, the accuracy of the structure can be evaluated by the following probe, and corrected if necessary. In most assessments, an extra problem $\mathbf{p}^*$ is added to the questioning, which is not used in the choice of the final knowledge state K representing the student. Using K, one can predict the student answer to $\mathbf{p}^*$ which should be correct if $\mathbf{p}^*$ is in K—except for careless errors—and false otherwise. In the knowledge structure for Beginning Algebra for example, as it is used by students today, the correlation between predicted and observed answers hovers between .7 and .85, depending on the sample of students. These high values actually *underestimate* the accuracy of the structure: a student having mastered some problem $\mathbf{p}^*$ contained in his or her knowledge state may nevertheless make a careless error in solving it. This correlation index is a powerful statistical tool continuously monitoring the validity of the knowledge structure, pointing to weaknesses, and evaluating the corrections prompted by some earlier analysis.

Uncovering a Knowledge State in a Knowledge Structure

Suppose that a satisfactory knowledge structure has been obtained. The task of the assessment is to uncover, by efficient questioning, the knowledge state of a particular student under examination. The situation is similar to that of *adaptive testing*—i.e. the computerized forms of the S.A.T. and the like—with the critical difference that the outcome of the assessment here is a knowledge state, rather than a numerical estimate of a student's competence in the topic.

The assessment procedures available all pertain to the scheme outlined in Figure 6.

In this article, we focus on one particular assessment procedure in which the plausibility of a state at any time in the course of an assessment is its current likelihood, based on all the information accumulated so far. At the outset of the assessment (trial 1 of the procedure), each of the knowledge states is assigned a certain *a priori* likelihood, which may depend upon the school year of the student if it is known, or some other information. The sum of these *a priori* likelihoods is equal to 1. They play no role in the final result of the assessment but may be helpful in shortening it. If no useful information is available, then all the states

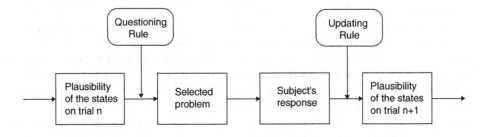

Fig. 6. Diagram of the transitions in an assessment procedure

are assigned the same likelihood. The first problem p_1 is chosen so as to be 'maximally informative.' This is interpreted to mean that, on the basis of the current likelihoods of the states, the student has about a 50% chance of knowing how to solve p_1. In other words, the sum of the likelihoods of all the states containing p_1 is as close to .5 as possible[10]. If several problem types are equally informative (as may happen at the beginning of an assessment) one of them is chosen at random. The student is then asked to solve an instance of that problem, also picked randomly. The student's answer is then checked by the system, and the likelihood of all the states are modified according to the following *updating rule*. If the student gave a correct answer to p_1, the likelihoods of all the states containing p_1 are increased and, correspondingly, the likelihoods of all the states *not* containing p_1 are decreased (so that the overall likelihood, summed over all the states, remains equal to 1). A false response given by the student has the opposite effect: the likelihoods of all the states *not* containing p_1 are increased, and that of the remaining states decreased[11] If the student does not know how to solve a problem, he or she can choose to answer "I don't know" instead of guessing. This results in a substantial increase in the likelihood of the states not containing p_1, thereby decreasing the total number of questions required to uncover the student's state. Problem p_2 is then chosen by a mechanism identical to that used for selecting p_1, and the likelihood values are increased or decreased according to the student's answer via the same updating rule. Further problems are dealt with similarly. In the course of the assessment, the likelihood of some states gradually increases. The assessment procedure stops when two criteria are fulfilled: (1) the entropy of the likelihood distribution, which measures the uncertainty of the assessment system regarding the student's state, reaches a critical low level, and (2) there is no longer any useful question to be asked (all the problems have either a very high or a very low probability of being responded to correctly). At that moment, a few likely states remain and the system selects the most likely one among them. Note that, because of the stochastic nature of the assessment procedure, the final state may very well contain a problem to which the student gave a false response. Such a response is thus regarded as due to a careless error. On the other hand, because all the problems have open-ended responses (no multiple choice), with a large number of possible solutions, the probability of lucky guesses is negligible.

To illustrate the evolution of an assessment, we use a graphic representation in the guise of the *likelihood map* of a knowledge structure. In principle, each point in the oval shape of Figure 7 represents one of the 57,147 states of the knowledge structure for the part of Arithmetic used for the graphs in this paper. (Because of graphics limitations, some grouping of similar states into a single point was necessary. To simplify the exposition, we suppose in the sequel that each point

---

[10] A different interpretation of 'maximally informative' was also investigated, based on the minimization of the expected entropy of the likelihood distribution. This method did not result in an improvement, and was computationally more demanding.

[11] The operator used to modify the likelihoods is Bayesian. This was shown by Mathieu Koppen (personal communication).

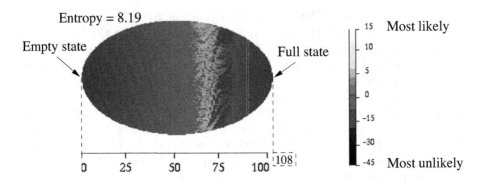

Fig. 7. Likelihood map of the Arithmetic knowledge structure whose precedence diagram was given at the top of Figure 3 (the blue points)

of the map represents one state.) The precedence diagram of this structure was given at the top of Figure 3 (appearing as blue points in the online version).

Knowledge states are sorted according to the number of problem types they contain, from 0 problems on the far left to 108 problems on the far right. The leftmost point stands for the empty knowledge state, which is that of a student knowing nothing at all in Arithmetic. The rightmost point represents the full knowledge state and corresponds to a student having mastered all the problems in Arithmetic. The points located on any vertical line within the oval represent knowledge states containing exactly the number of problems indicated on the abscissa.

The oval shape is chosen for esthetic reasons and reflects the fact that, by and large, there are many more states around the middle of the scale than around the edges. For instance, there are 1,668 states containing exactly 75 problems, but less than 100 states, in Arithmetic, containing more than 100 problems or less than 10 problems. The arrangement of the points on any vertical line is largely arbitrary.

The shading (or color when seen on line) of a point represents the likelihood of the corresponding state. A shading-coded logarithmic scale, pictured on the right of Figure 7, is used to represent the likelihood values with the lightest shades corresponding to the most likely states.

Figure 8 displays a sequence of likelihood maps describing the evolution of an assessment in Arithmetic from the very beginning, before the first problem, to the end, after the response to the last problem is recorded by the system and acted upon to compute the last map. The full assessment took 24 questions which is close to the average for Arithmetic. The initial map results from preliminary information obtained from that student. The vertical strip[12] of that map represents the a priori relatively high likelihood of the knowledge states containing between 58 and 75 problems: as a six grader, this student can be assumed to have mastered about two thirds of the Arithmetic curriculum.

---

[12] The strip is barely visible on the printed version and appears in bright red in the online version.

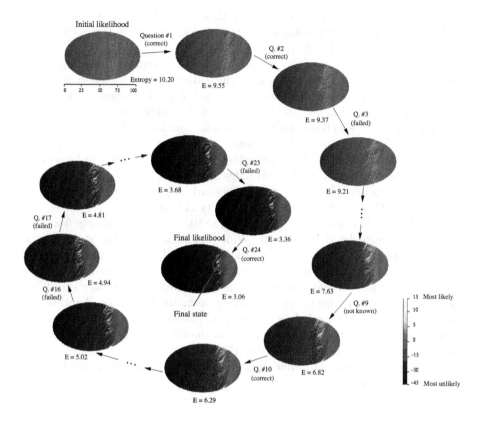

Fig. 8. Sequence of likelihood maps representing an assessment converging toward the student's knowledge state

Next to each map in Figure 8, we indicate the entropy of the corresponding likelihood distribution, and the student's response to the question (correct, false, or not known). Note that the initial entropy is 10.20, which is close to the theoretical maximum of 10.96 obtained for a uniform distribution on a set of 57,147 knowledge states. As more information is gathered by the system via the student's responses to the questions, the entropy decreases gradually. Eventually, after 24 questions have been answered a single very bright point remains among mostly dark (blue) points and a few bright points. This very bright point indicates the most likely knowledge state for that student, based on the answers to the problems. The assessment stops at that time because the entropy has reached a critical low level and the next 'best' problem to ask has only a 19% chance of being solved by the student, and so would not be very informative. In this particular case only 24 problems have sufficed to pinpoint the student's knowledge state among 57,147 possible ones. This striking efficiency is achieved by the numerous inferences implemented by the system in the course of the assessment.

The assessment procedure described in this article is the core engine of an Internet based, automated mathematics tutor which is used in several hundred colleges and school districts in the U.S.

In the U.S., the extensive research leading to this system has been supported since 1983 by various grants, mostly from the National Science Foundation. The first paper on this research topic, which is named 'Knowledge Spaces,' was published by Doignon and Falmagne [1985], two of the authors of this article. Important results have also been obtained by researchers from other countries, such as D. Albert, J. Heller and C. Hockemeyer (Austria), C. Dowling and R. Suck (Germany), and M. Koppen (The Netherlands). Most of the basic results are presented in a monograph entitled *'Knowledge Spaces,'* by Doignon and Falmagne [1999], but much work has been done afterward. A data base on this topic is maintained by C. Hockemeyer at the University of Graz.

Related research results, classified under the title 'Formal Concepts Analysis' have been obtained by B. Ganter, R. Wille and their team. The states in the sense of our paper correspond to the concepts as used in most of the papers of this volume. Most of the mathematical tools are similar. One critical difference between the two lines of research, however, is that axioms such as those specifying a learning space as not especially compelling in concept analysis. Another one lies in the role played by probability theory and stochastic processes as described in the latter part of this paper. For a study of the relationships between the two fields, see for instance Rusch and Wille [1996].

An exemplary sample of publications relevant to knowledge space theory is included in the references.

References

D. Albert, editor. *Knowledge Structures.* Springer Verlag, New York, 1994.

D. Albert and J. Lukas, editors. *Knowledge Spaces: Theories, Empirical Research, Applications.* Lawrence Erlbaum Associates, Mahwah, NJ, 1999.

G. Birkhoff. Rings of sets. *Duke Mathematical Journal,* 3:443–454, 1937.

E. Cosyn and H.B. Uzun. Axioms for Learning Spaces. *Journal of Mathematical Psychology,* 2005. To be submitted.

J.-P. Doignon and J.-Cl. Falmagne. Spaces for the assessment of knowledge. *International Journal of Man-Machine Studies,* 23:175–196, 1985.

J.-P. Doignon and J.-Cl. Falmagne. *Knowledge Spaces.* Springer, Berlin, 1999.

C. E. Dowling. Applying the basis of a knowledge space for controlling the questioning of an expert. *Journal of Mathematical Psychology,* 37:21–48, 1993a.

C. E. Dowling and C. Hockemeyer. Computing the intersection of knowledge spaces using only their basis. In Cornelia E. Dowling, Fred S. Roberts, and Peter Theuns, editors, *Recent Progress in Mathematical Psychology,* pages 133–141. Lawrence Erlbaum Associates Ltd., Mahwah, NJ, 1998.

C.E. Dowling. On the irredundant construction of knowledge spaces. *Journal of Mathematical Psychology,* 37:49–62, 1993b.

J.-Cl. Falmagne and J-P. Doignon. A class of stochastic procedures for the assessment of knowledge. *British Journal of Mathematical and Statistical Psychology,* 41:1–23, 1988a.

J.-Cl. Falmagne and J-P. Doignon. A Markovian procedure for assessing the state of a system. *Journal of Mathematical Psychology*, 32:232–258, 1988b.

B. Ganter and R. Wille. *Formal Concept Analysis*. Springer-Verlag, Berlin, 1999. Mathematical foundations, Translated from the 1996 German original by C. Franzke.

J. Heller. A formal framework for characterizing querying algorithms. *Journal of Mathematical Psychology*, 48:1–8, 2004.

W.T. Kelvin. *Popular Lectures and Addresses. Volume 1-3*. MacMillan, London, 1889. Volume 1: Constitution of Matter (Chapter: Electrical Units of Measurement).

M. Koppen. Extracting human expertise for constructing knowledge spaces: An algorithm. *Journal of Mathematical Psychology*, 37:1–20, 1993.

M. Koppen. The construction of knowledge spaces by querying experts. In Gerhard H. Fischer and Donald Laming, editors, *Contributions to Mathematical Psychology, Psychometrics, and Methodology*, pages 137–147. Springer–Verlag, New York, 1994.

K. Pearson. *The Life, Letters and Labours of Francis Galton*. Cambridge University Press, London, 1924. Volume 2: Researches of Middle Life.

F.S. Roberts. *Measurement Theory, with Applications to Decisionmaking, Utility, and the Social Sciences*. Addison-Wesley, Reading, MA, 1979.

A. Rusch and R. Wille. Knowledge spaces and formal concept analysis. In H.-H. Bock and W. Polasek, editors, *Data analysis and information systems*, pages 427–436, Heidelberg, 1996. Springer–Verlag.

R. Suck. The basis of a knowledge space and a generalized interval order. *Electronic Notes in Discrete Mathematics*, 2, 1999a. Abstract of a Talk presented at the OSDA98, Amherst, MA, September 1998.

R. Suck. A dimension–related metric on the lattice of knowledge spaces. *Journal of Mathematical Psychology*, 43:394–409, 1999b.

W.T. Trotter. *Combinatorics and Partially Ordered Sets: Dimension Theory*. The Johns Hopkins University Press, Baltimore and London, 1992.

The Basic Theorem on Preconcept Lattices

Christian Burgmann and Rudolf Wille

Technische Universität Darmstadt, Fachbereich Mathematik,
Schloßgartenstr., 7, D–64289 Darmstadt
wille@mathematik.tu-darmstadt.de

Abstract. *Preconcept Lattices* are identified to be (up to isomorphism) the completely distributive complete lattices in which the supremum of all atoms is equal or greater than the infimum of all coatoms. This is a consequence of the *Basic Theorem on Preconcept Lattices*, which also offers means for checking line diagrams of preconcept lattices.

Contents

1   Preconcept Lattices

The notion of a "formal preconcept" has been introduced into Formal Concept Analysis in [SW86] to mathematize the notion of a "preconcept" which is used in Piaget's cognitive psychology to explain the developmental stage between the stage of senso-motoric intelligence and the stage of operational intelligence (see [Pi73]). In Formal Concept Analysis [GW99], a *(formal) preconcept* of a formal context (G, M, I) is defined as a pair (A, B) with $A \subseteq G$, $B \subseteq M$, and $A \subseteq B'$ ($\Leftrightarrow A' \supseteq B$), which obviously generalizes the definition of a formal concept. The set of all preconcepts of a formal context (G, M, I) is denoted by $\mathfrak{V}(G, M, I)$. Preconcepts are naturally ordered by

$$(A_1, B_1) \leq (A_2, B_2) \iff A_1 \subseteq A_2 \text{ and } B_1 \supseteq B_2.$$

By Proposition 1 in [Wi04], the ordered set $\underline{\mathfrak{V}}(G, M, I) := (\mathfrak{V}(G, M, I), \leq)$ is even a completely distributive complete lattice with the following infima and suprema:

$$\bigwedge_{t \in T}(A_t, B_t) = (\bigcap_{t \in T} A_t, \bigcup_{t \in T} B_t) \text{ and } \bigvee_{t \in T}(A_t, B_t) = (\bigcup_{t \in T} A_t, \bigcap_{t \in T} B_t)$$

for all $(A_t, B_t) \in \mathfrak{V}(\mathbb{K})$ ($t \in T$). The lattice $\underline{\mathfrak{V}}(G, M, I)$ is called the *preconcept lattice* of the formal context (G, M, I).

A preconcept lattice $\underline{\mathfrak{V}}(G, M, I)$ can be understood as the concept lattice of the *derived context* $(G \dot\cup M, G \dot\cup M, I \cup (\neq \backslash G \times M))$ because, by the proof of Proposition 1 in [Wi04], the assignment

$$(A, B) \overset{\iota}{\mapsto} (A \cup (M \backslash B), (G \backslash A) \cup B)$$

R. Missaoui and J. Schmid (Eds.): ICFCA 2006, LNAI 3874, pp. 80–88, 2006.

is an isomorphism ι from the preconcept lattice $\mathfrak{V}(G, M, I)$ onto the concept lattice $\mathfrak{B}(G \dot{\cup} M, G \dot{\cup} M, I \cup (\neq \backslash G \times M))$. This connection shall be demonstrated by an example: We choose the formal context in Fig. 1, the concept lattice of which is represented in Fig. 2. The derived context of the context in Fig. 1 is shown in Fig. 3 and its concept lattice in Fig. 4.

The line diagram in Fig. 4 may also be considered as a representation of the preconcept lattice of the context in Fig. 1. The preconcepts can be identified in the line diagram by the following rule: an object (resp. attribute) belongs to a preconcept if and only if its name is attached to a circle representing a subpreconcept (resp. superpreconcept) of that preconcept. As atoms (resp. coatoms) of the preconcept lattice, we obtain the four (resp. three) preconcepts

$$
\begin{array}{ll}
(\emptyset, \{large, not-small, not-large\}), & (\{sun, earth\}, \emptyset), \\
(\emptyset, \{large, not-small, small\}), & (\{sun, moon\}, \emptyset), \\
(\emptyset, \{large, not-large, small\}), & (\{earth, moon\}, \emptyset). \\
(\emptyset, \{not-small, not-large, small\}). &
\end{array}
$$

In general, the *atoms* of a preconcept lattice $\mathfrak{V}(G, M, I)$ are just the preconcepts $(\emptyset, M \setminus \{m\})$ with $m \in M$ and the preconcepts $(\{g\}, M)$ with $g \in G$ and $\{g\}' = M$; the *coatoms* are just the preconcepts $(G \setminus \{g\}, \emptyset)$ with $g \in G$ and the preconcepts $(G, \{m\})$ with $m \in M$ and $\{m\}' = G$. The $\bigvee$-*irreducible preconcepts* which are not atoms are the preconcepts $(\{g\}, \{g\}')$ with $g \in G$ satisfying

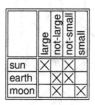

Fig. 1. A formal context of celestial bodies

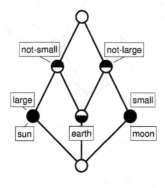

Fig. 2. Line diagram of the concept lattice of the formal context in Fig. 1

	/sun	/earth	/moon	large	not-large	not-small	small
sun		×	×	×		×	
earth	×		×		×	×	
moon	×	×			×		×
\large	×	×	×		×	×	×
\not-large	×	×	×	×		×	×
\not-small	×	×	×	×	×		×
\small	×	×	×	×	×		

Fig. 3. The derived context of the formal context in Fig. 1

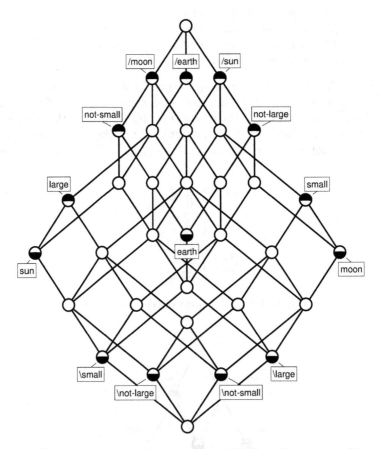

Fig. 4. Line diagram of the preconcept lattice of the formal context in Fig. 1 and of the concept lattice of the formal context in Fig. 3

$\{g\}' \neq M$, and the $\bigwedge$-*irreducible preconcepts* which are not coatoms are the preconcepts $(\{m\}', \{m\})$ with $m \in M$ satisfying $\{m\}' \neq G$. Specific preconcepts are the *smallest preconcept* $(\emptyset, M)$, the *largest preconcept* $(G, \emptyset)$, and the *central preconcept* $(\emptyset, \emptyset)$ which is a subpreconcept of the supremum of all atoms and a superconcept of the infimum of all coatoms.

2 $\bigvee$-Irreducible and $\bigwedge$-Irreducible Elements

Let us recall that an element x of a complete latice L is called $\bigvee$-*irreducible* (resp. $\bigwedge$-*irreducible*) if x is not the supremum (resp. infimum) of properly smaller (resp. larger) elements in L. The set of all $\bigvee$-irreducible (resp. $\bigwedge$-irreducible) elements in L is denoted by $\mathcal{J}(L)$ (resp. $\mathcal{M}(L)$). *Atoms*, the upper covers of the smallest element 0, are special $\bigvee$-irreducible elements; the set of all atoms of L is denoted by $\mathcal{A}(L)$. *Coatoms*, the lower covers of the greatest element 1, are special $\bigwedge$-irreducible elements; the set of all coatoms of L is denoted by $\mathcal{C}(L)$.

Proposition 1. *Let L be a completely distributive complete lattice. Then there is an isomorphism $\underline{\alpha}$ from $([0, \bigvee \mathcal{A}(L)], \leq)$ onto the ordered power set $(\mathfrak{P}(\mathcal{A}(L)), \subseteq)$ defined by $\underline{\alpha}(\underline{x}) := \{a \in \mathcal{A}(L) \mid a \leq \underline{x}\}$ and an isomorphism $\overline{\alpha}$ from $([\bigwedge \mathcal{C}(L), 1], \geq)$ onto the ordered power set $(\mathfrak{P}(\mathcal{C}(L)), \supseteq)$ defined by $\overline{\alpha}(\overline{x}) := \{c \in \mathcal{C}(L) \mid c \geq \overline{x}\}$; in particular, $\underline{x} = \bigvee\{a \in \mathcal{A}(L) \mid a \leq \underline{x}\}$ for all $\underline{x} \in [0, \bigvee \mathcal{A}(L)]$ and $\overline{x} = \bigwedge\{c \in \mathcal{C}(L) \mid c \geq \overline{x}\}$ for all $\overline{x} \in [\bigwedge \mathcal{C}(L), 1]$.*

Proof: Although the assertions of the proposition follow directly from wellknown results about completely distributive complete lattices, a short proof shall be given to support a better understanding of the assertions. This clarifies how the equality $\underline{x} = \bigvee\{a \in \mathcal{A}(L) \mid a \leq \underline{x}\}$ is a consequence of the following:

$$\underline{x} = \underline{x} \wedge \bigvee \mathcal{A}(L)$$
$$= \underline{x} \wedge \left(\bigvee\{a \in \mathcal{A}(L) \mid a \leq \underline{x}\} \vee \bigvee\{a \in \mathcal{A}(L) \mid a \not\leq \underline{x}\} \right)$$
$$= \bigvee\{\underline{x} \wedge a \in \mathcal{A}(L) \mid a \leq \underline{x}\} \vee \bigvee\{\underline{x} \wedge a \in \mathcal{A}(L) \mid a \not\leq \underline{x}\}$$
$$= \bigvee\{a \in \mathcal{A}(L) \mid a \leq \underline{x}\}$$

Mapping every subset B of $\mathcal{A}(L)$ to $\bigvee B$ yields the inverse map $\underline{\alpha}^{-1}$ because $a \leq \bigvee B$ implies $a = a \wedge \bigvee B = \bigvee\{a \wedge b \mid b \in B\}$, hence $a \in B$ and therefore $B = \underline{\alpha}(\bigvee B)$. Since $\underline{\alpha}$ and $\underline{\alpha}^{-1}$ are obviously order-preserving, $\underline{\alpha}$ is an isomorphism from $([0, \bigvee \mathcal{A}(L)], \leq)$ onto the Boolean lattice $(\mathfrak{P}(\mathcal{A}(L)), \subseteq)$. Dually, we get that $\overline{\alpha}$ is an isomorphism from $([\bigwedge \mathcal{C}(L), 1], \geq)$ onto the Boolean lattice $(\mathfrak{P}(\mathcal{C}(L)), \supseteq)$.

Since preconcept lattices are completely distributive complete lattices having *enough $\bigvee$-irreducible and $\bigwedge$-irreducible elements*, it is worthwhile to learn more about those lattices in general (see [BD74], Section XII.4). "Enough $\bigvee$-irreducible and $\bigwedge$-irreducible elements" in a complete lattice L usually means that every element of L is the supremum of $\bigvee$-irreducible elements and the infimum of $\bigwedge$-irreducible elements. A well-known theorem states that the completely distributive complete lattices with enough $\bigvee$-irreducible and $\bigwedge$-irreducible elements are up to isomorhism the complete sublattices of power set lattices (see [Ra52], [Ba54]). Most interesting for our investigation are completely distributive complete lattices L in which the supremum of all atoms is equal or greater than the infimum of all coatoms. Fortunately, those lattices have enough $\bigvee$-irreducible and $\bigwedge$-irreducible elements what results from the following proposition:

Proposition 2. *Let L be a completely distributive complete lattice in which the supremum of all atoms is equal or greater than the infimum of all coatoms, and let $p \in [\bigwedge \mathcal{C}(L), \bigvee \mathcal{A}(L)]$. Then, the $\bigvee$-irreducible elements which are not atoms of the interval $[0, p]$ are the infima*

$$\Downarrow d := \bigwedge \{x \in [0, d] \mid x \not\leq p\} \text{ where } d \text{ is an atom of the interval } [p, 1],$$

and the $\bigwedge$-irreducible elements which are not coatoms of the interval $[p, 1]$ are the suprema

$$\Uparrow b := \bigvee \{y \in [b, 1] \mid y \not\geq p\} \text{ where } b \text{ is a coatom of the interval } [0, p].$$

Let $\diagdown\!\!\Downarrow d := \bigvee \{x \in L \mid x \not\geq \Downarrow d\}$ and let $\diagup\!\!\Uparrow b := \bigwedge \{y \in L \mid y \not\leq \Uparrow b\}$. Then $\{\diagdown\!\!\Downarrow d \mid d \in \mathcal{A}([p, 1])\} = \mathcal{C}([p, 1])$ and $\{\diagup\!\!\Uparrow b \mid b \in \mathcal{C}([0, p])\} = \mathcal{A}([0, p])$; furthermore, for all $z \in L$,

$$z = \bigvee\{\Downarrow d \mid d \in \mathcal{A}([p, 1]), \Downarrow d \leq z\} \vee \bigvee\{\diagup\!\!\Uparrow b \mid b \in \mathcal{C}([0, p]), \Uparrow b \not\geq z\}$$

$$= \bigwedge\{\Uparrow b \mid b \in \mathcal{C}([0, p]), \Uparrow b \geq z\} \wedge \bigwedge\{\diagdown\!\!\Downarrow d \mid d \in \mathcal{A}([p, 1]), \Downarrow d \not\leq z\}.$$

This clarifies that each element of L is the supremum of $\bigvee$-irreducible elements and the infimum of $\bigwedge$-irreducible elements of L.

Proof: First, we notice that $[0, p]$ is a Boolean interval in $[0, \bigvee \mathcal{A}(L)]$ and $[p, 1]$ is a Boolean interval in $[\bigwedge \mathcal{C}(L), 1]$. Thus, we can confirm the existence of atoms in $[p, 1]$ and coatoms in $[0, p]$ as follows: For each coatom c of $[p, 1]$, the infimum of $\mathcal{C}([p, 1]) \setminus \{c\}$ is an atom of $[p, 1]$ by Proposition 1. Dually, we get for each atom a of $[0, p]$ that the supremum of $(\mathcal{A}([0, p]) \setminus \{a\}$ is a coatom of $[0, p]$. Clearly, all atoms of $[p, 1]$ and coatoms of $[0, p]$ can be obtained by the considered constructions.

Next, we identify the $\bigvee$-irreducible and $\bigwedge$-irreducible elements which are not atoms of $[0, p]$ and coatoms of $[p, 1]$, respectively. For each atom d of $[p, 1]$, $\Downarrow d$ is obviously an upper cover of $\Downarrow d \wedge p$ which is the supremum of all properly smaller elements of $\Downarrow d$; hence $\Downarrow d$ is $\bigvee$-irreducible. Dually, for each coatom b of $[0, p]$, $\Uparrow b$ is obviously the lower cover of $\Uparrow b \vee p$ which is the infimum of all properly larger elements of $\Uparrow b$; hence $\Uparrow d$ is $\bigwedge$-irreducible. For $z \in L \setminus ([0, p] \cup [p, 1])$, the interval $[z \wedge p, z \vee p]$ is isomorphic to the direct products $[z \wedge p, z] \times [z \wedge p, p]$ and $[z, z \vee p] \times [p, z \vee p]$ by the distributive laws (the isomorphisms are given by $x \mapsto (x \wedge z, x \wedge p)$ and $x \mapsto (x \vee z, x \vee p)$). Thus, z can only be $\bigvee$-irreducible if z is an upper cover of $z \wedge p$; hence $z = \Downarrow d$ for some $d \in \mathcal{A}([p, 1])$. Dually, we get that z is only $\bigwedge$-irreducible if $z = \Uparrow b$ for some $b \in \mathcal{C}([0, p])$.

Now, we prove $\{\searrow\Downarrow d \mid d \in \mathcal{A}([p,1])\} = \mathcal{C}([p,1])$. For $d \in \mathcal{A}([p,1])$, we obtain $\Downarrow d \wedge \bigvee\{x \in L \mid x \not\geq \Downarrow d\} = \bigvee\{\Downarrow d \wedge x \mid x \in L \text{ with } x \not\geq \Downarrow d\} < \Downarrow d$ because $\Downarrow d$ is $\bigvee$-irreducible. Therefore $\bigvee\{x \in L \mid x \not\geq \Downarrow d\} \not\geq \Downarrow d$ and hence $\searrow\Downarrow d = \bigvee\{x \in L \mid x \not\geq \Downarrow d\} = \bigvee \mathcal{A}([p,1]) \setminus \{d\} \in \mathcal{C}([p,1])$ which proves the desired equality. Dually, it follows $\{\nearrow\Uparrow b \mid b \in \mathcal{C}([0,p])\} = \mathcal{A}([0,p])$.

Finally, because of $[0,z] \cap (\mathcal{J}(L) \setminus \mathcal{A}([0,p])) = \{\Downarrow d \mid d \in \mathcal{A}([p,1]), \Downarrow d \leq z]\}$ and $[0,z] \cap \mathcal{A}([0,p]) = \{\nearrow\Uparrow b \mid b \in \mathcal{C}([0,p]), \Uparrow b \not\geq z\}$, we obtain

$$z \geq \underline{z} := \bigvee\{\Downarrow d \mid d \in \mathcal{A}([p,1]), \Downarrow d \leq z]\} \vee \bigvee\{\nearrow\Uparrow b \mid b \in \mathcal{C}([0,p]), \Uparrow b \not\geq z\},$$

and dually

$$z \leq \overline{z} := \bigwedge\{\Uparrow b \mid b \in \mathcal{C}([0,p]), \Uparrow b \geq z\} \wedge \bigwedge\{\searrow\Downarrow d \mid d \in \mathcal{A}([p,1]), \Downarrow d \not\leq z\}.$$

Since $\underline{z} \vee p$ is the infimum of the coatoms above it and $\overline{z} \wedge p$ is the supremum of the atoms below it, it follows that $\underline{z} \vee p = \bigwedge\{\searrow\Downarrow d \mid d \in \mathcal{A}([p,1]), \Downarrow d \not\leq z\}$ and $\overline{z} \wedge p = \bigvee\{\nearrow\Uparrow b \mid b \in \mathcal{C}([0,p]), \Uparrow b \not\geq z\}$, which forces $\underline{z} = z = \overline{z}$ by distributivity. Thus, the above inequalities are even equalities as claimed in the proposition.

3   The Basic Theorem

Now we are prepared to state and prove the basic theorem in which we use the following conventions: For subsets $A \subseteq \mathcal{J}(L)$ and $B \subseteq \mathcal{M}(L)$, where L is a completely distributive complete lattice, we define

$$\searrow A := \{\searrow a \mid a \in A\} \qquad (\searrow a := \bigvee\{x \in L \mid x \not\geq a\}),$$
$$\nearrow B := \{\nearrow b \mid b \in B\} \qquad (\nearrow b := \bigwedge\{x \in L \mid x \not\leq b\}).$$

The Basic Theorem on Preconcept Lattices.   *For a formal context (G, M, I), the preconcept lattice $\underline{\mathfrak{V}}(G, M, I)$ is a completely distributive complete lattice in which the supremum of all atoms is equal or greater than the infimum of all coatoms; in particular, the preconcept $(\emptyset, \emptyset)$ is a subpreconcept of the supremum of all atomic preconcepts and a superpreconcept of the infimum of all coatomic preconcepts. Arbitrary infima and suprema in $\underline{\mathfrak{V}}(G, M, I)$ are given by*

$$\bigwedge_{t \in T}(A_t, B_t) = (\bigcap_{t \in T} A_t, \bigcup_{t \in T} B_t) \text{ and } \bigvee_{t \in T}(A_t, B_t) = (\bigcup_{t \in T} A_t, \bigcap_{t \in T} B_t).$$

In general, a completely distributive complete lattice L, which has a singular element p below or equal the supremum of all atoms and above or equal the infimum of all coatoms, admits an isomorphism ψ from L to the preconcept lattice $\underline{\mathfrak{V}}(G, M, I)$ with $\psi(p) = (\emptyset, \emptyset)$ if and only if there are bijections $\tilde{\gamma}: G \longrightarrow \mathcal{J}(L) \setminus \mathcal{A}([0,p])$ and $\tilde{\mu}: M \longrightarrow \mathcal{M}(L) \setminus \mathcal{C}([p,1])$ such that

(1) $\tilde{\gamma}G \subseteq L \setminus [0,p]$ *and* $\tilde{\gamma}G \cup \nearrow\tilde{\mu}M$ *is $\bigvee$-dense in L*
    (i.e. $L = \{\bigvee X \mid X \subseteq \tilde{\gamma}G \cup \nearrow\tilde{\mu}M\}$),
(2) $\tilde{\mu}M \subseteq L \setminus [p,1]$ *and* $\tilde{\mu}M \cup \searrow\tilde{\gamma}G$ *is $\bigwedge$-dense in L*
    (i.e. $L = \{\bigwedge X \mid X \subseteq \tilde{\mu}M \cup \searrow\tilde{\gamma}G\}$),
(3) $gIm \Longleftrightarrow \tilde{\gamma}g \leq \tilde{\mu}m$ *for $g \in G$ and $m \in M$;*
in particular, $L \cong \underline{\mathfrak{V}}(\mathcal{J}(L) \setminus \mathcal{A}([0,p]), \mathcal{M}(L) \setminus \mathcal{C}([p,1]), \leq)$.

Proof: The results in Section 1 establish the properties of preconcept lattices listed in the first part of the Basic Theorem up to the equalities for the infima and suprema.

Now, let us assume that L is a completely distributive complete lattice with a singular element p satisfying $\bigvee \mathcal{A}(L) \geq p \geq \bigwedge \mathcal{C}(L)$. As in [GW99], p.21, we prove, first for the special case $L := \mathfrak{V}(G, M, I)$ and $p := (\emptyset, \emptyset)$, the existence of mappings $\tilde{\gamma}$ and $\tilde{\mu}$ with the required properties. We define $\tilde{\gamma}(g) := (\{g\}, \{g\}')$ for all $g \in G$ and $\tilde{\mu}(m) := (\{m\}', \{m\})$ for all $m \in M$. Obviously, $\tilde{\gamma}G \subseteq L \setminus [0, p]$ and $\tilde{\mu}M \subseteq L \setminus [p, 1]$. As claimed, we have $\tilde{\gamma}(g) \leq \tilde{\mu}(m) \iff \{g\} \subseteq \{m\}'$ and $\{g\}' \supseteq \{m\} \iff g \in \{m\}'$ and $\{g\}' \ni m \iff gIm$.

Furthermore, for each preconcept (A, B) of (G, M, I), the characterization of infima and suprema in $\mathfrak{V}(G, M, I)$ yields

$$(A, B) = (\bigcup_{g \in A} \{g\}, \bigcap_{m \in M \setminus B} M \setminus \{m\})$$

$$= (\bigcup_{g \in A} \{g\} \cup \emptyset, \bigcap_{g \in A} \{g\}' \cap \bigcap_{m \in M \setminus B} M \setminus \{m\})$$

$$= \bigvee_{g \in A} (\{g\}, \{g\}') \vee \bigvee_{m \in M \setminus B} (\emptyset, M \setminus \{m\})$$

$$= \bigvee_{g \in A} \tilde{\gamma}g \vee \bigvee_{m \in M \setminus B} \diagup \tilde{\mu}m$$

$$= \bigvee \tilde{\gamma}A \cup \diagup \tilde{\mu}(M \setminus B),$$

$$(A, B) = (\bigcap_{g \in G \setminus A} G \setminus \{g\}, \bigcup_{m \in M} \{m\})$$

$$= (\bigcap_{g \in G \setminus A} G \setminus \{g\} \cap \bigcap_{m \in B} \{m\}', \emptyset \cup \bigcup_{m \in B} \{m\})$$

$$= \bigwedge_{g \in G \setminus A} (G \setminus \{g\}, \emptyset) \wedge \bigwedge_{m \in M} (\{m\}', \{m\})$$

$$= \bigwedge_{g \in G \setminus A} \diagdown \tilde{\gamma}g \wedge \bigwedge_{m \in B} \tilde{\mu}m$$

$$= \bigwedge \diagdown \tilde{\gamma}(G \setminus A) \cup \tilde{\mu}(B).$$

Thus, $\tilde{\gamma}G \cup \diagup \tilde{\mu}M$ is $\bigvee$-dense and $\tilde{\mu}M \cup \diagdown \tilde{\gamma}G$ is $\bigwedge$-dense in $\mathfrak{V}(G, M, I)$.

More generally, if there exists an isomorphism $\psi : L \longrightarrow \mathfrak{V}(G, M, I)$ then it is natural to define the desired mappings by $\tilde{\gamma}g := \psi^{-1}(\{g\}, \{g\}')$ for $g \in G$ and $\tilde{\mu}m := \psi^{-1}(\{m\}', \{m\})$ for $m \in M$ which obviously have the claimed properties.

Conversely, let L be a completely distributive complete lattice with a singular element p satisfying $\bigvee \mathcal{A}(L) \geq p \geq \bigwedge \mathcal{C}(L)$, and let $\tilde{\gamma} : G \to L$ and $\tilde{\mu} : M \to L$ with the properties (1), (2), (3). Then a suitable mapping $\varphi : \mathfrak{V}(G, M, I) \to L$ can be defined by

$$\varphi(A, B) := \bigvee \tilde{\gamma}A \cup \diagup \tilde{\mu}(M \setminus B).$$

Clearly, φ is order-preserving. In order to prove that φ is an isomorphism, it has to be shown that φ^{-1} exists and is also order-preserving. Let $\psi : L \to \mathfrak{V}(G, M, I)$ be the mapping defined by

$$\psi x := (\{g \in G \mid \tilde{\gamma}g \leq x\}, \{m \in M \mid x \leq \tilde{\mu}m\}).$$

We have to prove that ψx is a preconcept of (G, M, I). Clearly,

$$h \in \{g \in G \mid \tilde{\gamma}g \leq x\} \Leftrightarrow \tilde{\gamma}h \leq x$$
$$\Rightarrow \tilde{\gamma}h \leq \tilde{\mu}n \text{ for all } n \in \{m \in M \mid x \leq \tilde{\mu}m\}$$
$$\Leftrightarrow hIn \text{ for all } n \in \{m \in M \mid x \leq \tilde{\mu}m\}$$
$$\Leftrightarrow h \in \{m \in M \mid x \leq \tilde{\mu}m\}'$$

and hence $\{g \in G \mid \tilde{\gamma}g \leq x\} \subseteq \{m \in M \mid x \leq \tilde{\mu}m\}'$; therefore ψx is a preconcept of (G, M, I). This yields that $\psi : L \to \mathfrak{V}(G, M, I)$ is indeed a mapping which is obviously order-preserving too; in particular, $\psi p = (\emptyset, \emptyset)$.

Now we prove that $\varphi^{-1} = \psi$. By the assumed $\bigvee$-density in (1), we get

$$\varphi\psi x = \bigvee \tilde{\gamma}\{g \in G \mid \tilde{\gamma}g \leq x\} \cup \diagup \tilde{\mu}(M \setminus \{m \in M \mid x \leq \tilde{\mu}m\}) = x.$$

Furthermore, by the assumed $\bigwedge$-density in (2), we get

$$\varphi(A, B) = \bigwedge \diagdown \tilde{\gamma}(G \setminus A) \cup \tilde{\mu}(B).$$

Consequently, we obtain

$$\psi\varphi(A, B) = \psi(\bigwedge \diagdown \tilde{\gamma}(G \setminus A) \cup \tilde{\mu}(B))$$
$$= (\{g \in G \mid \tilde{\gamma}g \leq \bigwedge \diagdown \tilde{\gamma}(G \setminus A) \cup \tilde{\mu}(B)\},$$
$$\{m \in M \mid \tilde{\mu}m \geq \bigvee \tilde{\gamma}A \cup \diagup \tilde{\mu}(M \setminus B)\})$$
$$= (\{g \in G \mid \tilde{\gamma}g \leq \tilde{\mu}m \text{ for all } \tilde{\mu}m \in \diagdown \tilde{\gamma}(G \setminus A) \cup \tilde{\mu}(B)\},$$
$$\{m \in M \mid \tilde{\mu}m \geq \tilde{\gamma}g \text{ for all } \tilde{\gamma}g \in \tilde{\gamma}A \cup \diagup \tilde{\mu}(M \setminus B)\})$$
$$= (\{g \in G \mid gIm \text{ for all } m \in M \text{ with } \tilde{\mu}m \in \diagdown \tilde{\gamma}(G \setminus A) \cup \tilde{\mu}(B)\},$$
$$\{m \in M \mid gIm \text{ for all } g \in G \text{ with } \tilde{\gamma}g \in \tilde{\gamma}A \cup \diagup \tilde{\mu}(M \setminus B)\})$$
$$= (A, B) \text{ by Proposition 2.}$$

Finally, for a completely distributive complete lattice L with a singular element p satisfying $\bigvee \mathcal{A}(L) \geq p \geq \bigwedge \mathcal{C}(L)$, we consider the formal context $(\mathcal{J}(L) \setminus \mathcal{A}([0, p]), \mathcal{M}(L) \setminus \mathcal{C}([p, 1]), \leq)$. For this context we define $\tilde{\gamma} : \mathcal{J}(L) \setminus \mathcal{A}([0, p]) \to L$ and $\tilde{\mu} : \mathcal{M}(L) \setminus \mathcal{C}([p, 1]) \to L$ as the respective identity mappings. Then, by Section 2, $\tilde{\gamma}(\mathcal{J}(L) \setminus \mathcal{A}([0, p])) \cup \diagup \tilde{\mu}(\mathcal{M}(L) \setminus \mathcal{C}([p, 1]))$ is $\bigvee$-dense in L and $\tilde{\mu}(\mathcal{M}(L) \setminus \mathcal{C}([p, 1])) \cup \diagdown \tilde{\gamma}(\mathcal{J}(L) \setminus \mathcal{A}([0, p]))$ is $\bigwedge$-dense in L, i.e. conditions (1) and (2) are satisfied; condition (3) is obviously true. Thus, $L \cong \mathfrak{V}(\mathcal{J}(L) \setminus \mathcal{A}([0, p]), \mathcal{M}(L) \setminus \mathcal{C}([p, 1]), \leq)$.

Corollary 1. *The preconcept lattices are (up to isomorphism) the completely distributive complete lattices in which the supremum of all atoms is equal or greater than the infimum of all coatoms.*

The Basic Theorem on Preconcept Lattices offers a useful method for checking line diagrams of preconcept lattices. This shall be demonstrated by the line diagram in Fig. 4 representing the preconcept lattice L of the formal context in Fig. 1 where L has the special property $\bigvee \mathcal{A}(L) = \bigwedge \mathcal{C}(L)$. The object names are attached excactly to the circles representing an element of $\mathcal{J}(L) \setminus \mathcal{A}(L)$ and the attribute names are attached exactly to the circles representing an element of $\mathcal{M}(L) \setminus \mathcal{C}(L)$; furthermore, the labels to the atom circles indicate the $\nearrow$-images of the attribute preconcepts and the labels to the coatom circles indicate the $\nwarrow$-images of the object preconcepts. Therefore, the circles of the $\bigvee$-irreducible elements and the $\bigwedge$-irreducible elements are properly labelled which confirms the conditions (1) and (2); checking condition (3) is also straightforward. Thus, the line diagram in Fig. 4 represents indeed the preconcept lattice of the formal context in Fig. 1.

References

[Ba54]   V. K. Balachandran: A characterization of $\Sigma\Delta$-rings of subsets. *Fundamenta Mathematica* **41** (1954), 38–41.

[BD74]   R. Balbes, Ph. Dwinger: *Distributive lattices.* University of Missouri Press, Columbia, Missouri 1974.

[GW99]   B. Ganter, R. Wille: *Formal Concept Analysis: mathematical foundations.* Springer, Heidelberg 1999; German version: Springer, Heidelberg 1996.

[Pi73]   J. Piaget: Einführung in die genetische Erkenntnistheorie. suhrkamp taschenbuch wissenschaft 6. Suhrkamp, Frankfurt 1973.

[Ra52]   G. N. Raney: Completely distributive lattices. Proc. Amer. Matm. Soc. **3** (1952), 677-680.

[SW86]   J. Stahl, R. Wille: Preconcepts and set representations of contexts. In: W. Gaul, M. Schader (eds.): *Classification as a tool of research.* North-Holland, Amsterdam 1986, 431–438.

[Wi04]   R. Wille: Preconcept algebras and generalized double Boolean algebras. In: P. Eklund (ed.): *Concept lattices.* LNAI **2961**. Springer, Heidelberg 2004, 1–13.

The Tensor Product as a Lattice of Regular Galois Connections

Markus Krötzsch[1] and Grit Malik[2]

[1] AIFB, Universität Karlsruhe, Germany
[2] Institut für Algebra, Technische Universität Dresden, Germany

Abstract. Galois connections between concept lattices can be represented as binary relations on the context level, known as *dual bonds*. The latter also appear as the elements of the *tensor product* of concept lattices, but it is known that not all dual bonds between two lattices can be represented in this way. In this work, we define *regular* Galois connections as those that are represented by a dual bond in a tensor product, and characterize them in terms of lattice theory. Regular Galois connections turn out to be much more common than irregular ones, and we identify many cases in which no irregular ones can be found at all. To this end, we demonstrate that irregularity of Galois connections on sublattices can be lifted to superlattices, and observe close relationships to various notions of distributivity. This is achieved by combining methods from algebraic order theory and FCA with recent results on dual bonds. Disjunctions in formal contexts play a prominent role in the proofs and add a logical flavor to our considerations. Hence it is not surprising that our studies allow us to derive corollaries on the contextual representation of deductive systems.

1 Introduction

From a mathematical perspective, Formal Concept Analysis (FCA) [1] is usually considered as a formalism for syntactically representing complete lattices by means of formal contexts. A closer look reveals that this representation hinges upon the fact that the well-known (context) derivation operations of FCA constitute *Galois connections*[1] between certain power set lattices. Thus formal contexts can equally well be described as convenient representations of such Galois connections.

With this in mind, it should not come as a surprise that Galois connections between concept lattices have also been studied extensively. On the level of formal contexts, such Galois connections can be described through suitable types of binary relations, called *dual bonds* in the literature, which turn out to be a very versatile tool for further studies. Dual bonds arguably constitute a fundamental notion for the study of interrelations and mappings between concept lattices. Indeed, many well-known morphisms of FCA, such as *infomorphisms* and *scale measures*, have recently been recognized as special types of dual bonds [2]. We review some of the relevant results on dual bonds in Sect. 3.

Now dual bonds themselves allow for a nice representation in terms of FCA: each extent of the *direct product* of two contexts is a dual bond between them. The concept

---

[1] In this work, we study Galois connections only in their classical *antitone* formulation, as is done in [1].

R. Missaoui and J. Schmid (Eds.): ICFCA 2006, LNAI 3874, pp. 89–104, 2006.

lattice of the direct product is known as the *tensor product* of FCA. However, this representation of dual bonds is usually not complete: the majority of dual bonds, herein called *regular* dual bonds, appears in the direct product, but there can also be irregular ones for which this is not the case.

Interestingly, this situation is also reflected by the corresponding Galois connections and we can distinguish regular and irregular Galois connections on purely lattice theoretical grounds. This allows us to give a lattice theoretical characterization of the tensor product in Sect. 4. Due to this relation to the tensor product, there are always plenty of regular Galois connections while irregular ones can be very rare. A major goal of this work is to further explore this situation, thus shedding new light on the structure of Galois connections between concept lattices and on the lattice theoretical relevance of the tensor product. In particular, we identify various cases for which only regular Galois connections exist, such that the tensor product yields a complete representation of the according function space.

In Sect. 5, we observe close relationships to the notion of complete distributivity, which will be a recurrent theme in this work. Disjunctions in formal contexts play a prominent role in Sect. 6 and add a logical flavor to our considerations. Hence it is not surprising that our studies allow us to derive corollaries on the contextual representation of deductive systems. Moreover, in Sect. 7, we demonstrate that irregularity of Galois connections on sublattices can be lifted to superlattices, which allows us to establish further relationships with distributivity. Finally, Sect. 8 summarizes our results and points to various open questions that need to be addressed in future research.

2   Preliminaries and Notation

Our notation basically follows [1], with a few exceptions to enhance readability for our purposes. Especially, we avoid the use of the symbol ′ to denote the operations that are induced by a context. We shortly review the main terminology using our notation, but we assume that the reader is familiar with the notation and terminology from [1]. Our treatment also requires some familiarity with general notions from order theory [3].

A *(formal) context* $\mathbb{K}$ is a triple (G, M, I) where G is a set of *objects*, M is a set of *attributes*, and $I \subseteq G \times M$ is an *incidence relation*. Given $O \subseteq G$ and $A \subseteq M$, we define:

$$O^I := \{m \in M \mid g \, I \, m \text{ for all } g \in O\}, \qquad I(O) := \{m \in M \mid g \, I \, m \text{ for some } g \in O\},$$
$$A^I := \{g \in G \mid g \, I \, m \text{ for all } m \in A\}, \qquad I^{-1}(A) := \{g \in G \mid g \, I \, m \text{ for some } m \in A\}.$$

By $\text{Ext}(\mathbb{K})$ and $\text{Int}(\mathbb{K})$ we denote the lattices of extents and intents of $\mathbb{K}$, respectively, ordered by subset inclusion.

The complement of a context $\mathbb{K}$ is defined as $\mathbb{K}^c = (G, M, \cancel{I})$ with $\cancel{I} := (G \times M) \setminus I$. We remark that $\text{Ext}(\mathbb{K}^c)$ in general has no simple relationship to $\text{Ext}(\mathbb{K})$: even if two contexts represent isomorphic concept lattices, this is not necessarily true for their complements.

Finally, an antitone Galois connection $\phi = (\vec{\phi}, \overleftarrow{\phi})$ between complete lattices K and L is a pair of functions $\vec{\phi} : K \to L$ and $\overleftarrow{\phi} : L \to K$ such that $k \le \overleftarrow{\phi}(l)$ iff $l \le \vec{\phi}(k)$, for all $k \in K, l \in L$. Each component of a Galois connection uniquely determines the other, so we will often work with only one of the two functions. For further details, see [1].

3 Dual Bonds and the Tensor Product

To represent Galois connections between concept lattices on the level of the respective contexts, one uses certain relations called *dual bonds*. In this section, we recount some results that are essential to our subsequent investigations. Details and further references can be found in [2].

Definition 1. *A dual bond between formal contexts $\mathbb{K} := (G, M, I)$ and $\mathbb{L} := (H, N, J)$ is a relation $R \subseteq G \times H$ for which the following hold:*

- *for every element $g \in G$, g^R (which is equal to $R(g)$) is an extent of $\mathbb{L}$ and*
- *for every element $h \in H$, h^R (which is equal to $R^{-1}(h)$) is an extent of $\mathbb{K}$.*

This definition is motivated by the following result.

Theorem 1 ([1–Theorem 53]). *Consider a dual bond R between contexts $\mathbb{K}$ and $\mathbb{L}$ as above. The mappings*

$$\vec{\phi}_R : \mathsf{Ext}(\mathbb{K}) \to \mathsf{Ext}(\mathbb{L}) : X \mapsto X^R \quad and \quad \overleftarrow{\phi}_R : \mathsf{Ext}(\mathbb{L}) \to \mathsf{Ext}(\mathbb{K}) : Y \mapsto Y^R$$

form an antitone Galois connection between the concept lattices of $\mathbb{K}$ and $\mathbb{L}$.
   Given such a Galois connection $(\vec{\phi}, \overleftarrow{\phi})$, the relation

$$R_{(\vec{\phi}, \overleftarrow{\phi})} := \left\{ (g, h) \mid h \in \vec{\phi}(g^{II}) \right\} = \left\{ (g, h) \mid g \in \overleftarrow{\phi}(h^{JJ}) \right\}$$

is a dual bond, and these constructions constitute a bijection between the dual bonds between $\mathbb{K}$ and $\mathbb{L}$, and the Galois connections between $\mathsf{Ext}(\mathbb{K})$ and $\mathsf{Ext}(\mathbb{L})$.

The previous result allows us to switch between different context representations of dual bonds in a canonical way. Indeed, consider contexts $\mathbb{K}$, $\mathbb{K}'$, $\mathbb{L}$, and $\mathbb{L}'$ such that $\mathsf{Ext}(\mathbb{K}) \cong \mathsf{Ext}(\mathbb{K}')$ and $\mathsf{Ext}(\mathbb{L}) \cong \mathsf{Ext}(\mathbb{L}')$. Then any dual bond R between $\mathbb{K}$ and $\mathbb{L}$ bijectively corresponds to a dual bond R' between $\mathbb{K}'$ and $\mathbb{L}'$ that represents the same Galois connection. We describe this situation by saying that R and R' are *equal up to isomorphism* (of concept lattices) or just *equivalent*.

Since extents are closed under intersections, the same is true for the set of all dual bonds between two contexts. Thus the dual bonds (and hence the respective Galois connections) form a closure system, and one might ask for a way to cast this into a formal context which has dual bonds as concepts. An immediate candidate for this purpose is the direct product. Given contexts $\mathbb{K} := (G, M, I)$ and $\mathbb{L} := (H, N, J)$, the *direct product* of $\mathbb{K}$ and $\mathbb{L}$ is the context $\mathbb{K} \times \mathbb{L} := (G \times H, M \times N, \nabla)$, where ∇ is defined by setting $(g, h) \nabla (m, n)$ iff $g \, I \, m$ or $h \, J \, n$.

The *tensor product* of two complete lattices is defined as the concept lattice of the direct product of their canonical contexts. As shown in [1–Theorem 26], the tensor product does not depend on using canonical contexts: taking the direct product of any other contexts that represent the factor lattices yields an isomorphic result.

Proposition 1 ([4]). *The extents of a direct product $\mathbb{K} \times \mathbb{L}$ are dual bonds between the contexts $\mathbb{K}$ and $\mathbb{L}$.*

However, it is known that the converse of this result is false, i.e. there are dual bonds which are not extents of the direct product. This motivates the following definition.

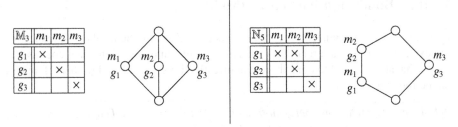

Fig. 1. The lattices M_3 and N_5 with their standard contexts

Definition 2. *A dual bond R between* $\mathbb{K}$ *and* $\mathbb{L}$ *is* regular *if it is an extent of* $\mathbb{K} \times \mathbb{L}$.

Example 1. For some prototypical examples of irregular dual bonds, consider the formal contexts in Fig. 1. Observe that for any complete lattice L, the identity function is a Galois connection between L and its dual order L^{op}. For M_3, this identity is represented by the dual bond $\{(g_1, m_1), (g_2, m_2), (g_3, m_3)\}$ (between the standard context and its dual). For N_5 it is given as $\{(g_1, m_2), (g_1, m_1), (g_2, m_2), (g_3, m_3)\}$. Some easy computations show that both of these dual bonds are irregular.

The next section is concerned with showing that regularity of dual bonds is equivalent to suitable properties of the induced Galois connection. For regularity of dual bonds, the following characterization will be very helpful.

Theorem 2 ([2–Theorem 3]). *Consider contexts* (G, M, I) *and* (H, N, J), *and a dual bond* $R \subseteq G \times H$. *Then R is regular iff* $R(g) = \bigcap_{m \in g^I} R(m^I)^{JJ}$ *for all* $g \in G$.

4   Regularity of Galois Connections

The previous section suggests to extend the notion of regularity from dual bonds to their respective Galois connections.

Definition 3. *A Galois connection* $\phi := (\vec{\phi}, \overleftarrow{\phi})$ *between lattices K and L is* regular *if its associated dual bond* R_ϕ *between the canonical contexts of K and L is regular.*

We know that the lattice structure of the regular Galois connections coincides with the tensor product of the respective lattices, and that the latter does not depend on using canonical contexts in the definition. Whatever contexts are chosen for representing the given complete lattices, the structure of their regular dual bonds is always the same. This, however, does not say that they always represent the same set of Galois connections. In order to obtain this, one needs to show that the isomorphisms used in [1–Theorem 26] for showing the context independence of the tensor product preserve the represented Galois connections. However, using our prior insights on the connections between dual bonds, Galois connections, and the direct product, we can produce an alternative proof which is more suggestive in the current setting.

Lemma 1. *Consider a dual bond R between* $\mathbb{K}$ *and* $\mathbb{L}$, *and a subset A of the set of attributes of* $\mathbb{K}$. *We find that* $A^{R^\triangledown} = \bigcap_{g \notin A^I} R(g)^J$.

Proof. We have $A^{R^\nabla} = \bigcap_{m \in A} R^\nabla(m)$ and, by [2–Lemma 3], this is equal to $\bigcap_{m \in A} R(m^I)^J$. Since $\cdot^J$ transforms unions into intersections, the latter equals $\bigcap_{m \in A} \bigcap_{g \in m^I} R(g)^J$. In other words, the expression is the intersection of the intents $R(g)^J$ for all g such that $g \not{I} m$ for some $m \in A$. But this is just $\bigcap_{g \notin A^I} R(g)^J$ as required.   □

The application of $\cdot^\nabla$ to a binary relation always yields a dual bond between the dual contexts, and thus a Galois connection between the dual concept lattices. Let us state the respective construction as a lattice theoretical operation on Galois connections.

Definition 4. *Consider a Galois connection* $\phi := (\vec{\phi}, \overleftarrow{\phi})$ *between complete lattices K and L. A pair of mappings* $\phi^\nabla := (\vec{\phi}^\nabla, \overleftarrow{\phi}^\nabla)$ *is defined as follows:*

$$\vec{\phi}^\nabla : K^{\mathrm{op}} \to L^{\mathrm{op}} : k \mapsto \bigvee \vec{\phi}(K \setminus {\downarrow}k) \qquad and \qquad \overleftarrow{\phi}^\nabla : L^{\mathrm{op}} \to K^{\mathrm{op}} : l \mapsto \bigvee \overleftarrow{\phi}(L \setminus {\downarrow}l),$$

where $\cdot^{\mathrm{op}}$ *denotes order duals, and* $\bigvee$ *and* ${\downarrow}$ *refer to K and L, not to K^{op} and L^{op}.*

Lemma 2. *Consider a dual bond R between $\mathbb{K}$ and $\mathbb{L}$, and let* $\phi := (\vec{\phi}, \overleftarrow{\phi}) : \mathsf{Ext}(\mathbb{K}) \to \mathsf{Ext}(\mathbb{L})$ *be the according Galois connection as in Theorem 1.*

Then $\phi^\nabla = (\vec{\phi}^\nabla, \overleftarrow{\phi}^\nabla)$ *is a Galois connection from $\mathsf{Ext}(\mathbb{K})^{\mathrm{op}}$ to $\mathsf{Ext}(\mathbb{L})^{\mathrm{op}}$. Up to the isomorphism between the dual lattices of extents and the lattices of intents, it is the Galois connection associated with R^∇.*

Proof. Let $\psi = (\vec{\psi}, \overleftarrow{\psi})$ denote the Galois connection associated with R^∇. Given any intent A of $\mathbb{K}$, we compute

$$\vec{\psi}(A) = A^{R^\nabla} = \bigcap_{g \notin A^I} R(g)^J = \bigcap_{g^{II} \not\subseteq A^I} \vec{\phi}(g^{II})^J = \left(\bigvee_{g^{II} \not\subseteq A^I} \vec{\phi}(g^{II}) \right)^J$$

where we used Lemma 1 for the second equality, and where $\bigvee$ refers to the supremum of extents in $\mathsf{Ext}(\mathbb{L})$. Now it is easy to see that $\bigvee\{\vec{\phi}(g^{II}) \mid g^{II} \not\subseteq A^I\} = \bigvee\{\vec{\phi}(O) \mid O = O^{II}, O \not\subseteq A^I\}$. Indeed, whenever $O = O^{II}$, $O \not\subseteq A^I$ we find some $g \in O$ with $g \notin A^I$. But then $\vec{\phi}(O) \subseteq \vec{\phi}(g^{II})$ which allows for the desired conclusion. With this we conclude that $\vec{\psi}(A) = \left(\bigvee\{\vec{\phi}(O) \mid O = O^{II}, O \not\subseteq A^I\} \right)^J = \left(\bigvee \vec{\phi}(\mathsf{Ext}(\mathbb{K}) \setminus {\downarrow}A^I) \right)^J$.

Now it is easy to see that $\vec{\psi}$ is just the composition of $\vec{\phi}^\nabla$ with the two lattice isomorphisms $\cdot^I : \mathsf{Int}(\mathbb{K}) \cong \mathsf{Ext}(\mathbb{K})^{\mathrm{op}}$ and $\cdot^J : \mathsf{Ext}(\mathbb{L})^{\mathrm{op}} \cong \mathsf{Int}(\mathbb{L})$. Since the property of being a Galois connection is invariant under isomorphism this establishes the claim.   □

The previous result was stated for Galois connections between concept lattices only, which simplified the notation that was needed in the proof. Yet it is easy to see that the result extends to arbitrary Galois connections, since the claimed properties are invariant under isomorphism.

Also observe that the proof of Lemma 2 does not require the fact that ϕ is a Galois connection. This should not come as a surprise, since the operator $\cdot^\nabla$ on binary relations always produces an intent of the direct product, even if the input is not a dual bond.

Now we can easily derive the independence of regularity of Galois connections from the choice of canonical contexts for the representation via dual bonds.

Proposition 2. *Consider a dual bond R and a Galois connection ϕ that is, up to isomorphism of complete lattices, equal to the Galois connection induced by R. Then R is regular iff ϕ is regular. Moreover, for any Galois connection ψ, ψ^∇ is regular.*

Proof. R is regular iff $R = R^{\nabla\nabla}$. By Lemma 2, $\phi^{\nabla\nabla}$ is thus equivalent to the Galois connection induced by R and thus to ϕ. By similar reasoning, given that S is the canonical dual bond for ϕ, the Galois connection induced by $S^{\nabla\nabla}$ is equivalent to ϕ. But this implies that $S = S^{\nabla\nabla}$ such that ϕ is regular. The other direction is shown similarly.

This shows that regularity of Galois connections can be established by considering any (possibly non-canonical) representation by dual bonds. Hence, let R be any dual bond for ψ. Then, by Lemma 2, ϕ^{∇} is represented (up to isomorphism) by R^{∇} (as a dual bond between the dual contexts). Regularity of ϕ^{∇} follows from regularity of R^{∇}.   □

The above theorem asserts that regularity of dual bonds reflects a property not only of a particular dual bond, but of a whole class of dual bonds that represent the same Galois connection. This invariance under change of syntactic representation allows us to choose arbitrary contexts for studying regularity of Galois connections. As a corollary, we obtain that the structure of (binary) tensor products is independent of context representations as well. We find that the study of regularity of Galois connections or dual bonds is synonymous with the study of the structure of the tensor product.

In the remainder of this section, we provide some basic characterizations of regular Galois connections. Since the dual adjoints of a Galois connection uniquely determine each other, we state the following result only for one part of a Galois connection.

Theorem 3. *Given a mapping $\phi : K \to L$ between complete lattices K and L, the following are equivalent:*

  (i) ϕ is part of a regular Galois connection between K and L,

  (ii) $\phi = \phi^{\nabla\nabla}$,

 (iii) For all $k \in K$, $\phi(k) = \bigwedge_{m \not\geq k} \bigvee_{m \not\geq g} \phi(g)$.

Proof. We first show that (i) is equivalent to (ii). By Proposition 2, every mapping of the form $\phi^{\nabla\nabla}$ is a part of a regular Galois connection. For the other direction, by Lemma 2, we find that $\phi^{\nabla\nabla}$ is equivalent to a part of the Galois connection associated with $R^{\nabla\nabla}$ where R is any dual bond for ϕ. The claim follows since $R^{\nabla\nabla} = R$ for regular Galois connections ϕ.

Equivalence of (ii) and (iii) is immediate by noting that $\phi^{\nabla\nabla}(k) = \bigwedge_{m \not\geq k} \bigvee_{m \not\geq g} \phi(g)$, where the second application of ∇ refers to the dual order, so that the order-related expressions have to be dualized.   □

Note that this result establishes a purely lattice theoretical description of the tensor product of FCA, based on the closure operator (in the general order-theoretic sense) $\cdot^{\nabla}$ on Galois connections. An alternative characterization has been derived in [5], where a lattice-theoretical description of the original closure operator $\cdot^{\nabla}$ on subsets of $L \times K$ was given. The latter formulation is substantially more complex and hinges upon certain sets of filters, called *T-carpets*. The advantage of this approach is that it generalizes to n-ary direct products, while our description is specific to the binary case.

5   Regularity for Completely Distributive Lattices

Though the above characterization of regularity is precise, its rather technical conditions are not fully satisfactory for understanding the notion. Especially, it does not significantly enhance our understanding of the sets of regular and irregular Galois connections as a whole. Next, we are going to explore several situations for which all Galois connections must be regular. In these cases, the representation of Galois connections through the direct product is exhaustive, and the tensor product is fully described as the *function space* of all Galois connections between two lattices.

Considering condition (iii) of Theorem 3, it should not come as a surprise that distributivity has an effect on regularity. The following characterization will be useful for formalizing this intuition.

Proposition 3. *A complete lattice K is completely distributive iff, for each pair of elements $g, m \in K$, if $m \not\geq g$, then there are elements $m', g' \in K$ such that:*

- *$m' \not\geq g$ and $m \not\geq g'$,*
- *$K = \uparrow g' \cup \downarrow m'$.*

Proof. For the proof, we use a result on completely distributive concept lattices. Clearly, K is completely distributive iff the concept lattice of its canonical context $\mathbb{K} = (K, K, \leq)$ is completely distributive. Now, according to [1–Theorem 40], the concept lattice of $\mathbb{K}$ is completely distributive iff for every pair $m \not\geq g$ there are elements m', g' such that $m' \not\geq g$, $m \not\geq g'$, and $g' \in k^{\leq\leq}$ for all $k \in K \setminus m'^{\leq}$. In the canonical context, the last condition is equivalent to saying that $g' \leq k$ for all $k \not\leq m'$. Thus $K = \uparrow g' \cup \downarrow m'$, as required.   □

Note that the above also implies that $m \leq m'$ and $g \geq g'$. Now we can apply this result to establish the following sufficient condition for regularity of all Galois connections.

Theorem 4. *Consider complete lattices K and L. If either K or L is completely distributive, then all Galois connections from K to L are regular. Especially, this is the case if K or L are distributive and finite.*

Proof. Consider a Galois connection $\phi := (\vec{\phi}, \overleftarrow{\phi}) : K \to L$. Assume that L is completely distributive. For a contradiction, assume that ϕ is not regular. By Theorem 3, there is an element $k \in K$ such that $\vec{\phi}(k) \neq \bigwedge_{m \not\geq k} \bigvee_{m \not\geq g} \vec{\phi}(g)$. For notational convenience, we define $n := \vec{\phi}(k)$ and $h := \bigwedge_{m \not\geq k} \bigvee_{m \not\geq g} \vec{\phi}(g)$. It is easy to see that $n \leq h$, so we conclude that $n < h$ and thus $n \not\geq h$. This inequality satisfies the conditions of Proposition 3 and we obtain elements $h', n' \in L$ such that $n' \not\geq h$, $n \not\geq h'$, and $L = \uparrow h' \cup \downarrow n'$.

Now suppose that $\overleftarrow{\phi}(h') \geq k$. By anti-monotonicity of $\vec{\phi}$, this implies $\vec{\phi}(\overleftarrow{\phi}(h')) \leq \vec{\phi}(k) = n$. Since $h' \leq \vec{\phi}(\overleftarrow{\phi}(h'))$ (see, e.g., [1]), this entails $h' \leq n$, which contradicts the above assumptions on h'. Thus $\overleftarrow{\phi}(h') \not\geq k$, and we conclude that $h \leq \bigvee_{\overleftarrow{\phi}(h') \not\geq g} \vec{\phi}(g)$.

Now for any $g \in K$, if $\vec{\phi}(g) \geq h'$ then $\overleftarrow{\phi}(\vec{\phi}(g)) \leq \overleftarrow{\phi}(h')$ and thus $g \leq \overleftarrow{\phi}(h')$. Thus, whenever $\overleftarrow{\phi}(h') \not\geq g$, we find that $\vec{\phi}(g) \not\geq h'$. By our assumptions on h' and n', this shows that $\vec{\phi}(g) \leq n'$, and consequently $\bigvee_{\overleftarrow{\phi}(h') \not\geq g} \vec{\phi}(g) \leq n'$.

The conclusions of the previous two paragraphs imply that $h \leq n'$, which yields the desired contradiction. The claim for the case where K is completely distributive follows by symmetry. The rest of the statements is immediate since a finite lattice is distributive iff it is completely distributive.   □

The above proof might seem surprisingly indirect, given that our first motivation for investigating distributivity possibly stems from the interleaved infimum and supremum operations of Theorem 3 (iii). Could a more direct proof just apply distributivity to exchange the position of these operations, thus enabling us to exploit the interplay of Galois connections and infima for further conclusions? The answer is a resounding "no": indeed, the infima and suprema from Theorem 3 distribute over each other in *any* finite lattice[2] but many finite lattices admit irregular dual bonds. Hence, just applying distributivity directly cannot suffice for a proof.

In the finite case, Theorem 4 shows that distributivity of lattices is ensures that only regular Galois connections exist. In Sect. 7 we will see that distributivity is necessary as well.

6   Disjunctions in Contexts

Further sufficient characterizations for regularity of dual bonds have been investigated in [2]. In this section, we combine these ideas with logical considerations along the lines of [7]. The results we obtain are specifically relevant for representations of logical and topological systems within FCA. The following properties of dual bonds constitute our starting point.

Definition 5. *Consider contexts* $\mathbb{K} := (G, M, I)$ *and* $\mathbb{L} := (H, N, J)$. *A relation* $R \subseteq G \times H$ *is* (extensionally) *closed if it preserves extents of* $\mathbb{K}$, *i.e. if for every extent* O *of* $\mathbb{K}$ *the image* $R(O)$ *is an extent of* $\mathbb{L}$. *R is* (extensionally) *continuous if its inverse is extensionally closed.*

In [2] continuity was used to establish relations to continuous functions between contexts, known as *scale measures*. Here, we are mostly interested in the following result, that is a corollary from [2–Theorem 4].

Proposition 4. *A dual bond R between $\mathbb{K}$ and $\mathbb{L}$ is regular, whenever it is closed as a relation from $\mathbb{K}^c$ to $\mathbb{L}$. By symmetry, the same conclusion follows if R is continuous as a relation from $\mathbb{K}$ to $\mathbb{L}^c$.*

Now continuity and closedness, while yielding sufficient conditions for a dual bond to be regular, are not particularly convenient as characterizations either. This is partially due to the fact that these properties, in contrast to regularity, are not independent from the contexts used in the presentation of a dual bond. Though this problem will generally persist throughout our subsequent considerations, the next lemma shows the way to a more convenient characterization.

Lemma 3. *Consider a dual bond R between contexts $\mathbb{K} := (G, M, I)$ and $\mathbb{L} := (H, N, J)$, and an extent $O \in \mathsf{Ext}(\mathbb{K}^c)$. The following are equivalent:*

(i) $R(O)$ *is an extent of* $\mathbb{L}$.

(ii) For every $h \in R(O)^{JJ}$, there is a set $X \subseteq O$ such that $h \in R(X)^{JJ}$ and there is $g \in O$ with $X \subseteq g^{XX}$.

Proof. Clearly, if $R(O)$ is an extent, then for every $h \in R(O)^{JJ}$, there is an element $g \in O$ such that $h \in R(g)$. Since $g \in g^{XX}$, this shows that (i) implies (ii).

---

[2] This is so because it holds for any continuous lattice since the sets $\{g' \mid m \not\geq g'\}$ are directed. See [6].

For the converse, let g be as in (ii). As observed in [2–Lemma 1], we find $g \in x^{II}$ for all $x \in X$. Given $x \in X$, $R^{-1}(y)$ is an extent of $\mathbb{K}$ for all $y \in R(x)$, such that we find $g \in R^{-1}(y)$ for all $y \in R(x)$, i.e. $R(x) \subseteq R(g)$. This yields $R(X) \subseteq R(g)$ which implies $h \in R(g)$ since the latter is an intent of $\mathbb{L}$. $\square$

Intuitively, the previous lemma states that the closure of the image of a set O can be reduced to the closures of the images of single elements g, which are asserted by the definition of dual bonds. As shown in the proof, if the closure is given, then the existence of a sufficient amount of such elements g is certain and the subsets X of (ii) can be chosen as singleton sets. On the other hand, a sufficient condition for showing closure is obtained by requiring the existence of suitable g for arbitrary sets X. The *disjunctions* of X turn out to be just what is needed.

Definition 6. *Consider a context* $\mathbb{K} := (G, M, I)$ *and a set* $X \subseteq G$. *An object* $g \in G$ *is the* disjunction *of* X, *if, for any* $m \in M$, *we have that*

$$g \ I \ m \ \text{iff there is some } x \in X \text{ such that } x \ I \ m.$$

Disjunctions in contexts introduce a logical flavor and have previously been studied in relation with the representation of deductive systems in FCA, e.g. in [8] or [7].

It is easy to see that X has a disjunction iff X^{II} is an object extent g^{II}. A little reflection shows that the existence of disjunctions still strongly depends on the particular context used to represent some complete lattice. Intuitively, this is due to the fact that the concept lattice of $\mathbb{K}^c$ is not fully determined by the concept lattice of $\mathbb{K}$, but depends on the particular representation of $\mathbb{K}$.

Our strategy for deducing regularity of Galois connections from the above observations is as follows: given a Galois connection between complete lattices K and L, we try to find a corresponding dual bond R between contexts, such that the context for K has a "sufficient" amount of disjunctions to show closedness of R. Lemma 3 implies that the existence of arbitrary disjunctions certainly is sufficient in this sense; but weaker assumptions turn out to be sufficient in some cases. The existence of such a closed dual bond R then implies regularity of the Galois connection, even though both closedness and disjunctions may not be given for other representations of the same Galois connection. We discover that typical situations where many disjunctions exist again are closely related to distributivity.

Hence, our task is to seek context representations with a maximal amount of disjunctions. Unfortunately, the canonical context turns out to be mostly useless for this purpose. Indeed, given a complete lattice K, it is easy to see that $X \subseteq K$ has a disjunction in the canonical context iff X has a least element in K. In order to find contexts with more disjunctions, we state the following lemma.

Lemma 4. *Consider a complete lattice* K *and subsets* $J, M \subseteq K$ *such that* M *is* $\bigwedge$-*dense and* J *is* $\bigvee$-*dense in* K. *Then* K *is isomorphic to the concept lattice of* $(J, M, \leq)$. *Furthermore, a subset* $X \subseteq J$ *has a disjunction in* $(J, M, \leq)$ *iff*

(i) $\bigwedge X \in J$ *and*

(ii) $\uparrow \bigwedge X \cap M = \bigcup_{x \in X} \uparrow x \cap M$.

The disjunction then is given by the element $\bigwedge X$.

Proof. The claimed isomorphism of K and the concept lattice is a basic result of FCA, see [1]. According to the definition, g is a disjunction for X precisely when we find that for all $m \in M$, $g \leq m$ iff $x \leq m$ for some $x \in X$. This in turn is equivalent to $\uparrow g \cap M = \bigcup_{x \in X} \uparrow x \cap M$. Obviously this also implies that $g \leq \bigwedge X$. If $g < \bigwedge X$ then, by $\bigwedge$-density of M, there is some $m \in M$ with $m \geq g$ but $m \not\geq \bigwedge X$. But then $\uparrow g \cap M \subset \bigcup_{x \in X} \uparrow x \cap M$ and g cannot be the disjunction. This shows that $g = \bigwedge X$ is the only possible disjunction for X, and that this is the case iff (i) and (ii) are satisfied. $\square$

This result guides our search for contexts with many disjunctions. Indeed, it is easy to see that for (i), it is desirable to have as many objects as possible, while (ii) is more likely to hold for a small set of attributes.

Note that Lemma 4 entails some notational inconveniences, since disjunctions are usually marked by the symbol $\bigvee$. Yet we obtain $\bigwedge$ since we work on disjunctions *of objects*. If one would prefer to do all calculations on attributes (which are often taken to represent formulae when modelling logical notions in FCA), one would obtain $\bigvee$ as expected.

In the finite case, finding a possibly small set of attributes for a given lattice is easy: the set of $\bigwedge$-irreducible elements is known to be the least $\bigwedge$-dense set. While this is also true for some infinite complete lattices, it is not required in this case.

Proposition 5. *Consider a complete lattice K such that the set of $\bigwedge$-irreducible elements $M(K)$ is $\bigwedge$-dense in K. For every set $X \subseteq K$, the following are equivalent:*

(i) *X has a disjunction in the context $(K, M(K), \leq)$,*

(ii) *$k \vee \bigwedge X = \bigwedge_{x \in X}(k \vee x)$, for any $k \in K$.*

In particular, the distributivity law $k \vee \bigwedge X = \bigwedge_{x \in X}(k \vee x)$ holds in K iff $(K, M(K), \leq)$ has all disjunctions.

Proof. For the forward implication, assume that (i) holds for some set $X \subseteq K$. For a contradiction, suppose that there is some $k \in K$ such that (ii) does not hold for k and X. Since $k \vee \bigwedge X \leq \bigwedge_{x \in X}(k \vee x)$ holds in any complete lattice, this says that $k \vee \bigwedge X < \bigwedge_{x \in X}(k \vee x)$. By $\bigwedge$-density of $M(K)$, there is some $m \in M(K)$ such that $k \vee \bigwedge X \leq m$ but $\bigwedge_{x \in X}(k \vee x) \not\leq m$. The former shows that $k \leq m$ and $\bigwedge X \leq m$. Now by (i) and Lemma 4, $x \leq m$ for some $x \in X$, such that $(x \vee k) \leq m$. But then $\bigwedge_{x \in X}(k \vee x) \leq m$, which yields a contradiction.

For other direction, assume that (ii) holds for some set $X \subseteq K$. For a contradiction, suppose that (i) is not true for X. By Lemma 4, we find some attribute $m \in M(K)$ such that $\bigwedge X \leq m$ but $x \not\leq m$ for all $x \in X$. We conclude that $m = m \vee \bigwedge X$. By (ii), this implies $m = \bigwedge_{x \in X}(m \vee x)$. Since m is $\bigwedge$-irreducible, there must be some $x \in X$ with $(m \vee x) = m$ and hence $x \leq m$; but this contradicts our assumptions. $\square$

Note that most parts of the above proof do not make use of the $\bigwedge$-irreducibility of the chosen attribute-set. In particular, the implication from (i) to (ii) holds for arbitrary $\bigwedge$-dense sets of attributes. In the rest of the proof, irreducibility is used only in connection with the set X. Thus we obtain the following corollary.

Corollary 1. *Consider a complete lattice K such that the set of $\wedge$-irreducible elements $M_{Fin}(K)$ is $\bigwedge$-dense in K. Then K is distributive iff $(K, M_{Fin}(K), \leq)$ has all finite disjunctions.*

The previous result is useful for the subsequent consideration of logical deductive systems. Concerning regularity of dual bonds between finite lattices, we will establish a stronger characterization in Theorem 7 later on. We are now ready to combine the above results to derive further sufficient conditions for regularity of Galois connections.

Theorem 5. *Consider a complete lattice K where $M(K)$ is $\bigwedge$-dense and such that the distributivity law $k \vee \bigwedge X = \bigwedge_{x \in X}(k \vee x)$ holds for arbitrary $k \in K$, $X \subseteq K$. Then all Galois connections from K to any other complete lattice are regular.*

Proof. By Lemma 4, the concept lattice of $\mathbb{K} := (K, M(K), \leq)$ is isomorphic to K and, by Proposition 5, $\mathbb{K}$ has all disjunctions. Now any Galois connection from K to some complete lattice L corresponds to a dual bond R from $\mathbb{K}$ to the canonical context $\mathbb{L} := (L, L, \leq)$ of L. Now consider any extent O of $\mathbb{K}^c$. Obviously, the conditions of Lemma 3 (ii) are satisfied, where $X = O$ and g is the disjunction of O. Thus $R(O)$ is an extent of $\mathbb{L}$. Since O was arbitrarily chosen, this shows that R is closed from which we conclude that R is regular, as required. □

Complete lattices for which finite infima distribute over arbitrary suprema are also known as *locales*, and are the subject of study in point-free topology [9]. The reason is that the lattice of open sets of any topological space forms a locale. Thus the previous theorem can be considered as a statement about certain lattices of *closed* sets of topological spaces. On the other hand, the condition that $\bigwedge$-irreducibles be $\bigwedge$-dense is rather severe in this setting. Especialy, it would be possible that the conjunction of these assumptions already implies complete distributivity – see Sect. 8 for some discussions.

A statement similar to Theorem 5 can be made when restricting to finite disjunctions.

Theorem 6. *Consider a distributive complete lattice K where $M_{Fin}(K)$ is $\bigwedge$-dense. Then all Galois connections from K to any other complete algebraic lattice are regular. Especially, this applies to Galois connections from K to finite lattices.*

Proof. The proof proceeds as in Theorem 5. However, to apply Lemma 3, we note that for any $l \in R(O)^{\leq\leq}$, there is a *finite* set $Y \subseteq R(O)$ such that $l \in Y^{\leq\leq}$. This follows directly from algebraicity of L, see e.g. [1]. Thus there is a finite set $X \subseteq O$ with $Y \subseteq R(X)$ and its disjunction g allows us to invoke Lemma 3 as desired. □

Again this theorem can be related to topology, but in a way that is quite different from Theorem 5 and which finally relates disjunctions in contexts to logical disjunctions. The key observation is that $\bigwedge$-density of $\wedge$-irreducibles is the characteristic property for the *open* set lattices of certain topological spaces, called *sober* spaces in the literature [9]. Being locales as all topologies, these spaces are finitely distributive as well. Thus the conditions on K given in Theorem 6 are satisfied by the open set lattice of any sober space. These structures are commonly known as *spacial locales*, and have been studied extensively in research on point-free topology. We obtain the following corollary.

Corollary 2. *Every Galois connection between a spacial locale and an algebraic lattice is regular. In particular, every Galois connection between algebraic spacial locales is regular.*

To see how these observations are connected to logic, we have a brief look at the presentation of deductive systems in formal concept analysis. In general, deductive systems are characterized by a semantic *consequence relation* ⊨ between models and formulae of some logic. For an example, consider the consequence relation between models of propositional logic and propositional formulae.

Now such binary relations are naturally represented as formal contexts. This idea has, with more or less explicit reference to FCA, been investigated within *Institution Theory* [10] and the theory of *Information Flow* [8]. At this point, it is not apparent how this relates to topology, locales, and algebraicity. This relation can be established on quite general grounds, but here we just sketch the situation for propositional logic as an exemplary case.

Thus consider a language of propositional logic as a set of objects. For attributes, consider any set of models of some propositional logic theory.[3] With semantic consequence as incidence relation, this yields a "logical" context that represents a given theory. It is easy to see that disjunctions of the logic correspond to object-disjunctions within this context. On the other hand, disjunctions in the *complemented* context correspond to *conjunctions* of the logic. The according concept lattice is the lattice of logical theories over this background knowledge. In particular, object extents represent knowledge that is given by single formulae, while their order in the concept lattice describes entailment. As is well-known, the sublattice of object extents is a Boolean algebra in the propositional case. Furthermore, the lattice of propositional theories is known to be algebraic: every logical consequence can be derived from only a finite set of assumptions (in other words, propositional statements cannot describe infinite information).

Moreover, the complement of a logical context represents another interesting concept lattice: it is isomorphic to the open set lattice of a topological space, the so-called *Stone space* of the aforementioned Boolean algebra.[4] Basically, this is just an FCA version of Stone's famous representation theorem for Boolean algebras (see [3] for an introduction). It is well-known that open set lattices of Stone spaces are algebraic spacial locales, so that we can immediately conclude from Corollary 2 that every dual bond between complements of logical contexts in the above sense is regular. However, another consequence of our observations is more interesting from a logical perspective.

Corollary 3. *Consider contexts $\mathbb{K}$ and $\mathbb{L}$ that represent theories of propositional logic as described above. Then any dual bond from $\mathbb{K}^c$ to $\mathbb{L}$ is regular. More specifically, it is closed from $\mathbb{K}$ to $\mathbb{L}$ and continuous from $\mathbb{K}^c$ to $\mathbb{L}^c$.*

Proof. Regularity follows immediately from Corollary 2 and the above remarks. For closedness, we can apply Lemma 3, using algebraicity of Ext($\mathbb{L}$) (the lattice of theories) and the availability of finite conjunctions in the propositional logic of $\mathbb{K}$. Likewise, for continuity, we combine algebraicity of Ext($\mathbb{K}^c$) (the open set lattice) with finite propositional disjunctions of $\mathbb{L}$.   □

---

[3] This differs from the more common approach where formulae or "properties" are usually taken as attributes. This deviation ensures compatibility to our object-centered treatment. Also note that we do in general not consider the set of *all* propositional models, since the context would not contain much information in this case.

[4] Note that it is not the open set lattice, since the latter is not a closure system. However, the order-dual lattice of intents is exactly the according lattice of closed sets.

The above formulation exhibits a seemingly strange twist in the dual bond, since we consider the complement of the context $\mathbb{K}$. However, this formulation fits well into our logical framework, since such dual bonds can be interpreted as proof theoretical *consequence relations* between two logical theories. To see this, note that a a logical implication $p \rightarrow q$ can be translated into $\neg p \vee q$. Based on this intuition, it makes perfect sense to interpret the dual bond of Corollary 3 as a set of logical implications. The defining conditions on dual bonds now state that the consequences of any single statement from $\mathbb{K}$ are deductively closed in $\mathbb{L}$, and that the sets of premises of a statement from $\mathbb{L}$ are deductively closed in $\mathbb{K}^c$. Observe how this justifies regularity of all such consequence relations: given any binary relation between the logical languages, we can always derive an adequate consequence relation by computing missing deductive inferences. In logic, this process is usually described by application of certain deductive rules, while in FCA it corresponds to the concept closure within the direct product.

The reason for emphasizing closedness and continuity in Corollary 3 is that these properties enable us to compose consequence relations in a very intuitive way. Indeed, if p implies q, and q implies r, then one is usually tempted to derive that p implies r. Using dual bonds to represent implication, such reasoning is described by taking the relational product. Continuity and closedness ensure that this construction does again yield a dual bond as in Corollary 3. Hence we obtain a *category* of logical theories and consequence relations, the sets of morphisms of which can be described by the tensor product of FCA. However, to the author's knowledge, the resulting categories have not yet been investigated with respect to their general properties or their relationship to other categories from algebra or order theory.

More details on deductive systems, consequence relations, and their contextual representation are given in [7]. In [11], consequence relations between separate logical theories (and languages) have been introduced proof-theoretically for positive logic (the logic of conjunctions and disjunctions), and the emerging categories were shown to be of topological and domain theoretical relevance. Much more general cases of *Stone duality* and their relation to FCA have been considered in [12].

7   Regularity for Sublattices

In this section, we show that the irregularity of a Galois connection between sublattices can be lifted to a Galois connection between their respective superlattices, which enables us to improve our characterization of the interplay between distributivity and regularity.

Proposition 6. *Consider complete lattices K, L, U, and V such that $U \subseteq K$ and $V \subseteq L$ and*

(i) *for any non-empty set $X \subseteq U$, we have $\bigwedge_U X = \bigwedge_K X$ and $\bigvee_U X = \bigvee_K X$,*

(ii) *for any non-empty set $Y \subseteq V$, we have $\bigwedge_V Y = \bigwedge_L Y$ and $\bigvee_V Y = \bigvee_L V$.*

Then any irregular Galois connection between U and V induces an irregular Galois connection between K and L.

Proof. We use $\bot_K = \bigwedge K$ and $\top_K = \bigvee K$ to denote the least and greatest elements of K, respectively. Similar notations are used for U and L. Let $\phi := (\vec{\phi}, \overleftarrow{\phi}) : U \to V$ be an irregular Galois connection. Define a mapping $\psi : K \to L$ by setting

$$\psi(k) := \begin{cases} \top_L & \text{if } k = \bot_K \\ \vec{\phi}(\bigwedge\{u \in U \mid k \leq u\}) & \text{if } \bot_K < k \leq \top_U \\ \bot_L & \text{if } k \not\leq \top_U \end{cases}$$

Note that we do not have to distinguish between infima in K and in U for the second case, since the considered set is always non-empty. We claim that ψ is one part of an irregular Galois connection from K to L.

Consider some set $X \subseteq K$. To see that ψ is a Galois connection, it suffices to show that $\psi(\bigvee X) = \bigwedge(\psi(X))$ (see [1–Proposition 7]). It $\bigvee X = \bot_K$ then X is either empty or contains only $\bot_K$. Both cases are easily seen to satisfy the claim. If $\bigvee X \not\leq \top_U$, then there is some $x \in X$ such that $x \not\leq \top_U$, i.e. $\psi(x) = \bot_L$. Again the claim is obvious.

It remains to consider the case where $\bot_K \leq \bigvee X \leq \top_U$. To this end, first note that $\bigwedge\{u \in U \mid x \leq u \text{ for all } x \in X\} = \bigvee_{x \in X} \bigwedge\{u \in U \mid x \leq u\}$ (∗). Indeed, the left hand side (lhs) is greater-or-equal than the right hand side (rhs). Assuming that it is strictly greater, the rhs is not among the u on the left, i.e. there is $x \in X$ with $x \not\leq$ rhs. Since $x \leq$ rhs $\leq$ lhs, this yields a contradiction.

Furthermore, we can assume without loss of generality that $\bot_K \notin X$, since there is certainly some greater element in X as well, making $\bot_K$ redundant in all considered operations. We compute:

$$\psi(\bigvee X) = \vec{\phi}(\bigwedge\{u \in U \mid \bigvee X \leq u\}) \overset{*}{=} \vec{\phi}(\bigvee_{x \in X} \bigwedge\{u \in U \mid x \leq u\})$$
$$= \bigwedge_{x \in X} \vec{\phi}(\bigwedge\{u \in U \mid x \leq u\}) = \bigwedge_{x \in X} \psi(x).$$

This finishes our proof that ψ is part of a Galois connection. To see that it is irregular, we use condition (iii) of Theorem 3. By the assumption that ψ is irregular, there is some $u \in U$ such that $\vec{\phi}(u) \neq \bigwedge_{m \in U,\, m \not\geq u} \bigvee_{g \in U,\, m \not\geq g} \vec{\phi}(g)$. Since the left hand side is always smaller-or-equal to the right hand side, this inequality is in fact strict. We have to show that $\psi(u) \neq \bigwedge_{m \not\geq u} \bigvee_{m \not\geq g} \psi(g)$. Since $\vec{\phi}(u) = \psi(u)$, this follows by showing that $\bigwedge_{m \in U,\, m \not\geq u} \bigvee_{g \in U,\, m \not\geq g} \vec{\phi}(g) \leq \bigwedge_{m \not\geq u} \bigvee_{m \not\geq g} \psi(g)$. To obtain the latter, we observe that, for any $m \not\geq u$, there is some $n \in U$ with $n \not\geq u$ and $\bigvee_{g \in U,\, n \not\geq g} \vec{\phi}(g) \leq \bigvee_{m \not\geq g} \psi(g)$. Indeed, consider $m \not\geq u$ and set $n := \bigvee\{u \in U \mid u \leq m\}$. We claim that, for any $v \in U$ with $n \not\geq v$, $\vec{\phi}(v) \leq \bigvee_{m \not\geq g} \psi(g)$. But this is obvious since $\vec{\phi}(v) = \psi(v)$ and $n \not\leq v$ implies $m \not\leq v$. By what was said before, this finishes the proof of irregularity of ψ. □

As a corollary to this result, we find that distributivity is necessary to assert that only regular Galois connections exist for some complete lattice.

Corollary 4. *If a complete lattice K has only regular Galois connections to any other lattice, then it is distributive.*

Proof. For a contradiction, assume that L is not distributive. Then L has either M_3 or N_5 as a sublattice. We have seen in Example 1 that both of these have an irregular Galois connection to some other lattice, so Proposition 6 yields the required contradiction. □

Summarizing our results, we obtain a satisfactory characterization of regularity for doubly-founded complete lattices.[5]

Theorem 7. *Given a doubly-founded complete lattice L, the following are equivalent:*

(i) L is distributive,

(ii) L has all disjunctions,

(iii) L has only regular Galois connections to any other complete lattice.

Proof. Recall that since L is doubly-founded, $M(L)$ is $\bigwedge$-dense in L, and that distributivity is equivalent to complete distributivity in this case [1–Theorem 41]. Thus (i) is equivalent to (ii) by Proposition 5. The implication from (i) to (iii) was stated in Theorem 4. The other direction follows from Corollary 4.   □

8   Summary and Outlook

In this work, we identified a novel property of Galois connections, dubbed regularity, which describes whether a Galois connection between two complete lattices is represented in their FCA tensor product. We characterized this property and identified several cases for which only regular Galois connections exist. These cases are of particular interest, since they enable us to represent the function space of all Galois connections by means of the tensor product, thus providing a lattice theoretical motivation for this construction.

Though we applied rather diverse proof strategies based on ideas from FCA, order algebra, and logic, many results expose relationships to notions of distributivity. It is known from Theorem 4 that complete distributivity of a lattice disallows irregular Galois connections to any other lattice, but a full characterization of this situation was only established for complete lattices that are doubly-founded (Theorem 7). We conjecture that a similar result holds for the general case, i.e. that a complete lattice admits only regular Galois connections to *any* other lattice iff it is completely distributive. Theorem 5 described other, seemingly weaker, conditions for enforcing regularity, but it is conceivable that these assumptions entail complete distributivity as well. Confirming or refuting these conjectures remains the subject of future work.

Apart form this immediate question, the present work shows many other directions for future research. First and foremost, we have concentrated on characterizations that refer to only one lattice at a time. This allowed us to identify specific situations where regularity is ubiquitous, but it also neglects the fact that in general both lattices contribute to regularity. Future investigations should take this into account, for example by studying appropriate sublattices. Proposition 6 provides a theoretical foundation for this approach.

Considering mainly situations where no irregular Galois connections exist at all, we evaded the question for the role of irregular elements within the complete lattice of all Galois connections. Can the irregular elements be described lattice theoretically within this setting? We think that our results constitute the first steps towards such studies.

---

[5] Recall that every finite lattice is doubly-founded [1].

Last but not least, a completely different field of further questions was highlighted in Sect. 6, where we sketched fresh categories of deductive systems that use dual bonds as their morphisms. The study of these categories and their relevance in the field of logic/topology/domain theory remains open.

References

1. Ganter, B., Wille, R.: Formal Concept Analysis: Mathematical Foundations. Springer (1999)
2. Krötzsch, M., Hitzler, P., Zhang, G.Q.: Morphisms in context. In: Conceptual Structures: Common Semantics for Sharing Knowledge. Proceedings of the 13th International Conference on Conceptual Structures, ICCS '05, Kassel, Germany. (2005) Extended version available at www.aifb.uni-karlsruhe.de/WBS/phi/pub/KHZ05tr.pdf.
3. Davey, B.A., Priestley, H.A.: Introduction to Lattices and Order. second edn. Cambridge University Press (2002)
4. Ganter, B.: Relational Galois connections. *Unpublished manuscript* (2004)
5. Wille, R.: Tensor products of complete lattices as closure systems. Contributions to General Algebra **7** (1991) 381–386
6. Gierz, G., Hofmann, K.H., Keimel, K., Lawson, J.D., Mislove, M., Scott, D.S.: Continuous Lattices and Domains. Volume 93 of Encyclopedia of Mathematics and its Applications. Cambridge University Press (2003)
7. Krötzsch, M.: Morphisms in logic, topology, and formal concept analysis. Master's thesis, Technische Universität Dresden (2005)
8. Barwise, J., Seligman, J.: Information flow: the logic of distributed systems. Volume 44 of Cambridge tracts in theoretical computer science. Cambridge University Press (1997)
9. Johnstone, P.T.: Stone spaces. Cambridge University Press (1982)
10. Goguen, J., Burstall, R.: Institutions: abstract model theory for specification and programming. Journal of the ACM **39** (1992)
11. Jung, A., Kegelmann, M., Moshier, M.A.: Multi lingual sequent calculus and coherent spaces. Fundamenta Informaticae **XX** (1999) 1–42
12. Erné, M.: General Stone duality. Topology and its Applications **137** (2004) 125–158

Two Instances of Peirce's Reduction Thesis

Frithjof Dau and Joachim Hereth Correia

Technische Universität Dresden, Institut für Algebra,
D-01062 Dresden
{dau, heco}@math.tu-dresden.de

Abstract. A main goal of Formal Concept Analysis (FCA) from its very beginning has been the support of rational communication by formalizing and visualizing concepts. In the last years, this approach has been extended to traditional logic based on the doctrines of concepts, judgements and conclusions, leading to a framework called *Contextual Logic*. Much of the work on Contextual Logic has been inspired by the Existential Graphs invented by Charles S. Peirce at the end of the 19th century. While his graphical logic system is generally believed to be equivalent to first order logic, a proof in the strict mathematical sense cannot be given, as Peirce's description of Existential Graphs is vague and does not suit the requirements of contemporary mathematics.

In his book 'A Peircean Reduction Thesis: The Foundations of topological Logic', Robert Burch presents the results of his project to reconstruct in an algebraic precise manner Peirce's logic system. The resulting system is called Peircean Algebraic Logic (PAL). He also provides a proof of the Peircean Reduction Thesis which states that all relations can be constructed from ternary relations in PAL, but not from unary and binary relations alone.

Burch's proof relies on a major restriction on the allowed construction of graphs. Removing this restriction renders the proof much more complicated. In this paper, a new approach to represent an arbitrary graph by a relational normal form is introduced. This representation is then used to prove the thesis for infinite and two-element domains.

1   Introduction

From its very beginning, FCA was not only understood as an approach to restructure lattice theory (see [Wil82]) but also as a method to support rational communication among humans and as a concept-oriented knowledge representation. While FCA supports communication and argumentation on a concept level, an extended approach was needed to also support the representation of judgments and conclusions. This led to the development of contextual logic (see [DK03, Wil00]).

Work on contextual logic has been influenced by the Conceptual Graphs invented by John Sowa (see [Sow84, Sow92]). These graphs are in turn inspired by the Existential Graphs from Charles S. Peirce. In Peirce's opinion the main purpose of logic as a mathematical discipline is to analyze and display reasoning

R. Missaoui and J. Schmid (Eds.): ICFCA 2006, LNAI 3874, pp. 105–118, 2006.

in an easily understandable fashion. While he also contributed substantially to the development of the linear notation of formal logic, he considered the later developed Existential Graphs as superior notation (see [PS00, Pei35a]).

Intuitively, the system of Existential Graphs seems equivalent to first order logic. However, a proof in the strict mathematical sense cannot be given based on Peirce's work. His description of Existential Graphs is too vague to suit the requirements of contemporary mathematics.

To solve this problem, Robert Burch studied the large range of Peirce's philosophical work and presented in [Bur91] his results on attempting an algebraization of Peirce's logic system. This algebraic logic is called *Peirce's Algebraic Logic*. He uses this logic system to prove Peirce's reduction thesis, namely, that ternary relations suffice to construct arbitrary relations, but that not all relations can be constructed from unary and binary relations alone. While this thesis is not stated explicitly in Peirce's work [Pei35b], this idea appears repeatedly.

Burch's proof depends on a restriction on the constructions allowed in PAL: the juxtaposition of disjoint graphs is only allowed as last or second-last operation. While Burch proves that the expressivity is still the same, this restriction is a major difference to the original system of Existential Graphs. Removing this restriction make the PAL-system more similar to both the system of Existential Graphs and to the system of relational algebra. The equivalence of this restriction-free PAL and relational algebra has been shown in [HCP04]. The proof of Peirce's Reduction Thesis however is more complicated if we cannot rely on this restriction.

In this paper we provide the first steps toward the proof, concentrating on the special cases of a domain with only two elements and of domains with infinitely many elements. To achieve this, we define representations of the constructed relations similar to the disjunctive normal form (DNF) known from first-order propositional logic. Taking advantage of some properties the relations in the DNF have, we can then prove the reduction thesis for the two special cases.

Organization of This Paper

In the following section we provide the basic definitions used in this paper. To simplify notation in the later parts, we in particular introduce a slight generalization of *relation* in Def. 1. Together with the definition of the PAL-graph we then introduce the disjunctive normal form in Section 3. In the following sections, we prove Peirce's Reduction Thesis for infinite and two-element domains. We conclude the paper with an outlook on further research in this area.

2   Basic Definitions

Relations in the classical sense are sets of tuples, that is relations are subsets of A^n where A is an arbitrary set (in the following called *domain*) and n is a natural number, which is the arity of the relation. However, this definition leads to unnecessary complications when discussing the interpretation of the algebraically defined PAL-graphs. While the elements of a tuple are clearly ordered, the same cannot be said about the arcs and nodes of a graph. Consequently, it is difficult and leads to cumbersome notations if we force such an order onto the interpretation of the graphs.

For this reason we introduce the following generalization, we consider relations where the places of the relations are indicated by natural numbers.

Definition 1 (Relations). *Let $I \subseteq \mathbb{N}$ be finite. A I-*ARY RELATION OVER* A is a set $\varrho \subseteq A^I$, i. e. a set of mappings from I to A.*

While looking slightly more complicated at first, this definition is compatible with the usual one. Any n-tuple can be interpreted as a mapping from the set $\{1, \ldots, n\}$ to A. Instead of downsets of the natural numbers as domain of the mapping, we now allow arbitrary but finite subsets of $\mathbb{N}$.

If R is an I-ary relation over A and if $J \supseteq I$, then R can be canonically extended to a J-ary relation R' by $R' := \{f : J \to A \mid f|_I \in R\}$. In this work, we use the implicit convention that all relations are extended if needed. To provide an example, let R be an I-ary relation and S be a J-ary relation. With $R \cap S$ we denote the $I \cup J$-ary relation $R' \cap S'$, where R' is R extended to $I \cup J$ and S' is S extended to $I \cup J$.

We will use the following notations to denote the arity of a relation: usually we will append the arity as lower index to the relation name. Thus R_I denotes an I-ary relation. In Sec. 5, it is convenient to append the elements of I as lower indices to R. For example, both $R_{i,j}$ and $R_{j,i}$ are names for an $\{i, j\}$-ary relation. The elements of a relation will be noted in the usual tuple-notion with round brackets, where we use the order of the lower indices. For example, both $R_{i,j} := \{(a, b), (b, b)\}$ and $R_{j,i} := \{(b, a), (b, b)\}$ denote the same relation, namely the relation $\{f_1, f_2\}$ with $f_1(i) = a$, $f_1(j) = b$ and $f_2(i) = b$, $f_2(j) = b$.

Note that $\emptyset$-ary relations are allowed. There are exactly two $\emptyset$-ary relation, namely $\emptyset$ and $\{\emptyset\}$.

From given relations, we can construct new relations. In mathematics, this usually refers to relational algebra. In this paper, we use the PAL-operations as introduced by Burch in [Bur91]. While the operations from relational algebra provide the same expressive power (see [HCP04]), the PAL operations concentrate on a different aspect. The *teridentity* is the three-place equality, that is (in the notation of standard mathematical relations as used in [HCP04]) the relation $\doteq_3 := \{(a, a, a) \mid a \in A\}$. It plays a crucial role for Peirce and also in Burch's book. The core of the Peircean Reduction Thesis is that with the teridentity any relation can be constructed from the unary and binary (or the ternary) relations, but from unary and binary relations alone one cannot construct the teridentity. This means that the teridentity would be somehow hidden in the operations from relational algebra. As the operations have the same expressivity, we can define each operation of one system by operations from the other. The operations of relational algebra can easily be expressed in PAL using the teridentity, but this is at least difficult for the identification of the first two coordinates ([HCP04], Def. 2,R3), that is $\Delta(\varrho) := \{(a_1, \ldots, a_{m-1}) \mid (a_1, a_1, \ldots, a_{m-1}) \in \varrho\}$, and for the union of relations without teridentity. Proving the Peircean Reduction Thesis will show that this is not only difficult but impossible.

In relational algebra, we can construct the teridentity in relational algebra using product, cyclic shift ζ (the tuples are rotated: see [HCP04], Def. 2,R2a) and identification of the first two coordinates from the binary identity: $\doteq_3 = \Delta(\zeta(= \times =))$. As product and cyclic shift are also in PAL, we could deduce after a final proof, that teridentity is indeed involved in the identification of the first two coordinates.

The PAL-operations found by Burch also have an easy graphical interpretation as shown in [HCP04]. We will use this notatition (see Def. 3).

1. **Negation:** If R is an I-ary relation, then

$$\neg R := A^I \backslash R$$

2. **Product:** If R is an I-ary relation, S is an J-ary relation, and we have $I \cap J = \emptyset$, then

$$R \times S := \{f : I \cup J \to A \mid (f|_I \in R) \wedge (f|_J \in S)\}$$

3. **Join:** If R is an I-ary relation with $i, j \in I$, $i \neq j$, then

$$\delta^{i,j}(R) := \{f : I\backslash\{i,j\} \to A \mid \exists F \in R : (F|_{I\backslash\{i,j\}} = f) \wedge (F(i) = F(j))\}$$

We need two further technical operations which do not belong to PAL (but they can be constructed within PAL), but which are needed in the ongoing proofs:

1. **Projection:** Let $I := \{i, j\}$ and R be an I-ary relation. Then

$$\pi_i(R) = \{(f(i)) \mid f \in R\} \quad \text{and} \quad \pi_j(R) = \{(f(j)) \mid f \in R\}$$

2. **Renaming:** If R is an I-ary relation with $i \in I$ and $j \notin I$, we set

$$\sigma^{i \to j}(R) := \{f|_{I\backslash\{i\}} \cup \{(j, f(i))\} \mid f \in R\}$$

Finally, for a given domain A, we need names for some special relations. With $\doteq_I$ we denote the I-ary identity relation, i.e. $\{f : I \to A \mid \exists a \in A \forall i \in I : f(i) = a\}$. For three-element sets I, this identity is called the I-ary TERIDENTITY. We will write $\doteq_I$ to emphasize this. With $\neq_I$ we denote the complement of the teridentity. With A_I or A_I^n (we assume $|I| = n$) we denote the I-ary universal relation A^I.

After the neccessary definitions for relations, we can now define PAL-graphs over I. They are basically mathematical graphs (multi-hypergraphs), enriched with an additional structure describing the cuts. The vertices are either labelled with an element of I (then such a vertex is a free place of the graph), or with an additional sign '$*$' (in this case, the vertex denotes an unqualified, existentially quantifed object).

Definition 2 (PAL-Graphs). *For* $I \subseteq \mathbb{N}$*, a structure* $(V, E, \nu, \top, Cut, area, \kappa, \varrho)$ *is called an* I-ARY PAL-GRAPH OVER A *iff*

1. *V, E and Cut are pairwise disjoint, finite sets whose elements are called* VERTICES, EDGES *and* CUTS, *respectively,*

2. $\nu : E \to \bigcup_{k \in \mathbb{N}} V^k$ *is a mapping,*[1]

3. $\top$ *is a single element with* $\top \notin V \cup E \cup Cut$, *called the* SHEET OF ASSERTION,

4. $area : Cut \cup \{\top\} \to \mathfrak{P}(V \cup E \cup Cut)$ *is a mapping such that*
   a) $c_1 \neq c_2 \Rightarrow area(c_1) \cap area(c_2) = \emptyset$,
   b) $V \cup E \cup Cut = \bigcup_{d \in Cut \cup \{\top\}} area(d)$,
   c) $c \notin area^n(c)$ *for each* $c \in Cut \cup \{\top\}$ *and* $n \in \mathbb{N}$ *(with* $area^0(c) := \{c\}$ *and* $area^{n+1}(c) := \bigcup \{area(d) \mid d \in area^n(c)\}$*).*

5. $\kappa : E \to \bigcup_{n \in \mathbb{N}} \mathfrak{P}(A^n)$ *is a mapping with* $\kappa(e) \subseteq A^n$ *for* $|e| = n$ *(see below for the notion of* $|e|$*),*

6. $\varrho : V \to I \cup \{*\}$ *is a mapping such that for each* $i \in I$, *there is exactly one vertex* v_i *with* $\varrho(v_i) = i$, *this vertex is incident with exactly one edge and we have* $v_i \in area(\top)$, *and*

7. $\mathfrak{G}$ *has* DOMINATING NODES, *i.e., for each edge* $e = (v_1, \dots, v_k)$ *and each incident vertex* $v_i \in \{v_1,, \dots, v_k\}$, *there is* $e \in area^n(cut(v_i))$ *for an* $n \geq 1$ *(see below for.the notions of* $e = (v_1, \dots, v_k)$ *and* $cut(v_i)$*).*

For an edge $e \in E$ *with* $\nu(e) = (v_1, \dots, v_k)$ *we set* $|e| := k$ *and* $\nu(e)\big|_i := v_i$. *Sometimes, we also write* $e\big|_i$ *instead of* $\nu(e)\big|_i$, *and* $e = (v_1, \dots, v_k)$ *instead of* $\nu(e) = (v_1, \dots, v_k)$. *We set* $E^{(k)} := \{e \in E \mid |e| = k\}$.

As for every $x \in V \cup E \cup Cut$ *there is exactly one context* $c \in Cut \cup \{\top\}$ *with* $x \in area(c)$, *we can write* $c = area^{-1}(x)$ *for every* $x \in area(c)$, *or even more simple and suggestive:* $c = cut(x)$.

We set $V^* := \{v \in V \mid \varrho(v) = *\}$ *and* $V^? := \{v \in V \mid \varrho(v) \in \mathbb{N}\}$, *and we set* $FP(\mathfrak{G}) := I$ *('FP' stands for 'free places').*

In the following, PAL-graphs will be abbreviated by PG.

An example for this definition is the following PG:

$$\mathfrak{G} := (\ \{v_1, v_2, v_3, v_4\}, \{e_1, e_2, e_3\}, \{(e_1, (v_1, v_2)), (e_2, (v_2, v_3)), (e_3, (v_3, v_4))\},$$
$$\top, \{c_1, c_2\}, \{(\top, \{v_1, v_2, e_1, c_1\}), (c_1, \{v_3, v_4, e_3, c_2\}), (c_2, \{e_2\})\},$$
$$\{(e_1, \text{emp}), (e_2, \text{work}), (e_3, \text{proj})\}, \{(v_1, 1), (v_2, 2), (v_3, *), (v_4, *)\}\ )$$

Below, the left diagram is a possible representation of $\mathfrak{G}$. In the right diagram, we have sketched furthermore assignments of the elements (the vertices, edges, and cuts) of the $\mathfrak{G}$ to the graphical elements of the diagram. The precise conventions on how the graphs are diagrammatically represented will be given in Def. 3.

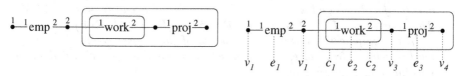

(This is a standard example for querying relational databases. If emp relates names of employees and their ids, proj relates description of projects and their ids, and work is a relation between employee ids and project describing which

---

[1] We set $\mathbb{N} := \{1, 2, 3, \dots\}$ and $\mathbb{N}_0 := \mathbb{N} \cup \{0\}$.

employee works in which project, then this graph retrieves all employees who work in all projects.)

A PG $\mathfrak{G}$ with $\mathrm{FP}(\mathfrak{G}) = I$ describes the I-ary relation of all tuples $(a_1, \ldots, a_n)$ such that when the free places of $\mathrm{FP}(\mathfrak{G})$ are replaced by $a_1, \ldots, a_n$, we obtain a graph which evaluates to true. Following the approach of [Dau03] and [Dau04], PGs have been defined in one step, and the evaluation of graphs could be defined analogously to the evaluation of concept/query graphs with cuts, which is done over the tree of contexts $Cut \cup \{\top\}$. In this paper, we follow a different approach.

PGs can be defined inductively as well, such that the inductive construction of PGs corresponds to the operations on relations. In the following, this inductive construction of PGs is introduced, and we define the semantics of the graphs along their inductive construction. Moreover, a graphical representation of PGs is provided as well.

Definition 3 (Inductive Definition of PGs, Semantics, Graphical Representation).

1. **Atomar graphs:** Let R be an I-ary relation with $I = \{i_1, \ldots, i_n\} \neq \emptyset$. Let $R' := \{(f(i_1), \ldots, f(i_n)) \mid f \in R\}$ be the corresponding 'ordinary' n-ary relation over the domain A. The graph

$$(\{v_1, \ldots, v_n\}, \{e\}, \{(e, (v_1, \ldots, v_n))\}, \top, \emptyset, \emptyset, \{(e, R')\}, \{(v_1, i_1), \ldots, (v_n, i_n)\})$$

is the atomic PG corresponding to R. If this graph is named $\mathfrak{G}$, we see that $\mathfrak{G}$ is an I-ary PG. We set $\mathcal{R}(\mathfrak{G}) := R$.

Graphically, a vertex v of $\mathfrak{G}$ with $\varrho(v) = *$ is depicted as bold spot $\bullet$, and a vertex v with $\varrho(v) = i$ is labelled with i. The edge $e = (v_1, \ldots, v_n)$ is depicted by its label $R := \kappa(e)$, which is linked for each vertex v_i, $i = 1, \ldots, n$ to its representing sign. This line is labelled with i. For example, the following diagrams depict the same $\{1, 3, 5, 8\}$-ary relation R:

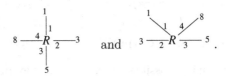

2. **Cut Enclosure:** Let $\mathfrak{G} := (V, E, \nu, \top, Cut, area, \kappa, \varrho)$ be an I-ary PAL-graph. Let c be a fresh cut (i.e., $c \notin E \cup V \cup Cut \cup \{\top\}$). Then let $\neg\mathfrak{G}$ be the PG defined by $(V, E, \nu, \top, Cut', area', \kappa, \varrho)$ with $Cut' := Cut \cup \{c\}$, $area'(d) := area(d)$ for $d \neq c$ and $d \neq \top$, $area'(\top) := V^?$ and $area'(c) := area(\top) \backslash V^?$. This graph is an I-ary PG. We set $\mathcal{R}(\neg\mathfrak{G}) := (R)^c := A^I \backslash \mathcal{R}(\mathfrak{G})$.

In the graphical notation, all elements of the graph, except the vertices labelled with a free place, are enclosed by a finely drawn, closed line, the cut-line of c. For example,

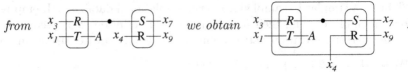

3. **Juxtaposition:** Let $\mathfrak{G}_1 := (V_1, E_1, \nu_1, \top_1, Cut_1, area_1, \kappa_1, \varrho_1)$ be an I-ary PG and let $\mathfrak{G}_2 := (V_2, E_2, \nu_2, \top_2, Cut_2, area_2, \kappa_2, \varrho_2)$ be a J-ary PG such that $\mathfrak{G}_1$ and $\mathfrak{G}_2$ are disjoint, and I and J are disjoint. The JUXTAPOSITION OF $\mathfrak{G}_1$ AND $\mathfrak{G}_2$ is defined to be the PG $\mathfrak{G} := (V, E, \nu, \top, Cut, area, \kappa, \varrho)$:

$$\mathfrak{G}_1 \ \mathfrak{G}_2 := (V_1 \cup V_2, E_1 \cup E_2, \nu_1 \cup \nu_2, \top, Cut_1 \cup Cut_2, area, \kappa_1 \cup \kappa_2, \varrho_1 \cup \varrho_2)$$

where $\top$ is a fresh sheet of assertion (part. $\top \neq \top_1, \top_2$), and we set $area(c) := area_i(c)$ for $c \in Cut_i$, $i = 1, 2$, and $area(\top) := area_1(\top_1) \cup area_2(\top_2)$. This graph is an $I \cup J$-ary PG. We set $\mathcal{R}(\mathfrak{G}_1 \ \mathfrak{G}_2) := \mathcal{R}(\mathfrak{G}_1) \times \mathcal{R}(\mathfrak{G}_2)$.

In the graphical notation, the juxtaposition of $\mathfrak{G}_1$ and $\mathfrak{G}_2$ is simply noted by writing the graphs next to each other, i.e. we write: $\mathfrak{G}_1 \ \mathfrak{G}_2$.

4. **Join:** Let $\mathfrak{G} := (V, E, \nu, \top, Cut, area, \kappa, \varrho)$ be an I-ary PG, and let $i, j \in I$ with $i \neq j$. Let v_i, v_j be the vertices with $\varrho(v_i) = i$ and $\varrho(v_j) = j$. Let v be a fresh vertex. Then the JOIN OF i AND j FROM $\mathfrak{G}$ is

$$\delta^{i,j}(\mathfrak{G}) := (V', E, \nu', \top, Cut, area', \kappa, \varrho')$$

with $V' := V \setminus \{v_i, v_i\} \cup \{v\}$, ν' satisfies $\nu'(e)|_k := \nu(e)|_k$ for $\nu(e)|_k \neq v_i, v_j$ and $\nu'(e)|_k := v$ otherwise, $area'(c) := area(c)$ for $c \in Cut$ and $area'(\top) := area(\top) \setminus \{v_i, v_i\} \cup \{v\}$, and $\varrho'(w) := \varrho(w)$ for $w \neq v$ and $\varrho'(v) := *$. This graph is an $I \setminus \{i, j\}$-ary PG. We set $\mathcal{R}(\delta^{i,j}(\mathfrak{G})) := \delta^{i,j}(\mathcal{R}(\mathfrak{G}))$.

In the graphical notation, the vertices v_i, v_j are both replaced by the same, heavily drawn dot, which stands for an existential quantified object. For example, with joining the vertices with 2 and 8,

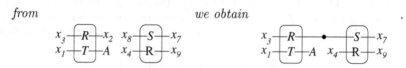

We have seen in the definition that all inductively constructed graphs are PGs. On the other hand, for a given PG $\mathfrak{G} := (V, E, \nu, \top, Cut, area, \kappa, \varrho)$, it can easily be shown by induction over the tree of contexts $Cut \cup \{\top\}$ that $\mathfrak{G}$ can be constructed with the above PAL-operations, and that different inductive constructions of $\mathfrak{G}$ yield the same semantics and the same graphical representation. Thus for each PG $\mathfrak{G}$, we have a well-defined meaning $\mathcal{R}(\mathfrak{G})$ and a well-defined graphical representation of $\mathfrak{G}$.

Graphs similar to PGs have already been studied by one of the authors in [Dau03] and [Dau04]. In [Dau03], *concept graphs with cuts*, which are based on Peirce's Existential Graphs and which, roughly speaking, correspond to closed formulas of first order logic, have been investigated. In [Dau04], concept graphs with cuts are syntactically extended to *query graphs with cuts* by adding labelled query markers to their alphabet, so query graphs with cuts are evaluated to *relations* in models. Both [Dau03] and [Dau04] focus on providing sound and complete calculi for the systems. This is done as common in mathematical logic, that is, graphs are defined as purely syntactical structures, built over an alphabet of names, which gain their meaning when their alphabet are interpreted in models.

Both query graphs with cuts and PGs are graphs which describe relations. The main difference between these types of graphs is as follows: PGs are *semantical* structures, that is, we directly assign relations to the edges of PGs, instead of assigning relation names, which then would have to be interpreted in models. Moreover, in query graphs with cuts, object names may appear, objects are classified by *types*, and we have orders on the set of types and relation names. From this point of view, PGs can be considered to be restrictions of query graphs with cuts, but this restriction is only a minor one.

3   Disjunctive Normal Form for PGs

Let $\mathfrak{G} := (V, E, \nu, \top, Cut, area, \kappa, \varrho)$ be a PG. Let $\sim$ be the smallest equivalence relation on V such that for all $e = (v_1, \ldots, v_n)$, there is $v_1 \sim v_2 \sim \ldots \sim v_n$, and for $v \sim v'$, we say that v and v' are CONNECTED. As for each free place $i \in \mathrm{FP}(\mathfrak{G})$ there exists a uniquely given vertex $w_i \in V$ with $\varrho(w_i) = i$, this equivalence relation is transferred to $\mathrm{FP}(\mathfrak{G})$ by setting $i \sim j :\Leftrightarrow w_i \sim w_j$. Finally we set

$$P(\mathfrak{G}) := \{[i]_\sim \mid i \in \mathrm{FP}(\mathfrak{G})\} \cup \{\emptyset\} \quad .$$

$P(\mathfrak{G})$ is simply the set of all equivalence classes, together with the empty set $\emptyset$. Next we show that for a PG $\mathfrak{G}$, the relation $\mathcal{R}(\mathfrak{G})$ can be described as a union of intersections of I-ary relations with $I \in P(\mathfrak{G})$. In the proof, we may obtain $\emptyset$-ary relations , that is why we have to add $\emptyset$ to $P(\mathfrak{G})$.

Theorem 1 (Disjunctive Normal Form (DNF) for Relations described by PGs). *Let $\mathfrak{G}$ be a PG. Then there is a $n \in \mathbb{N}$, and for each $m \in \{1, \ldots, n\}$ and for each class $p \in P(\mathfrak{G})$ there is a p-ary relation R_p^m, such that we have*

$$\mathcal{R}(\mathfrak{G}) = \bigcup_{m \in \{1,\ldots,n\}} \bigcap_{p \in P(\mathfrak{G})} R_p^m$$

The relations R_p^m shall be called GROUND RELATIONS OF $\mathfrak{G}$.

Proof: The proof is done by induction over the construction of PGs.

1. Atomar graphs: If R is an relation and $\mathfrak{G}_R$ be the corresponding atomar graph, it is easy to see that the theorem holds for $\mathfrak{G}_R$ by setting $n := 1$ and $R_p^1 := R$.
2. Juxtaposition: Let $\mathfrak{G}_1, \mathfrak{G}_2$ be two PGs with $\mathrm{N}(\mathfrak{G}_1) \cap \mathrm{N}(\mathfrak{G}_2) = \emptyset$. If we use the letter R to denote the relations of $\mathfrak{G}_1$ and the letter S to denote the relations of $\mathfrak{G}_2$, we have

$$\mathcal{R}(\mathfrak{G}_1) = \bigcup_{m \in \{1,\ldots,n_1\}} \bigcap_{p \in P(\mathfrak{G}_1)} R_p^m \quad \text{and} \quad \mathcal{R}(\mathfrak{G}_2) = \bigcup_{m \in \{1,\ldots,n_2\}} \bigcap_{p \in P(\mathfrak{G}_2)} S_p^m$$

Thus we have with the canonical extension of the ground relations

$$\mathcal{R}(\mathfrak{G}) = \left(\bigcup_{m \in \{1,\ldots,n_1\}} \bigcap_{p \in P(\mathfrak{G}_1)} R_p^m \right) \cap \left(\bigcup_{m \in \{1,\ldots,n_2\}} \bigcap_{p \in P(\mathfrak{G}_2)} S_p^m \right)$$

Now an application of the distributive law, using $P(\mathfrak{G}) = P(\mathfrak{G}_1) \cup P(\mathfrak{G}_2)$ and $n := n_1 + n_2$, yields the theorem for $\mathfrak{G}$.

3. Cut enclosure: We consider $\neg\mathfrak{G}$. Due to the induction hypothesis, we have

$$\mathcal{R}(\mathfrak{G}) = \bigcup_{m\in\{1,\dots,n\}} \bigcap_{p\in P(\mathfrak{G})} R_p^m$$

Thus, using De Morgan's law, we have

$$\mathcal{R}(\neg\mathfrak{G}) = \Big(\bigcup_{m\in\{1,\dots,n\}} \bigcap_{p\in P(\mathfrak{G})} R_p^m \Big)^c = \bigcap_{m\in\{1,\dots,n\}} \bigcup_{p\in P(\mathfrak{G})} (R_p^m)^c$$

Similar to the last case, we apply the distributive law to obtain a union of intersections of relations. Due to the distributive law, given a class $p \in P(\mathfrak{G})$, the p-ary ground relations of $\neg\mathfrak{G}$ are intersections of 0 up to d relations $(R_p^m)^c$, and these intersections are relations over p, too. Thus the theorem holds for $\neg\mathfrak{G}$ as well.

4. Join: We consider $\mathfrak{G}$ and two distinct free places $i, j \in \mathbb{N}(\mathfrak{G})$. With $q :=$ $([i]_\sim \cup [j]_\sim)\backslash\{i,j\}$, we have $P(\delta^{i,j}(\mathfrak{G})) = P(\mathfrak{G})\backslash\{[i]_\sim, [j]_\sim\} \cup \{q\}$. Now we conclude

$$\delta^{j,k}(\mathcal{R}(\mathfrak{G})) = \delta^{j,k}\left(\bigcup_{m\in\{1,\dots,n\}} \bigcap_{p\in P(\mathfrak{G})} R_p^m \right)$$

$$= \bigcup_{m\in\{1,\dots,n\}} \delta^{j,k}\left(\bigcap_{p\in P(\mathfrak{G})} R_p^m \right)$$

$$= \bigcup_{m\in\{1,\dots,n\}} \bigcap_{p\in P(\mathfrak{G}), j,k\notin p} \left(R_p^m \cap \delta^{j,k}\left(R_{[i]_\sim}^e \cap R_{[j]_\sim}^e \right) \right)$$

As $\delta^{j,k}(R_{[i]_\sim}^e \cap R_{[j]_\sim}^e)$ is a q-ary relation, we are done. $\square$

4   Proof of the Peircean Reduction Thesis for Infinite Domains

Using the theorem from the last section, the first instance of the Peircean Reduction Thesis can easily be shown as a corollary. Before that, some observations about the theorem and its proof are provided.

For a given PG $\mathfrak{G} := (V, E, \nu, \top, Cut, area, \kappa, \varrho)$, the relations R_p^m in Thm. 1 depend on the relations which appear in $\mathfrak{G}$, i.e., they depend on κ, but the proof of Thm. 1 yields that the number n of disjuncts $\bigcap_{p\in P(\mathfrak{G})} R_p^m$ does not depend on κ. That is, if we denote n by $n(\mathfrak{G})$, two PGs $\mathfrak{G}_1$, $\mathfrak{G}_2$ which differ only in κ, i.e., $\mathfrak{G}_1 = (V, E, \nu, \top, Cut, area, \kappa_1, \varrho)$ and $\mathfrak{G}_2 = (V, E, \nu, \top, Cut, area, \kappa_2, \varrho)$), satisfy $n(\mathfrak{G}_1) = n(\mathfrak{G}_2)$.

Now we are prepared to prove the reduction thesis for infinite domains with a simple counting argument.

Corollary 1 (Reduction Thesis for infinite Domains). *Let $\mathfrak{G}$ be an I-ary PG over a domain A with $|I| = 3$, and let each relation in $\mathfrak{G}$ have an arity ≤ 2. If we have $|A| > n(\mathfrak{G})$, then $\mathcal{R}(\mathfrak{G}) \neq \dot{=}_I$. Particularly, for an infinite set A, there exists no PG which evaluates to the teridentity on A.*

Proof: W.l.o.g. let $\mathrm{FP}(\mathfrak{G}) = \{1, 2, 3\}$. As each relation of $\mathfrak{G}$ has an arity ≤ 2, we cannot have $1 \sim 2 \sim 3$. For the proof, we assume that we have two equivalent free places (the case $P(\mathfrak{G}) = \{\{1\}, \{2\}, \{3\}, \emptyset\}$ can be proven analogously), and w.l.o.g. let $2 \sim 3$. Now Thm. 1 yields

$$\mathcal{R}(\mathfrak{G}) = \bigcup_{m \in \{1, \ldots, n\}} R_\emptyset^m \cap R_1^m \cap R_{2,3}^m$$

Now let A be a domain with $|A| > n(\mathfrak{G})$. Assume $\mathcal{R}(\mathfrak{G}) = \dot{=}_{1,2,3}$. Then there exists an $m \leq n$ and distinct $a, b \in A$ with $(a, a, a), (b, b, b) \in R_\emptyset^m \cap R_1^m \cap R_{2,3}^m$. We obtain $R_\emptyset^m = \{\emptyset\}$, $(a), (b) \in R_1^m$ and $(a, a), (b, b) \in R_{2,3}^m$, thus we have $(a, b, b), (b, a, a) \in R_\emptyset^m \cap R_1^m \cap R_{2,3}^m$, too, which is a contradiction. □

5   Peirce's Reduction Thesis for Two-Element Domains

In the last section, we have proven Peirce's reduction thesis with a counting argument. But this argument does not apply to finite domains. For example, if $A = \{a_1, \ldots, a_n\}$ is an n-element domain, one might think that we can construct a PG such that its DNF has n disjuncts, each of them evaluating to exactly one triple $\{(a_i, a_i, a_i)\}$. In this section, we show that for two-element domains $A = \{a, b\}$, there is no PG $\mathfrak{G}$ with $\mathcal{R}(\mathfrak{G}) = \dot{=}_3$. This is done by classifying the relations over A into classes such that no class is suited to describe (a, a, a) in one disjunct and (b, b, b) in another disjunct, and by proving that the operations on relations 'respect' the classes.

For a relation $R_{i,j}$, we set $\gamma_x^i(R_{i,j}) := \{y \mid (x, y) \in R_{i,j}\}$. Now we define the following classes:[2]

$$C_a^{i,j} := \{R_{i,j} \mid \gamma_a^i(R_{i,j}) \supseteq \gamma_b^i(R_{i,j})\} \qquad \text{and} \qquad C_a^i := \{\emptyset, \{a\}, \{a, b\}\}$$
$$C_b^{i,j} := \{R_{i,j} \mid \gamma_b^i(R_{i,j}) \supseteq \gamma_a^i(R_{i,j})\} \qquad \text{and} \qquad C_b^i := \{\emptyset, \{b\}, \{a, b\}\}$$
$$C_{\dot{=}}^{i,j} := \{\emptyset_{i,j}^2, \dot{=}_{i,j}, A_{i,j}^2\}, \quad C_{\neq}^{i,j} := \{\emptyset_{i,j}^2, \neq_{i,j}, A_{i,j}^2\} \quad \text{and} \quad C_{a,b}^i := \{\{a, b\}\}$$

For our purpose, the intuition behind this definition is as follows: $R_{i,j} \in C_a^{i,j}$ means that b cannot be separated (in position i) resp. $R_{i,j} \in C_b^{i,j}$ means that a cannot be separated (in position i).

PGs are built up inductively with the construction steps juxtaposition, cut enclosure, and join. The next three lemmata show how the classes are respected by the corresponding operations for relations.

---

[2] Recall the notion $R_{i,j}$ for an $\{i, j\}$-ary relation, and recall that both $R_{i,j}$ and $R_{j,i}$ denote the same relation. But for the definition of γ and the classes $C_a^{i,j}$, $C_b^{i,j}$, the order of the indices is important. For example, given a relation $R_{i,j}$ $(= R_{j,i})$, it might happen that $R_{i,j} \in C_a^{i,j}$ and $R_{j,i} \in C_b^{j,i}$.

To ease the notation, we abbreviate the composition of product and join. So let $\mathfrak{G}$ be an PG and let $i, j \in \mathrm{FP}(\mathfrak{G})$ with $i \not\sim j$. Then we write

$$R_{[i]_\sim} \circ_{j,k} R_{[j]_\sim} := \delta^{j,k}\left(R_{[i]_\sim} \cap R_{[j]_\sim}\right) \quad \left(= \delta^{j,k}\left(R_{[i]_\sim} \times R_{[j]_\sim}\right)\right)$$

Next we investigate how these classes are respected by the operations on relations. We start with the classes $C_a^{i,j}$ and $C_b^{i,j}$.

Lemma 1 (Class-Inheritance for $C_a^{i,j}$ and $C_b^{i,j}$). *Let $R_{i,j} \in C_a^{i,j}$. Then:*

1. *If $S_{k,l}$ is arbitrary, then $R_{i,j} \circ_{j,k} S_{k,l} \in C_a^{i,l}$*
2. *If S_k is arbitrary, then $R_{i,j} \circ_{j,k} S_k \in C_a^i$*
3. *$\neg(R_{i,j}) \in C_b^{i,j}$*
4. *$C_a^{i,j}$ is closed under (possibly empty) finite intersections (with $\bigcap \emptyset = A_{i,j}^2$).*

The analogous propositions hold for $R_{i,j} \in C_b^{i,j}$ as well.

Proof:

1. Let $(b, y) \in R_{i,j} \circ_{j,k} S_{k,l} \in C_a^{i,l}$. Then there exists x with $(b, x) \in R_{i,j}$ and $(x, y) \in S_{k,l}$. From $R_{i,j} \in C_a^{i,j}$ we obtain $(a, x) \in R_{i,j}$, thus we have $(a, y) \in R_{i,j} \circ_{j,k} S_{k,l}$ as well. So we conclude $R_{i,j} \circ_{j,k} S_{k,l} \in C_a^{i,l}$.
2. Done analogously to the last case.
3. We have $\gamma_x(\neg R_{i,j}) = (\gamma_x(R_{i,j}))^c$ for $x \in \{a, b\}$. So we get $R_{i,j} \in C_a^{i,j} \Leftrightarrow \gamma_a(R_{i,j}) \supseteq \gamma_b(R_{i,j}) \Leftrightarrow (\gamma_a(R_{i,j}))^c \subseteq (\gamma_b(R_{i,j}))^c \Leftrightarrow \gamma_a(\neg R_{i,j}) \subseteq \gamma_b(\neg R_{i,j}) \Leftrightarrow \neg R_{i,j} \in C_b^{i,j}$
4. If $R_{i,j}^n, n \in \mathbb{N}$ are arbitrary relations, we have $\gamma_x(\bigcap_{n \in N} R_{i,j}^n) = \bigcap_{n \in N} \gamma_x(R_{i,j}^n)$, which immediately yields this proposition. $\square$

The next lemma corresponds to Lem. 1, now for the class $C_{\underset{=}{}}^{i,j}$.

Lemma 2 (Class-Inheritance for $C_{\underset{=}{}}^{i,j}$). *Let $R_{i,j} \in C_{\underset{=}{}}^{i,j}$. Then:*

1. *If $S_{k,l} \in C_{\underset{=}{}}^{k,l}$, then $R_{i,j} \circ_{j,k} S_{k,l} \in C_{\underset{=}{}}^{i,l}$.*
   If $S_{k,l} \in C_{\neq}^{k,l}$, then $R_{i,j} \circ_{j,k} S_{k,l} \in C_{\neq}^{i,l}$.
   If $S_{k,l} \in C_a^{k,l}$, then $R_{i,j} \circ_{j,k} S_{k,l} \in C_a^{i,l}$.
   If $S_{k,l} \in C_b^{k,l}$, then $R_{i,j} \circ_{j,k} S_{k,l} \in C_b^{i,l}$.
2. *If $S_k \in C_{a,b}^k$, then $R_{i,j} \circ_{j,k} S_k \in C_{a,b}^i$.*
   If $S_k \in C_a^k$, then $R_{i,j} \circ_{j,k} S_k \in C_a^i$.
   If $S_k \in C_b^k$, then $R_{i,j} \circ_{j,k} S_k \in C_b^i$.
3. *$\neg(R_{i,j}) \in C_{\neq}^{i,l}$.*
4. *$C_{\underset{=}{}}^{i,j}$ is closed under (possibly empty) finite intersections.*

Proof:

1. For each relation $R_{k,l}$ we have $\doteq_{i,j} \circ_{j,k} R_{k,l} = \sigma^{k \to i}(R_{k,l})$.
   For each relation $R_{k,l}$ we have $A_{i,j}^2 \circ_{j,k} R_{k,l} = A_i \times \pi_l(R_{k,l})$. Particularly, for each relation $R_{i,j}$, we have both $A_{i,j}^2 \circ_{j,k} R_{k,l} \in C_a^{i,l}$ and $A_{i,j}^2 \circ_{j,k} R_{i,j} \in C_b^{i,l}$.
   Moreover, for $R_{k,l} \in C_{\underset{=}{}}^{k,l}$ or $R_{k,l} \in C_{\neq}^{k,l}$, we have $A_{i,j}^2 \circ_{j,k} R_{k,l} = A_{i,l}^2$.
   From these obervations we conclude this proposition.

2. For each relation R_k we have $\doteq_{i,j} \circ_{j,k} R_k = \sigma^{k \to i}(R_k)$.
   For each relation $R_k \neq \emptyset$ we have $A^2_{i,j} \circ_{j,k} R_k = A_i$, for $R_k = \emptyset$ we have $A^2_{i,j} \circ_{j,k} R_k = \emptyset$.
   From these obervations we conclude this proposition.
3. Trivial.
4. Trivial. $\qquad\square$

Of course, we have an analogous lemma for the class $C^{i,j}_{\neq}$. The proof is analogous to the last proof and henceforth omitted.

Lemma 3 (Class-Inheritance for $C^{i,j}_{\neq}$).

Let $R_{i,j} \in C^{i,j}_{\neq}$. Then we have:

1. If $S_{k,l} \in C^{k,l}_{\doteq}$, then $R_{i,j} \circ_{j,k} S_{k,l} \in C^{i,l}_{\doteq}$.
   If $S_{k,l} \in C^{k,l}_{\neq}$, then $R_{i,j} \circ_{j,k} S_{k,l} \in C^{i,l}_{\neq}$.
   If $S_{k,l} \in C^{k,l}_a$, then $R_{i,j} \circ_{j,k} S_{k,l} \in C^{i,l}_a$.
   If $S_{k,l} \in C^{k,l}_b$, then $R_{i,j} \circ_{j,k} S_{k,l} \in C^{i,l}_b$.
2. If $S_k \in C^k_{a,b}$, then $R_{i,j} \circ_{j,k} S_k \in C^i_{a,b}$.
   If $S_k \in C^k_a$, then $R_{i,j} \circ_{j,k} S_k \in C^i_b$.
   If $S_k \in C^k_b$, then $R_{i,j} \circ_{j,k} S_k \in C^i_a$.
3. $\neg(R_{i,j}) \in C^{i,l}_{\doteq}$.
4. $C^{i,j}_{\neq}$ is closed under (possibly empty) finite intersections.

Theorem 2 (Properties of the relations in the DNF for PGs). Let $\mathfrak{G}$ a PG. Let $i \in FP(\mathfrak{G})$ with $\{i\} \in P(\mathfrak{G})$.

Then one of the following properties holds:

1. $R^m_i \in C^i_a$ for all $m \in \{1, \ldots, n\}$
2. $R^m_i \in C^i_b$ for all $m \in \{1, \ldots, n\}$
3. $R^m_i \in C^i_{a,b}$ for all $m \in \{1, \ldots, n\}$

Let $i, j \in FP(\mathfrak{G})$ with $i \sim j$. Then one of the following properties holds:

1. $R^m_{i,j} \in C^{i,j}_{\doteq}$ for all $m \in \{1, \ldots, n\}$
2. $R^m_{i,j} \in C^{i,j}_{\neq}$ for all $m \in \{1, \ldots, n\}$
3. $R^m_{i,j} \in C^{i,j}_a$ for all $m \in \{1, \ldots, n\}$
4. $R^m_{i,j} \in C^{i,j}_b$ for all $m \in \{1, \ldots, n\}$

Proof: The proof is done by induction over the construction of PAL-graphs.

Atomar graphs: For each relation $R_{i,j}$ we have $R_{i,j} \in C^{i,j}_a \cup C^{i,j}_b \cup \{\doteq, \neq\}$. Thus it is easy to see that the theorem holds for atomar graphs.

Juxtaposition: If we consider the juxtaposition of two graphs $\mathfrak{G}_1$, $\mathfrak{G}_2$, then the ground relations of the juxtaposition are the ground relations of $\mathfrak{G}_1$ and the ground relations of $\mathfrak{G}_2$.

Cut enclosure: As said in the proof of Thm.1, given a class $p \in P(\mathfrak{G})$, the p-ary ground relations of $\neg\mathfrak{G}$ are intersections of 0 up to d relations $(R^m_p)^c$, where the relations R_p^m are the p-ary ground relations of $\mathfrak{G}$. First of all, due to Lem. 1.3., 2.3., 3.3., the set of all complements of the ground relations fulfill the property of this theorem. Moreover, due to Lem. 1.4., 2.4., 3.4., all classes $C_{\doteq}^{i,j}$, $C_{\neq}^{i,j}$, $C_a^{i,j}$, $C_b^{i,j}$ are closed under (possibly empty) intersections. Thus the theorem holds for $\neg\mathfrak{G}$ as well.

Join: We consider $\delta^{j,k}(\mathfrak{G})$. We have $\mathbb{N}(\delta^{j,k}(\mathfrak{G})) = \mathrm{FP}(\mathfrak{G})\backslash\{j,k\}$. Due to the proof of Thm. 1, we have to show that the proposition holds for the new ground relations $R_{[j]_\sim}^m \circ_{j,k} R_{[k]_\sim}^m = \delta^{j,k}(R_{[j]_\sim}^m \cap R_{[k]_\sim}^m)$.

First we consider the case that $\{j\},\{k\} \in P(\mathfrak{G})$. We have $P(\delta^{j,k}(\mathfrak{G})) = P(\mathfrak{G})\backslash\{\{j\},\{k\}\}$. For $p \neq \{j\},\{k\}$, the ground relations of $\mathfrak{G}$ and of $\delta^{j,k}(\mathfrak{G})$ which are not over j or k (or over $\emptyset$) are the same, thus we are done. The case $j \sim k$, i.e. $\{j,k\} \in P(\mathfrak{G})$, can be handled analogously.

Next we consider the case that there is an i with $i \sim j$, but there is no l with $k \sim l$. We have $P(\delta^{j,k}(\mathfrak{G})) = P(\mathfrak{G})\backslash\{\{i,j\},\{k\}\} \cup \{\{i\}\}$. The new ground relations are of the form $R_{i,j}^m \circ_{j,k} R_k^m$. We have to do a case distinction, both for $R_{i,j}^m$ and R_k^m.

Assume for example $R_{k,l}^m \in C_{\doteq}^{i,j}$ for all $m \leq n$ and $R_k^m \in C_{a,b}^{i,j}$, then $R_{i,j} \circ_{j,k} R_k \in C_{a,b}^{i,j}$ due to Lem. 2.2. All other cases are proven analogously with Lem. 1.2., 2.2., 3.2..

The case when there is an l with $k \sim l$, but there is no i with $i \sim j$ can be done analogously to the last case (now looking which properties $R_{l,k}^m$ has. Note that we have to consider $R_{l,k}^m$ instead of $R_{k,l}^m$).

Now we finally consider the case that there are i,k with $i \sim j$ and $k \sim l$. Then $P(\delta^{j,k}(\mathfrak{G})) = P(\mathfrak{G})\backslash\{\{i,j\},\{k,l\}\} \cup \{\{i,l\}\}$. The new ground relations we obtain are $R_{i,j}^m \circ_{j,k} R_{k,l}^m$ with $m \leq n$. Again, we have to do a case distinction, both for the classes $\{i,j\}$ and $\{k,l\}$. This case distinction is done analogously to the last case one, now using Lem. 1.2., 2.2. and 3.2. $\qquad\square$

Corollary 2 (Reduction Thesis for two-element Domains). *Let A be a domain with $|A| = 2$, and let $\mathfrak{G}$ be a ternary PG over A where each relation has an arity ≤ 2. Then $\mathcal{R}(\mathfrak{G}) \neq \doteq_3$.*

Like in the proof of Cor. 1, let $\mathrm{FP}(\mathfrak{G}) = \{1,2,3\}$, and let

$$\mathcal{R}(\mathfrak{G}) = \bigcup_{m\in\{1,\dots,n\}} R_\emptyset^m \cap R_1^m \cap R_{2,3}^m$$

be a DNF for $\mathcal{R}(\mathfrak{G})$.

Assume $\mathcal{R}(\mathfrak{G}) = \doteq_{1,2,3}$. Due to $R_1^m \cap R_{2,3}^m = R_1^m \times R_{2,3}^m$, each relation R_1^m contains at most one element (a) or (b). On the other hand, there must then be an m with $R_1^m = (a)$ and an f with $R_1^f = (b)$. However, one of the three classes C_a^i, C_b^i, $C_{a,b}^i$ contains the relations R_1^m and R_1^f, but none of the three classes contains both $\{(a)\}$ and $\{(b)\}$. Contradiction. $\qquad\square$

6   Further Research

The methods and ideas presented in this paper will be continued to a complete proof of Peirce's Reduction Thesis. The main structure will be similar to the second proof presented here, but the necessary generalizations still pose problems in some details.

Acknowledgments

We want to thank Reinhard Pöschel from Technische Universität Dresden for his valuable input and conttributions in many discussions. Moreover, we want to thank the anonymous referees for proofreading this paper very carefully and their valuable hints to make it more readable.

References

[Bur91]   Robert W. Burch. *A Peircean Reduction Thesis: The Foundation of Topological Logic*. Texas Tech. University Press, 1991.

[Dau03]   Frithjof Dau. *The Logic System of Concept Graphs with Negations and its Relationship to Predicate Logic*, volume 2892 of *LNAI*. Springer, Berlin – Heidelberg – New York, November 2003.

[Dau04]   Frithjof Dau. Query graphs with cuts: Mathematical foundations. In Alan Blackwell, Kim Marriott, and Atsushi Shimojima, editors, *Diagrams*, volume 2980 of *LNAI*, pages 32–50. Springer, Berlin – Heidelberg – New York, 2004.

[DK03]   Frithjof Dau and J. Klinger. From formal concept analysis to contextual logic. In Bernhard Ganter, Gerd Stumme, and Rudolf Wille, editors, *Formal Concept Analysis: The State of the Art*. Springer, Berlin – Heidelberg – New York, 2003.

[HCP04]   Joachim Hereth Correia and Reinhard Pöschel. The Power of Peircean Algebraic Logic (PAL). In Peter W. Eklund, editor, *ICFCA*, volume 2961 of *Lecture Notes in Computer Science*, pages 337–351. Springer, 2004.

[Pap83]   H. Pape. *Charles S. Peirce: Phänomen und Logik der Zeichen*. Suhrkamp Verlag Wissenschaft, 1983. German translation of Peirce's Syllabus of Certain Topics of Logic.

[Pei35a]   C. S. Peirce. *MS 478: Existential Graphs*. Harvard University Press, 1931–1935. Partly published in of [Pei35b] (4.394-417). Complete german translation in [Pap83].

[Pei35b]   Charles Sanders Peirce. *Collected Papers*. Harvard University Press, 1931–1935.

[PS00]   Charles Sanders Peirce and John F. Sowa. Existential Graphs: MS 514 by Charles Sanders Peirce with commentary by John Sowa, 1908, 2000. Available at: http://www.jfsowa.com/peirce/ms514.htm.

[Sow84]   John F. Sowa. *Conceptual structures: information processing in mind and machine*. Addison-Wesley, 1984.

[Sow92]   John F. Sowa. Conceptual graphs summary. In T. E. Nagle, J. A. Nagle, L. L. Gerholz, and P. W. Eklund, editors, *Conceptual Structures: current research and practice*, pages 3–51. Ellis Horwood, 1992.

[Wil82]   Rudolf Wille. Restructuring lattice theory: an approach based on hierarchies of concepts. In Ivan Rival, editor, *Ordered sets*, pages 445–470, Dordrecht–Boston, 1982. Reidel.

[Wil00]   Rudolf Wille. Contextual logic summary. In Gerd Stumme, editor, *Working with Conceptual Structures. Contributions to ICCS 2000*, pages 265–276. Shaker Verlag, Aachen, 2000.

Very Fast Instances for Concept Generation

Anne Berry[1], Ross M. McConnell[2], Alain Sigayret[1], and Jeremy P. Spinrad[3]

[1] LIMOS (CNRS UMR 6158), Université Clermont-Ferrand II,
Ensemble scientifique des Cézeaux, 63177 Aubière Cedex, France
`berry@isima.fr, sigayret@isima.fr`
[2] Computer Science Department, Colorado State University,
Fort Collins, CO 80523-1873, USA
`rmm@cs.colostate.edu`
[3] EECS Department, Vanderbilt University,
Nashville, TN 37235, USA
`spin@vuse.vanderbilt.edu`

Abstract. Computing the maximal bicliques of a bipartite graph is equivalent to generating the concepts of the binary relation defined by the matrix of this graph. We study this problem for special classes of input relations for which concepts can be generated much more efficiently than in the general case; in some special cases, we can even say that the number of concepts is polynomially bounded, and all concepts can be generated particularly quickly.

1 Introduction

One of the important current directions of research related to Formal Concept Analysis deals with the generation of item sets, whether these are defined as 'frequent item sets' or using other more complex criteria. These problems are closely related to concept generation, which has given rise to recent publications (see e.g. [1]).

The problem of concept generation has been shown to be equivalent to various graph problems: computing the maximal transversals of a hypergraph or finding the maximal bicliques of a bipartite graph ([17]). More recently, [4] showed that concept generation is equivalent to generating the minimal separators of a co-bipartite graph. On all three of these problems there exist publications which may well yield algorithmic improvements for concept generation.

In this paper, we aim to use graph results which are related to the form of the matrix representing the relation defined by a context. In some cases we require the matrix to be input in a certain form, in other cases there are good graph algorithms which re-order the rows and columns of the matrix so that the result is in the desired form when the input relation permits it.

We address the issue of generating concepts more quickly than in the general case on special binary matrices defined by a context $(\mathcal{O}, \mathcal{P}, \mathcal{R})$, with the requirement that only polynomial space is used to encounter all the concepts, whether or not their number is polynomial. The best current complexity for the general

R. Missaoui and J. Schmid (Eds.): ICFCA 2006, LNAI 3874, pp. 119–129, 2006.
© Springer-Verlag Berlin Heidelberg 2006

case of generating all concepts using only polynomial space is of $O(|\mathcal{R}|\cdot|\mathcal{P}|)$ per concept using Ganter's algorithm [9], or $O(|\mathcal{P}|^{\alpha})$ per concept using the version of Bordat's algorithm introduced by Berry, Bordat and Sigayret which uses only polynomial space [3], where n^{α} is the time required to perform matrix multiplication, currently $\alpha = 2.376$ ([7]).

We start with the case in which the relation has the *consecutive ones property*: the columns of the matrix representation can be permuted so that in every row, the ones form a consecutive block (such a permutation is called a *consecutive ones arrangement*). We show that in this case, the number of concepts is $O(|\mathcal{R}|)$, and all these concepts can be found in global $O(|\mathcal{R}|)$ time. We generalize from this starting point in several ways.

One form of generalization is to use the decomposition of a general matrix into a PQR-tree ([13]), which efficiently finds submatrices which have the consecutive ones property.

Other forms of generalization come from using natural extensions of the consecutive ones property. Perhaps the most natural is the *circular ones property*. Although the number of concepts can become exponential in this case, we show that the concepts can be generated in $O(|\mathcal{P}|)$ time per concept, which is optimal, since it takes $\Theta(|\mathcal{P}|)$ space to represent a concept in this case. There also are fast algorithms for permuting rows and columns to obtain a circular ones ordering, if such an ordering exists.

Finally, we generalize to orderings in which the number of blocks of ones is at most constant for every row. If we are given an ordering of this type, we can still generate the set of concepts in $O(|\mathcal{P}|)$ time per concept, although no polynomial algorithm is known for finding such an ordering when it is not given as part of the input.

In these time analyses, as in all the rest of the discussions in this paper, $\mathcal{P}$ and $\mathcal{O}$ can be freely interchanged by duality of rows and columns.

2   Background and Previous Results

In this paper, we consider contexts $(\mathcal{O}, \mathcal{P}, \mathcal{R})$, where $\mathcal{O}$ is the set of objects, $\mathcal{P}$ is the set of properties, and both $\mathcal{O}$ and $\mathcal{P}$ are finite. We will work on the 0-1 matrix of $\mathcal{R}$; we will refer to the elements of the relation as *ones*, and to the non-elements as *zeroes*. We will use the following classical set notations: $+$ denotes the union of two disjoint sets, $-$ denotes set difference.

A *concept* or *closed set*, also called a *maximal rectangle* of $\mathcal{R}$, is a sub-product $A \times B \subseteq \mathcal{R}$ such that $\forall x \in \mathcal{O} - A, \exists y \in B \,|\, (x,y) \notin \mathcal{R}$, and $\forall y \in \mathcal{P} - B, \exists x \in A \,|\, (x,y) \notin \mathcal{R}$. Given a subset $\mathcal{P}_1$ of $\mathcal{P}$ and a subset $\mathcal{O}_1$ of $\mathcal{O}$, we will say that the set $\mathcal{R} \cap (\mathcal{O}_1 \times \mathcal{P}_1)$ is a *sub-relation* of $\mathcal{R}$ which we will denote $\mathcal{R}(\mathcal{O}_1, \mathcal{P}_1)$.

We will also use finite undirected graphs. Such graphs are classically denoted $G = (V, E)$, where V is the vertex set and E is the edge set. A *bipartite graph* is a graph $G = (V_1 + V_2, E)$ where V_1 and V_2 induce edgeless subgraphs (i.e. V_1 and V_2 are independent sets). A *maximal biclique* (or *maximal complete bipartite subgraph*) is a subgraph $H = (W_1 + W_2, F)$, with $W_1 \subseteq V_1, W_2 \subseteq V_2$, such that all edges are present between any vertex of W_1 and any vertex of W_2, and which is maximal for this property. A graph $G = (V, E)$ is said to be an *interval graph* if there exists a one-to-one mapping I from V to a family of intervals of the real line such that $xy \in E$ iff the corresponding intervals $I(x)$ and $I(y)$ are intersecting.

Example 1. $\mathcal{O} = \{1, 2, 3, 4, 5, 6\}$, $\mathcal{P} = \{a, b, c, d, e, f\}$. Relation $\mathcal{R}$ is presented below.

$\mathcal{R}$	a	b	c	d	e	f
1		×	×	×	×	
2	×	×	×			
3	×	×				×
4				×	×	
5			×	×		
6	×					

The maximal rectangles/maximal bicliques/concepts are:
$\mathcal{O} \times \emptyset$, $\{2, 3, 6\} \times \{a\}$, $\{1, 2, 3\} \times \{b\}$, $\{1, 2, 5\} \times \{c\}$, $\{1, 4, 5\} \times \{d\}$, $\{2, 3\} \times \{a, b\}$, $\{1, 2\} \times \{b, c\}$, $\{1, 5\} \times \{c, d\}$, $\{1, 4\} \times \{d, e\}$, $\{3\} \times \{a, b, f\}$, $\{2\} \times \{a, b, c\}$, $\{1\} \times \{b, c, d, e\}$, $\emptyset \times \mathcal{P}$.

For example, $\{1, 2\} \times \{b, c\}$ is a maximal rectangle of the matrix given above, or, equivalently, a maximal biclique of the graph of Figure 1.

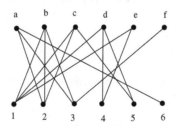

Fig. 1. $\{1, 2\} \times \{b, c\}$ is a maximal biclique of graph G associated with relation $\mathcal{R}$ of Example 1

3   Consecutive Ones Matrices

The starting point for this research is the *consecutive ones property*. A matrix is said to have the consecutive ones property if its columns can be permuted so that in each row all the ones are consecutive. This property has been studied heavily in the context of graph theory: Fulkerson and Gross ([8]) showed that a graph is an interval graph iff the vertex-clique incidence matrix can be permuted to have the consecutive ones property (the vertex-clique incidence matrix shows the maximal cliques of the graph which each vertex belongs to). Booth and Lueker ([6]) gave a linear-time algorithm for finding a consecutive ones arrangement of a given matrix if such an arrangement exists, as part of their interval graph recognition algorithm.

Let us assume that we have a context $(\mathcal{O}, \mathcal{P}, \mathcal{R})$ whose 0-1 matrix has the consecutive ones property. It takes $O(|\mathcal{P}| + |\mathcal{O}| + |\mathcal{R}|)$ time to find a consecutive ones ordering of the columns ([14]). We claim that every concept of this relation will have a column set that is consecutive in this ordering. Therefore, we can associate with each concept a unique *starting column*, namely, the leftmost of the concept's columns in the ordering.

In this section we will use the common notations for 0-1 matrices; thus R will denote a set of rows in a matrix, C will denote a set of columns. m will denote the number of ones in the matrix.

Example 2. $\mathcal{O} = \{1, 2, 3, 4, 5, 6, 7\}$, $\mathcal{P} = \{a, b, c, d, e, f\}$. Matrix $\mathcal{M}$, presented below, has the consecutive ones property.

	a	b	c	d	e	f
1	×	×	×			
2	×	×	×	×		
3	×	×	×	×	×	×
4		×				
5		×	×	×		
6		×	×	×	×	
7				×	×	

The corresponding concept lattice is presented in Figure 2.

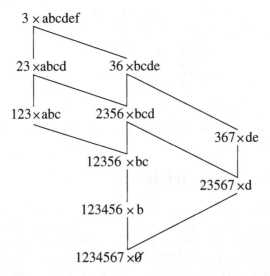

Fig. 2. Concept lattice associated with the consecutive ones matrix of Example 2. The set notations have been simplified for more clarity.

We may delete rows and columns that are all zeros or all ones, compute the lattice for the remaining submatrix, and then correct the elements of this lattice in a trivial way. Henceforth, we will assume that the matrix has no rows that are all zeros or all ones. This implies that the minimal element of the lattice is $\mathcal{O} \times \emptyset$ and the maximal element is $\emptyset \times \mathcal{P}$. We give a recursive algorithm that finds the remaining concepts.

Let $(c_1, c_2, ..., c_p)$ be the order of the columns in a consecutive ones arrangement. The input to each call is a submatrix of the initial matrix given by $\{c_i, c_{i+1}, ..., c_p\}$, the rows that have at least one 1 in any of these columns. In addition, there is a mark on each row that has a 1 preceding column c_i in the original matrix. The purpose of the marks on the rows is to allow the recursive call to avoid returning concepts of the submatrix that are not maximal in the original matrix.

The algorithm finds all concepts of the original matrix that have i as their starting column, and makes a recursive call on columns c_{i+1} through c_p to find the concepts that have their starting column anywhere in $\{c_{i+1}, c_{i+2}, ..., c_p\}$.

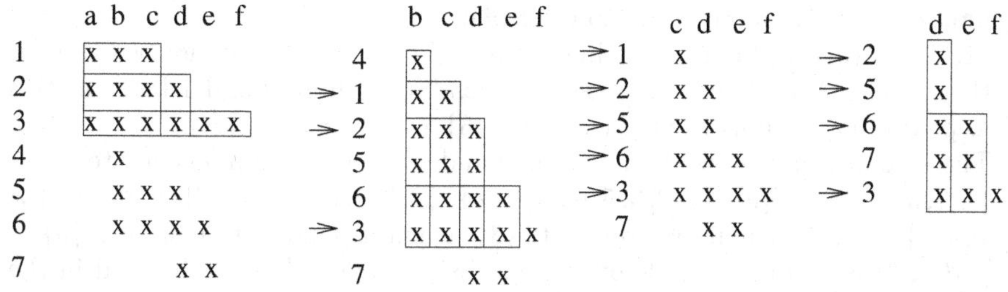

First submatrix, columns a b c d e f:

	a	b	c	d	e	f
1	X	X	X			
2	X	X	X	X		
3	X	X	X	X	X	X
4			X			
5		X	X	X		
6		X	X	X	X	
7					X	X

Second submatrix, columns b c d e f:

	b	c	d	e	f
4	X				
1	X	X			
2	X	X	X		
5	X	X	X		
6	X	X	X	X	
3	X	X	X	X	X
7				X	X

Third submatrix, columns c d e f:

	c	d	e	f
→1	X			
→2	X	X		
→5	X	X		
→6	X	X	X	
→3	X	X	X	X
7		X	X	

Fourth submatrix, columns d e f:

	d	e	f
→2	X		
→5	X		
→6	X	X	
7	X	X	
→3	X	X	X

Fig. 3. Finding the concepts in the consecutive ones matrix of Example 3. The corresponding lattice is presented in Figure 2.

Example 3. In Figure 3, the lefthand table corresponds to matrix $\mathcal{M}$ of Example 2 passed to the initial call of the algorithm, with starting column a. The initial matrix does not have any marked row. The next table shows the submatrix, $\mathcal{M}$ minus the first column, passed to the second call (starting column b). In this second call, rows 1, 2 and 3 are marked as their starting column is a. The third table corresponds to a submatrix with starting column c, all rows but 7 are marked. The fourth table corresponds to a submatrix with starting column d; row 1 has disappeared as it has no one in columns d,e,f. There will be no more recursive call.

Let the *ending column* of a row be the rightmost column where the row has a 1. In the recursive call on columns $\{c_i, ..., c_p\}$, let R_i be the set of rows that have a 1 in column i. We permute the rows so that the rows in R_i appear in ascending order of ending column. When the rows of a set R_i are tied for ending column, we place the marked ones before the unmarked ones.

Example 4. In Figure 3, $R_1 = \{1, 2, 3\}$, as these objects have the same starting column a, and these rows appear in ascending order of ending column (respectively c, d, f). In the second call, on columns $\{b, c, d, e, f\}$, we will have

$R_2 = \{4, 1, 2, 5, 6, 3\}$. The members of R_2 appear in ascending order of ending column, and the tie between rows 2 and 5 (ending column c) has been broken in favor of 2, since row 2 is marked and row 5 is not.

Let $(r_1, r_2, ..., r_q)$ be the resulting ordering of R_i. The ones in row r_j of the submatrix extend from c_i to c_k for some $k \geq i$. Because of the way R_i has been ordered, every row after r_j in the ordering also has ones in every column in $\{c_i, ..., c_k\}$. Therefore, $\{r_j, ..., r_q\} \times \{c_i, ..., c_k\}$ is a rectangle. It is easy to see that this rectangle is maximal in the submatrix, hence is a concept of the submatrix, if and only if $j = 1$ or r_j extends farther to the right than its predecessor, r_{j-1}.

A concept of the submatrix can fail to be a concept of the original matrix if and only if it can be extended to a larger rectangle in the whole matrix by adding column c_{i-1}. This is easy to detect: a concept of the submatrix fails to be a concept in the original matrix if and only if all of its rows are marked.

Example 5. In Figure 3, the initial call on rows $(1, 2, 3, 4, 5, 6, 7)$ and starting column a gives rise successively to concepts:

$\{1, 2, 3\} \times \{a, b, c\}$ (as 1 has ending column c, this concept can not be extented to the right), $\{2, 3\} \times \{a, b, c, d\}$ (as 2 has ending column d and 1 has not, this concept can not be upper extended), and $\{3\} \times \{a, b, c, d, e, f\}$.

The second call on rows $(4, 1, 2, 5, 6, 3)$ and starting column b generates:

$\{1, 2, 3, 4, 5, 6\} \times \{b\}$, $\{1, 2, 3, 5, 6\} \times \{b, c\}$, $\{2, 3, 5, 6\} \times \{b, c, d\}$, and $\{3, 6\} \times \{b, c, d, e\}$. As 3 is marked, row set $\{3\}$ generates no new concept: $\{3\} \times \{b, c, d, e, f\}$ is a sub-rectangle of concept $\{3\} \times \{a, b, c, d, e, f\}$ obtained in the initial step.

The third call, on starting column c, generates no concept, as all the rows are marked.

The fourth call, on starting column d, generates $\{2, 3, 5, 6, 7\} \times \{d\}$, and $\{3, 6, 7\} \times \{d, e\}$, as at least one object (7) is unmarked in row set $\{2, 3, 5, 6, 7\}$ and in row set $\{3, 6, 7\}$. $\{3\}$ fails to contain an unmarked row and gives rise to no new concept.

The corresponding lattice is presented in Figure 2. Concepts are generated in this figure from left to right and from bottom to top.

Note that, in handling R_i, the algorithm never generates two concepts with the same upper-left hand corner. This gives a one-to-one mapping from the concepts the algorithm generates to the ones in the matrix, so that there can be at most m of these concepts. The algorithm can only fail to generate directly the top and bottom elements of the lattice, so we get the following bound:

Theorem 1. *If a matrix with m ones has the consecutive ones property, then the corresponding relation has at most $m + 2$ concepts.*

That the bound is tight is illustrated by the identity matrix, where each 1 is itself a concept, and $\mathcal{O} \times \emptyset$ and $\emptyset \times \mathcal{P}$ are also concepts. The number of concepts can be $\Omega(m)$ in dense matrices, as it can be seen by running the algorithm on an $(n/2) \times n$ matrix where row i has ones in columns i through $i + n/2$: half of the ones are the upper-left corner of a concept.

The proof of correctness is elementary, given the foregoing observations. In order to obtain an $O(m)$ time bound, it suffices to spend time proportional to $(\sum_{i=1}^{p} R_i) \in O(m)$. This is also easy to accomplish by elementary techniques. In particular, each concept can be represented by giving its column set as an interval of $(c_1, ..., c_p)$, and giving its row set as an interval on the ordering of R_i computed in the recursive call on columns $(c_i, ..., c_p)$.

4   Matrices with Bounded PQR Diameter

In this section, we will show that the number of concepts may be bounded by the size of the corresponding relation, provided some property on the PQR-decomposition of its matrix.

The *PQ-tree* of a consecutive ones matrix is a way of representing all consecutive-ones orderings of the columns ([6]). It is a tree whose leaves are the columns of the matrix. The tree has two kinds of internal nodes: Q-nodes, whose children are ordered left-to-right, and P nodes, whose children are not ordered. Assigning a left-to-right ordering to children of each P node and reversing the left-to-right ordering of children of any subset of the Q nodes imposes a leaf order on the leaves, hence an order on columns of the matrix. Such an ordering is always a consecutive ones ordering. Conversely, all consecutive ones orderings can be obtained in this way.

[13] provides a generalization of PQ-trees to arbitrary matrices, called the *PQR-tree*, as illustrated in Figure 4. The third types of nodes, *R nodes*, appear if and only if the matrix does not have the consecutive ones property, otherwise the PQR-tree is a PQ-tree. The PQR-tree can be constructed in time proportional to the number of ones of the matrix. One of the interesting aspects of PQR-trees is that is gives a compact representation of all possible PQR arrangements of a matrix.

The PQR-tree has a set-theoretic definition. Let us consider the leaves to be a set V of 'properties', and let us consider each internal node to represent a subset of V, namely the set of leaf descendants of the node. Such a subset corresponds to a row (an 'object') of the matrix, more precisely to the property set associated with this object. We say that two subsets X and Y of V *overlap* if they intersect, but neither contains the other. Let $\mathcal{F}$ be the family of nonempty subsets of V that do not overlap with any row, and let $\mathcal{F}_0$ be the set of members of $\mathcal{F}$ that do not overlap with any other member of $\mathcal{F}$: then $\mathcal{F}_0$ is the set of nodes of the PQ-tree.

As a direct consequence of this definition, every row of a matrix is a union of one or more children of the node of the PQR-tree, as a row that is not such a union must overlap some member of $\mathcal{F}_0$, hence of $\mathcal{F}$, a contradiction. To each internal node of the PQR-tree, we may assign a *quotient matrix*, as follows. Let $ch(X)$ denote the children of X in the tree, let $Row(X)$ denote the set of rows that are given by the union of more than one member of $ch(X)$, and let $Col(X)$ denote a set of columns with one representative column from each member of $ch(X)$. The quotient at X is the submatrix induced by rows $Row(X)$ and columns $Col(X)$.

The exact choice of representatives from $ch(X)$ to obtain $Col(X)$ is irrelevant, since all choices yield isomorphic submatrices. It is possible that $Row(X)$ is empty, but the family of sets given by $\{Row(X) \neq \emptyset \mid X \text{ is an internal node}\}$ is a partition of the rows.

Example 6. Figure 4 shows a PQR-decomposition of a matrix. Traditionally, the Q nodes are drawn as horizontal bars, the P nodes are drawn as points, and R nodes are drawn as filled ovals.

In this decomposition, the children of $V = \{a, b, c, d, e, f, g, h, i, j, k, l\}$ are $W = \{a, b, c, d, e\}$, $\{f\}$ and $Y = \{g, h, i, j, k, l\}$ (Q decomposition), rows 1 and 2 are the rows that are unions of more than one of these children, and these rows, together with a selection of one column in each of W, $\{f\}$ and Y, yield the two-row quotient whose rows are $(1, 1, 0)$ and $(0, 1, 1)$.

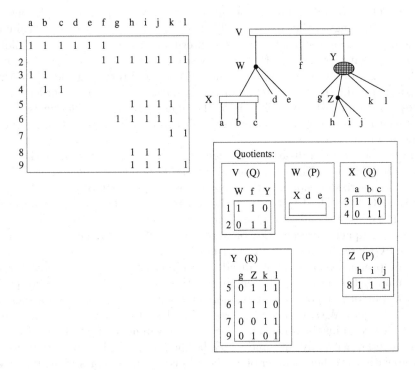

Fig. 4. The matrix of Example 6 and the corresponding PQR-tree and quotients

Lemma 1. *In a PQR-tree, a node is a P node if and only if its quotient matrix is a rectangle of ones.*

The matrix can be uniquely reconstructed from the quotients by inverting this process. We now illustrate a similar operation by which the concepts of the original matrix can be obtained from the concepts that occur in the quotients. In Example 6, rows $\{5, 6\}$ and columns $\{Z, k\}$ are a concept in the quotient at Y. Substituting $\{h, i, j\}$ for Z in the column set, and adding rows labeled Y in Y's parent V yields the concept with rows $\{2, 5, 6\}$ and columns $\{h, i, j, k\}$.

In general, each concept $A_1 \times B_1$ in the original matrix is obtained from a concept $A_0 \times B_0$ in a quotient at a node X, by expanding the sets that appear as column labels in B_0 and adding rows to A_0 that have a 1 in a column labeled with an ancestor of X in a quotient at an ancestor of X.

As in the case of consecutive ones matrices, we may use a compact representation of each concept found. From results shown in [13] we can derive that, on the path from the quotient of $A_0 \times B_0$ to the root, at least every other quotient has at least one 1 in every column. Each of these contributes at least one row to A_1. The length of the path to the root is $O(|A_1|)$, and the time to list out the elements of A_1 and B_1, given $A_0 \times B_0$, is $O(|A_1| + |B_1|)$. Therefore, we may let $A_0 \times B_0$ serve to represent $A_1 \times B_1$.

Definition 1. *The decomposition diameter of a binary matrix M is the maximum number of children of an R node, or 1 if there are no R nodes.*

The following is a consequence of the foregoing observations:

Theorem 2. *If a matrix has a bounded decomposition diameter, the corresponding relation $\mathcal{R}$ has $O(|\mathcal{R}|)$ concepts, and they can be found in $O(|\mathcal{R}|)$ time.*

5   Circular Ones Property

Perhaps the more natural generalization of the consecutive ones property is the circular ones property. When we first considered this generalization, we were discouraged to find that the number of concepts becomes exponential, for example, a matrix with only a diagonal of zeroes, which clearly has the circular ones property, has a lattice isomorphic to that of the power set of $\mathcal{P}$ or $\mathcal{O}$.

To generate concepts efficiently in this case, we will use the concept generation process described in [3]: the algorithm starts with the minimum element of the lattice, and recursively processes the direct successors of each concept. Since in a concept lattice, all the concepts containing a given property form a sub-lattice, we can store in the recursive stack information necessary to avoid re-processing a concept. Another useful feature is that each concept $A \times B$, $(A \subseteq \mathcal{O}, B \subseteq \mathcal{P})$ is the minimal element of a sub-lattice described by sub-relation $\mathcal{R}(A, \mathcal{P} - B)$. Finally, the direct successors of concept $A \times B$ are described by the properties X which in $\mathcal{R}(A, \mathcal{P} - B)$ are not properly contained in another; the elements of X which have identical columns are then grouped together to be added to A to form one of the successors of $A \times B$. The bottleneck of this algorithm is computing the containments; this requires $O(|\mathcal{P}|^2 \cdot |\mathcal{O}|)$ in general.

However, we will show that the time for generating all concepts can be significantly reduced if the matrix has a circular ones ordering.

First, we note that although PQ-trees are generally viewed as a tool for finding a consecutive ones ordering, they can also be used to find a circular ordering in $O(|\mathcal{P}| + |\mathcal{O}| + |\mathcal{R}|)$ time. Thus we may assume that we have a circular arrangement of objects, and for each property, we are given the circular ordering of its objects in concise form. For example, if there are 100 objects, the objects of a property p_1 may be given as $o_{19} - o_{54}$, while those of p_2 may be given as $o_1 - o_{23}$ and $o_{93} - o_{100}$.

Given this form of storage, for each property p_i, it is easy to determine whether p_i contains property p_j in constant time. This simple observation, together with the fact that an arrangement remains circular as objects are deleted, is the key to reducing the time for concept generation using polynomial space, from $O(|\mathcal{P}|^2 \cdot |\mathcal{O}|)$ time per generated concept to $O(|\mathcal{P}|)$ per generated concept.

6   Constant Number of Blocks

Our last extension of the consecutive ones property is to relations given as a matrix in which the number of blocks of ones entries in every row is bounded by a constant.

For each property, we maintain a concise representation of the ones in the property; that is, for a property p_i with k_i blocks, the starting column number and the ending column number of each block: $j_1 - j_2, j_3 - j_4, ..., j_{2k_i-1} - j_{2k_i}$.

As long as the number of blocks is constant, it is still possible to test containment between properties in constant time. It is also clear that properties continue to have at most a fixed number of blocks of ones as objects are deleted from the current universe of discourse. These observations allow us to generate the concepts in $O(|\mathcal{P}|)$ time per concept, using the same strategy as in the circular ones case.

In one sense, going to a constant number k of consecutive blocks seems to be a huge generalization of the previous algorithms: the consecutive ones property corresponds to $k = 1$, while circular ones are a special case of $k = 2$. One key advantage of the earlier cases discussed in this paper is the existence of polynomial time algorithms to permute rows and columns of a matrix to obtain a matrix with the consecutive/circular ones property, if such a permutation exists. No such algorithm is known for k blocks of consecutive ones. An obvious open question is to investigate whether this problem is polynomial or NP-complete. However, in applications such as social science yes/no questionnaires ([2]), this kind of arrangement may often arise naturally, for example when some questions are logically related to others.

7   Conclusion

This paper shows that for a number of special classes of matrices, the matrix properties can be used to design efficient algorithms for concept generation. Indeed, even in the general case, if we use a recursive concept-generation algorithm such as the one described in Section 5, we can afford to check on each sub-relation encountered whether the matrix has the consecutive ones property, or the circular ones property, and in this case speed up the rest of the remaining recursive call from $O(|\mathcal{P}|^2 \cdot |\mathcal{O}|)$ time to $O(|\mathcal{P}| \cdot |\mathcal{O}|)$, $O(|\mathcal{P}| + |\mathcal{O}|)$ or even constant time per generated concept.

These results show that for special classes of input, the number of concepts may be much less than exponential, notwithstanding the matrix density.

One of the questions which arises is how to embed a relation within one of our special classes, while adding or removing a small or inclusion-minimal set of ones.

There are many other classes of matrices and bipartite graphs to consider for this problem. One example which springs to the mind is that of 'Gamma-free matrices' which are obtainable if, and only if, the corresponding bipartite graph is weakly chordal and bipartite (also called 'chordal bipartite'). In this case, [11] showed that all the maximal bicliques can be identified in $\min(|\mathcal{R}| \cdot \log(|\mathcal{O}| + |\mathcal{P}|), (|\mathcal{O}| + |\mathcal{P}|)^2)$ time. Other results may later appear using this relationship between graphs and lattices.

References

1. Alexe G., Alexe S., Crama Y., Foldes S., Hammer P.L., Simeone B.: Consensus algorithm for the generation of all maximal bicliques. *Discrete Applied Mathematics*, **145** (2004), 11–21.

2. Barbut M., Monjardet B.: *Ordre et classification*. Classiques Hachette, (1970).

3. Berry A., Bordat J-P., Sigayret A.: Concepts can't afford to stammer. *INRIA Proc. International Conference "Journées de l'Informatique Messine" (JIM'03)*, Metz (France), (Sept. 2003). Submitted as 'A local approach to concept generation.'

4. Berry A., Sigayret A.: Representing a concept lattice by a graph. *Discrete Applied Mathematics*, **144(1-2)** (2004) 27–42.

5. Bordat J-P.: Calcul pratique du treillis de Galois d'une correspondance. *Mathématiques, Informatique et Sciences Humaines*, **96** (1986) 31–47.

6. Booth S., Lueker S.: Testing for the consecutive ones property, interval graphs, and graph planarity using PQ-tree algorithms. *J. Comput. Syst. Sci.*, **13** (1976) 335–379.

7. Coppersmith D., Winograd S.: On the Asymptotic Complexity of Matrix Multiplication. *SIAM J. Comput.*, **11**:3 (1982) 472–492.

8. Fulkerson D.R., Gross O.A.: Incidence matrices and interval graphs. Pacific J. Math. **15** (1965) 835–855.

9. Ganter B.: Two basic algorithms in concept analysis. *Preprint 831, Technische Hochschule Darmstadt*, (1984).

10. Ganter B., Wille R.: *Formal Concept Analysis*. Springer, (1999).

11. Kloks T., Kratsch D.: Computing a perfect edge without vertex elimination ordering of a chordal bipartite graph. Information Processing Letter **55** (1995) 11–16.

12. Kuznetsov S. O., Obiedkov S. A.: Comparing performance of algorithms for generating concept lattices. *Journal for Experimental and Theoretical Artificial Intelligence (JETAI)*, **14:2-3** (2002) 189–216.

13. McConnell R. M.: A certifying algorithm for the consecutive ones property. *Proceedings of the 15th Annual ACM-SIAM Symposium on Discrete Algorithms (SODA04)*, **15**, (2004) 761–770.

14. Paige R., Tarjan R. E.: Three Partition Refinement Algorithms. *SIAM Journal on Computing*, **16** (1987) 973–989.

15. Spinrad J. P.: *Efficient Graph Representation*. Fields Institue Monographs; 19. American Mathematical Society, Providence (RI, USA), (2003).

16. Spinrad J. P.: Doubly Lexical Orderings of Dense 0-1 Matrices. *Information Processing Letters*, **45** (1993) 229–235.

17. Zaki M. J., Parthasarathy S., Ogihara M., Li W.: New Algorithms for Fast Discovery of Association Rules. *Proceedings of 3rd Int. Conf. on Database Systems for Advanced Applications*, (April 1997).

Negation, Opposition, and Possibility
in Logical Concept Analysis

Sébastien Ferré

Irisa/Université de Rennes 1, Campus de Beaulieu,
35042 Rennes cedex, France
ferre@irisa.fr

Abstract. We introduce the *epistemic extension*, a logic transformation based on the modal logic AIK (All I Know) for use in the framework of Logical Concept Analysis (LCA). The aim is to allow for the
distinction between negation, opposition, and possibility in a unique formalism. The difference between negation and opposition is examplified
by the difference between "young/not young" and "young/old". The difference between negation and possibility is examplified by the difference
between "(certainly) not young" and "possibly not young". Furthermore
this epistemic extension entails no loss of genericity in LCA.

1 Introduction

Many have felt the need to extend Formal Concept Analysis in various
ways: valued attributes, partially ordered attributes, graph patterns, 3-valued
contexts, distinction between negation and opposition, etc. Often a new and
specific solution was proposed: conceptual and logical scales [Pre97], 3-valued
contexts [Obi02], two different negation operators [Wil00, Kan05], etc.

We have introduced *Logical Concept Analysis* [FR00] in order to have a framework that could cover as many extensions as possible by simply changing the logic
used to describe objects. Other authors have proposed similar frameworks where
they talk about partial orderings of patterns instead of logics [CM00, GK01]. Indeed our logics can be seen as partial orderings, but we emphasize the use of
the term logic as this brings useful notions like the distinction between syntax
and semantics, reasoning through entailment or subsumption, and important
properties like consistency and completeness. Moreover this makes available the
numerous existing logics, and the theory and practice that comes with them.

We have already applied LCA to various logics for querying and navigating
in logical information systems [FR04] (e.g., combinations of string patterns, intervals over numbers and dates, custom taxonomies, functional programming
types). We further support its capabilities by showing how an existing modal
logic, AIK, can be used to represent at the same time *complete* and *incomplete* knowledge (Closed World Assumption and distinction beteeen *certain* and
possible facts), and the distinction between *negation* and *opposition*. The result is something more expressive than existing extensions of FCA, because for
each object one can express certain and possible facts, these facts being expressed in an almost arbitrary logic whose negation plays the role of opposition;

R. Missaoui and J. Schmid (Eds.): ICFCA 2006, LNAI 3874, pp. 130–145, 2006.

and three levels of negation allow for distinguishing certainly/possibly true/false normal/opposite properties of objects. For example it becomes possible to distinguish between "possibly not young" (possibility), "certainly not young" (usual negation), and "old", i.e., "opposite of young" (opposition). All these distinctions are quite important when querying a context.

Section 2 presents useful preliminaries about Logical Concept Analysis (LCA), and the modal logic All I Know (AIK). Section 3 explains the ambiguity that lies in the interpretation of negation. Section 4 then gives a solution based on the logic AIK for distinguishing (usual) negation and opposition. Section 5 goes further in the use of the logic AIK in order to distinguish certainty and possibility. In Section 6 the logic AIK is replaced by a logic transformation in order to retain the genericity of LCA, and to simplify the use of AIK. Finally we compare our solution to related works in Section 7, before concluding.

2 Preliminaries

This paper is based on Logical Concept Analysis (LCA) where the logic is a modal logic called AIK (All I Know). As preliminaries, we first recall some definitions of LCA, and differences to FCA. Then we introduce the logic AIK with an emphasis on its semantics as it is crucial in the rest of the paper.

2.1 Logical Concept Analysis

Logical Concept Analysis (LCA) [FR00] has been introduced in order to allow for richer object descriptions, and to bring in automated reasoning when deciding whether an object belongs to a concept or not. The principle is to replace sets of attributes (object descriptions and intents) by the formulas of an "almost arbitrary logic". The kind of logics we consider is defined as follows.

Definition 1 (logic). *A logic (in LCA) is made of*

- *a* syntax *or* language, *i.e., a set L of formulas,*
- *a set of* operations *like conjunction ($\sqcap$, binary), disjunction ($\sqcup$, binary), negation ($\neg$, unary), tautology ($\top$, nullary), and contradiction ($\bot$, nullary),*
- *a* semantics, *i.e., a set of interpretations I, and a binary relation $\models$ ("is a model of") between interpretations and formulas,*
- *a* subsumption *relation $\sqsubseteq$, which decides whether a formula is subsumed ("is more specific/less general") than another formula. Its intended meaning is that $f \sqsubseteq g$ iff every model of f is also a model of g.*

We also define $M(f)$ as a shorthand for $\{i \in I \mid i \models f\}$, i.e., the set of models of the formula f.

This definition is quite large, and covers most existing languages and semantics in logic. There are two differences with the usual presentation of logics. Firstly, logical operations (e.g., conjunction, negation) are not necessarily connectors in the language, which makes them more general. For example, if formulas are intervals, then the conjunction is the intersection on intervals. Secondly, the left argument of the entailment relation (subsumption) is restricted to a single formula instead of a set of formulas. This makes subsumption a generalization ordering, hence its name, which is crucial for use in concept analysis.

In LCA it is possible to look at logics only as partial orderings, and to forget about semantics. However semantics plays an important role when defining and reasoning about logics, like in this paper. This is why we introduce it above, and in the following we define the relation we usually expect between logical operations, subsumption, and semantics, i.e. consistency and completeness.

Definition 2 (consistency and completeness). *An operation of some logic $\mathcal{L}$ is* consistent and complete *if its expected meaning agrees with semantics, i.e., for all formulas $f, g \in \mathcal{L}$:*

1. $M(f \sqcap g) \quad = \quad M(f) \cap M(g)$
2. $M(f \sqcup g) \quad = \quad M(f) \cup M(g)$
3. $M(\neg f) \quad = \quad I \setminus M(f)$
4. $M(\top) \quad = \quad I$
5. $M(\bot) \quad = \quad \emptyset$
6. $f \sqsubseteq g \quad \Longleftrightarrow \quad M(f) \subseteq M(g)$

From these definitions a *logical context* can now be defined, where a logic replaces a set of attributes. However remind that logical formulas replace sets of attributes, and not single attributes, as is demonstrated by the mapping from objects to formulas that replaces the binary relation.

Definition 3 (logical context). *A* logical context *is a triple $K = (\mathcal{O}, \mathcal{L}, d)$, where:*

- $\mathcal{O}$ *is a finite set of objects,*
- $\mathcal{L}$ *is a logic equiped with at least disjunction $\sqcup$, contradiction $\bot$, and subsumption $\sqsubseteq$ as consistent and complete operations. Because these 3 operations are consistent and complete, this logic can equivalently be seen as a 0-sup-semilattice, whose ordering is $\sqsubseteq$, bottom is $\bot$, and join operation is $\sqcup$.*
- $d \in \mathcal{O} \to L$ *maps each object $o \in \mathcal{O}$ to a single formula $d(o) \in \mathcal{L}$, its logical description.*

This definition is a slight weakening of the original definition [FR00] (conjunction and tautology operations are no more required as they have no consequence on the concept lattice); and it is similar to the definition of a *pattern structure* [GK01] (except disjunction is denoted by $\sqcap$ instead of $\sqcup$, and no logical interpretation is given[1]). The disjunction $\sqcup$ plays the role of set intersection $\cap$ in FCA, which is normal as a consistent and complete disjunction returns the most specific common subsumer of two formulas. So, in the case formulas are sets of attributes, disjunction is defined as $\cap$; and in the case formulas are Prolog terms, disjunction is defined as anti-unification [Plo70].

---

[1] In logic, disjunction is usually denoted by $\vee$ or $\sqcup$.

Galois connections can then be defined on logical contexts, enabling us to go from formulas to sets of objects (their *extent*), and from sets of objects to formulas (their *intent*).

Lemma 1. *Let $K = (\mathcal{O}, \mathcal{L}, d)$ be a logical context. The pair (ext, int), defined by*

$$ext(f) = \{o \in \mathcal{O} \mid d(o) \sqsubseteq f\} \quad \text{for every } f \in L$$
$$int(O) = \bigsqcup\{d(o) \mid o \in O\} \quad \text{for every } O \subseteq \mathcal{O}$$

is a Galois connection between $\mathcal{P}(\mathcal{O})$ and L: $O \subseteq ext(f) \Longleftrightarrow int(O) \sqsubseteq f$.

From there, concepts, the concept lattice, and its labelling are defined as usual, only replacing sets of attributes by formulas where necessary. However, we do not detail them in this paper as we here focus on expressing and distinguishing negation, opposition, and possibility in formulas and in the subsumption relation, which determines extents, which determine in turn the concept lattice.

2.2   The Logic All I Know

In computer science, epistemic aspects in logic have often been discussed under the term *Closed World Assumption* (CWA). CWA says that every fact that cannot be deduced from a knowledge base can be considered as false, contrary to the open world assumption, which says that such a fact is neither true, nor false. This often leads to the notion of *non-monotonous reasoning*, i.e., the addition of knowledge can make false a fact that was previously true by CWA.

There exists several formalisms for handling CWA, but all of them except the logic All I Know (AIK) are non-monotonous [Lif91, DNR97, Moo85, McC86, Lev90]. However logics to be used in the framework of Logical Concept Analysis (LCA) must form at least a partial ordering, and so, must have a monotonous subsumption relation. In fact, CWA should be applied locally on the formulas used to describe objects rather than globally on the knowledge base. The logic AIK precisely defines such an operation, a modal operator O. Moreover it has been established that this logic covers all non-monotonous formalisms cited above (see [Che94] for correspondences between these formalisms), and a proof method exists for its propositional version [Ros00].

The logic AIK [Lev90] is essentially a modal logic [Bow79], to which the two modal operators N and O have been added[2]. The formula language of *AIK* is defined as the propositional language *Prop*, whose atomic propositions belong to an infinite set *Atom*, and connectors are 1 (tautology), 0 (contradiction), $\wedge$ (conjunction), $\vee$ (disjunction), and $\neg$ (negation). It is extended by the modal operators K, N and O. The five logical operations ($\sqcap$, $\sqcup$, $\neg$, $\top$, $\bot$) are simply realized by the connectors of the language (resp. $\wedge$, $\vee$, $\neg$, 1, 0).

A Kripke semantics is given to AIK. Given a set of *worlds* W, interpretations are couples (w, R), where $w \in W$ is a world, and $R \subseteq W \times W$ is an *accessibility*

---

[2] The basic modal operator is K. Sometimes modal logics are presented as having the two modal operators $\square$ and $\lozenge$, which are in fact equivalent to respectively K and $\neg K \neg$.

relation between worlds. More precisely, each world defines a valuation of atoms in *Atom* by boolean values $\{TRUE, FALSE\}$. The "is a model of" relation between interpretations and formulas is then defined as follows.

Definition 4 (semantics). *Let w be a world, and let R be an accessibility relation that is both* transitive[3] *and* euclidean[4]. *A Kripke structure (w, R) is a model of a formula $\phi \in AIK$, which is denoted by $(w, R) \models \phi$, iff the following conditions are satisfied ($R(w)$ denotes the set of successor worlds of w through R):*

1. $(w, R) \models a$ *iff* $w(a) = TRUE$, where $a \in Atom$;
2. $(w, R) \models 1$ *iff*   *true, i.e., for every (w, R)* ;
3. $(w, R) \models 0$ *iff*   *false, i.e., for no (w, R)* ;
4. $(w, R) \models \neg\phi_1$ *iff* $(w, R) \not\models \phi_1$;
5. $(w, R) \models \phi_1 \wedge \phi_2$ *iff* $(w, R) \models \phi_1$ *and* $(w, R) \models \phi_2$;
6. $(w, R) \models \phi_1 \vee \phi_2$ *iff* $(w, R) \models \phi_1$ *or* $(w, R) \models \phi_2$;
7. $(w, R) \models K\phi_1$ *iff*   *for every $w' \in R(w)$, $(w', R) \models \phi_1$* ;
8. $(w, R) \models N\phi_1$ *iff*   *for every $w' \notin R(w)$, $(w', R) \models \phi_1$* ;
9. $(w, R) \models O\phi_1$ *iff*   *for every w', $w' \in R(w)$ iff $(w', R) \models \phi_1$.*

The logic AIK can be equiped with a subsumption relation, which enables us to compare object descriptions and queries, for instance in the definition of extents. An axiomatization of AIK [Lev90] provides a consistent and complete algorithm for computing the subsumption relation $\sqsubseteq_{AIK}$. The logic $\mathcal{L}_{AIK} = (AIK, \wedge, \vee, \neg, 1, 0, \sqsubseteq_{AIK})$ can then be defined according to Definition 1. Given its semantics defined above, this logic also forms a complete lattice, whose ordering is $\sqsubseteq_{AIK}$, meet and join operations are respectively $\wedge$ and $\vee$, and top and bottom are respectively 1 and 0. This makes it applicable to LCA. In order to ease the understanding of modal operators, we provide the following lemma (a proof is available in [Fer02], p. 86).

Lemma 2. *If $\phi \in AIK$ and $W_R(\phi) = \{w | (w, R) \models \phi\}$ is the set of worlds where ϕ is true, then for every structure (w, R):*
1. $(w, R) \models K\phi$ *iff* $R(w) \subseteq W_R(\phi)$;
2. $(w, R) \models N\neg\phi$ *iff* $R(w) \supseteq W_R(\phi)$;
3. $(w, R) \models O\phi$ *iff* $R(w) = W_R(\phi)$.

This lemma shows that what counts in a model (w, R) of a modal formula is neither the initial world w, nor the accessibility relation itself, but the set of successor worlds $R(w)$. So modal formulas $M\phi$ ($M \in \{K, N\neg, O\}$) can be interpreted as sets of models of ϕ, rather than individual models of ϕ. For instance, the modal formula $K\phi$ represents some subsets of $W_R(\phi)$, in which at least ϕ, but not only ϕ is always true: $K\phi$ can be read "at least ϕ". Dually, the modal formula $N\neg\phi$ can be read "at most ϕ", and the modal formula $O\phi$, which is semantically equivalent to $K\phi \wedge N\neg\phi$ from definition 4, can be read "exactly ϕ" or "all I know is ϕ" (hence the name AIK [Lev90]).

---

[3] A relation R is transitive iff $\forall w, w', w'' : wRw'$ and $w'Rw''$ implies wRw''.
[4] A relation R is euclidean iff $\forall w, w', w'' : wRw'$ and wRw'' implies $w'Rw''$.

3 Ambiguity in the Interpretation of Negation

In this section we exhibit an ambiguity that lies in the interpretation of negation, when it is available in the logic. Practically, in information retrieval, the problem is that we get unsatisfactory answers (i.e., extents) to queries with negation. In order to show the problem more concretely, let us consider a logical context $K = (\mathcal{O}, \mathcal{L}_{Prop}, d)$, where $\mathcal{L}_{Prop}$ is the propositional logic (the same as AIK but with no modal operators). Let say objects are persons, and atoms are properties of these persons (e.g., young, rich). Describing objects then consists in expressing the knowledge one has about person properties. Let say one person of the context, Alice, is young, unhappy, and rich or smart. This can be represented by giving to object Alice the description $young \wedge \neg happy \wedge (rich \vee smart)$. Such a context can exhibit two "anomalies" w.r.t. our intuition, for every queries $q, q' \in Prop$:

1. $ext(q) \cup ext(\neg q) \subsetneq \mathcal{O}$: an object can satisfy neither a query, nor its negation. Example: Alice is not an answer of any of the queries $tall$ and $\neg tall$: so one cannot retrieve persons who are not known as tall.
2. $ext(q \vee q') \supsetneq ext(q) \cup ext(q')$: an object can satisfy $q \vee q'$ while satisfying neither q, nor q'.
   Example: Alice is an answer to the query $rich \vee smart$, but neither to the query $rich$, nor $smart$: so one cannot retrieve persons who are either known as rich or known as smart.

These anomalies come from the fact that formulas, rather than interpretations, are used to describe objects. In relational databases and object-oriented databases, these anomalies do not exist because objects are described by sets of valued attributes letting no ambiguity in their interpretation. On the contrary, formulas generally do not have a unique interpretation but a set of models. When an object description has several models, there are always at least 2 of them that disagree w.r.t. a query q: one satisfies q while the other satisfies $\neg q$, which causes the anomaly 1. Similarly, even if all models satisfy $q \vee q'$, some may satisfy $\neg q$, and others $\neg q'$, which causes the anomaly 2. But interpretations cannot be used in practice because they are in general infinite, and we do not want to use them because they are not flexible enough. For example, in databases, they tend to force users to give a value to every attribute. With formulas, a finite set of relevant descriptors can be chosen among a large or infinite set (open language).

In fact, the expected interpretation of boolean operators in queries is generally *extensional*, i.e., logical operations on formulas (conjunction, disjunction, and negation) are expected to match set operations on extents (intersection, union, and complement); while they are fundamentally *intensional* in descriptions. In the intensional interpretation, negation can be understood as *opposition*, like the opposition that exists between "male" and "female": some things are neither male nor female. Now disjunction should be understood as *undetermination*, like knowing that something has a sex; one can know that something has a sex without knowing whether it is male or female.

Both kinds of negation are present in natural languages. In English, the grammatical word "not" is the extensional negation. Indeed everything is either happy or not happy, hot or not hot. The intensional negation is not so obvious in English as it can be realized either by various prefixes (as in happy/unhappy, legal/illegal), or by a totally different word (hot/cold, tall/small). However there are languages, like Esperanto [JH93], where a unique prefix (in Esperanto, mal-) is used to build all opposites, and thus becomes a grammatical element similar to a logical connector (e.g., varma/malvarma, alta/malalta). In the following, when translating formulas in English, we use this prefix "mal-" to build the opposites instead of the normal english word in order to make opposition more visible (as it is in the logical language with negation): e.g., tall/mal-tall.

Our objective is not to choose between extensional and intensional interpretations, but to combine both in a same formalism. This requires distinguishing occurences of negation as extensional or intensional. Considering a single property "young" and only negation, we obtain 4 different queries: "young", "not young", "mal-young" (i.e., "old"), and "not mal-young" (i.e., "not old"). The logic should also recognize the two subsumption relations that exist between these formulas: "young" entails "not mal-young", and "mal-young" entails "not young" (non-contradiction law). Finally, only opposition is relevant in descriptions because it should be enough not to say a person is young so as to retrieve this person from the query "not young". This is known as the *Closed World Assumption* (CWA).

4   Distinguishing Negation and Opposition

As said in the introduction, the logic AIK (Section 2.2) enables us to apply the Closed World Assumption (CWA) on individual object descriptions through the modality O ("exactly"). The principle is to apply this modality on each object description of a context $K = (\mathcal{O}, \mathcal{L}_{Prop}, d)$, so as to obtain the context

$$K^1 =_{def} (\mathcal{O}, \mathcal{L}_{AIK}, d^1), \text{such that } d^1(o) =_{def} O(d(o)), \text{ for all } o \in \mathcal{O}.$$

For example the description of Alice in previous section becomes $O(young \wedge \neg happy \wedge (rich \vee smart))$.

In queries we propose to use the modality K ("at least"), and we claim that negation has a different interpretation whether it is inside or outside the scope of this modality: $\neg K(young)$ means "is not young", while $K(\neg young)$ means "is mal-young". So boolean operations are extensional outside modalities, and intensional inside modalities. We do not consider the use of the modality O in queries as it is unlikely that a user would ask for an object being "exactly q".

We now show that the anomalies exhibited in Section 3 are solved, i.e., extensional operations match set operations on extents (see proof in [Fer02], p. 88).

Theorem 1. *Let $q, q' \in Prop$ be propositional formulas.*

1. $ext_{K^1}(\neg K(q)) = \mathcal{O} \setminus ext_{K^1}(K(q))$;
2. $ext_{K^1}(K(q) \vee K(q')) = ext_{K^1}(K(q)) \cup ext_{K^1}(K(q'))$;
3. $ext_{K^1}(K(q) \wedge K(q')) = ext_{K^1}(K(q)) \cap ext_{K^1}(K(q'))$;

These equalities are not verified when negation and disjunction appear inside modalities. For example Alice is neither in the extent of $K(tall)$, nor in the extent of $K(\neg tall)$, but she is in the extent of both $\neg K(tall)$ and $\neg K(\neg tall)$ ("Alice is neither tall nor small"). Furthermore, Alice is in the extent of $K(rich \vee smart)$ but neither in the extent of $K(rich)$ nor in the extent of $K(smart)$ ("Alice is rich or smart but we do not know which"). This confirms that operations inside modalities are intensional, whereas operations outside are extensional. Finally it can be proved in AIK that for every propositional formula q:

- $K(q) \sqsubseteq_{AIK} \neg K(\neg q)$ ("is q" entails "is not mal-q"),
- $K(\neg q) \sqsubseteq_{AIK} \neg K(q)$ ("is mal-q" entails "is not q").

This shows there is a hierarchy between both negations; opposition is more specific than extensional negation.

5 Distinguishing Certainty and Possibility

In previous section we applied the Closed World Assumption (CWA) on object descriptions so that everything not true in a description is considered as false. For example, Alice is now considered as "not tall" ($d^1(Alice) \sqsubseteq_{AIK} \neg K(tall)$) because her description is not subsumed by "tall" ($d(Alice) \not\sqsubseteq tall$). This assumes we have a *complete knowledge* about objects. However we sometimes have only *incomplete knowledge* about objects. In this case, some property that is not true in a description may still be *possible*. In the example, as the description of Alice is subsumed neither by "tall" nor by "small" ("mal-tall"), Alice may be everything among "tall", "mal-tall", and "neither tall nor mal-tall".

Between these two extreme positions, a range of 5 intermediate positions can be imagined, from the most incomplete to the most complete:

1. Alice is young and unhappy, and may have any other property;
2. Bob is young and unhappy, and may be rich and smart;
3. Charlie is young and unhappy, and may be either rich or smart but not both;
4. David is young and unhappy, and may be rich;
5. Edward is young and unhappy (and has no other property).

Edward corresponds to previous section, where a property is either true or false. In other cases, we want to distinguish certain properties and possible properties. For example David is certainly young, possibly rich, and certainly not tall; whereas Edward is certainly not rich. In the rest of this section, we adapt the logic AIK and its use in order to distinguish and represent "(certainly) true", "possibly true", "possibly false", and "(certainly) false".

5.1 First Solution with AIK

In a combination of certain and possible facts certain facts represent some kind of *minimum* of what is true, while possible facts represent a kind of *maximum*. Now Lemma 2 shows that the modalities K and $N\neg$ in AIK can be read respectively "at least" and "at most".

So we propose the following representation of above descriptions in AIK, before showing they are not fully satisfying:

1. $d(Alice) = K(young \land \neg happy) \land N\neg(0)$: at least young and unhappy, and at most everything;
2. $d(Bob) = K(young \land \neg happy) \land N\neg(young \land \neg happy \land rich \land smart)$: at least young and unhappy, and at most young, unhappy, rich, and smart;
3. $d(Charlie) = K(young \land \neg happy) \land (N\neg(young \land \neg happy \land rich) \lor N\neg(young \land \neg happy \land smart))$: at least young and unhappy, and at most young, unhappy, and rich, or young, unhappy, and smart;
4. $d(David) = K(young \land \neg happy) \land N\neg(young \land \neg happy \land rich)$: at least young and unhappy, and at most young, unhappy, and rich;
5. $d(Edward) = K(young \land \neg happy) \land N\neg(young \land \neg happy)$: at least and at most young and unhappy, i.e., exactly young and unhappy (what can be represented equivalently by $O(young \land \neg happy)$).

In order to verify whether these descriptions correspond to what we want, we draw the following table where the lines are the 5 descriptions, the columns are different queries q, and a cell is marked $+$ if the description is subsumed by Kq, by $-$ if it is subsumed by $\neg Kq$, and empty otherwise.

$\sqsubseteq_{AIK}$	young	¬young	¬happy	rich	smart	rich ∧ smart	tall
$d(Alice)$	+		+				
$d(Bob)$	+	−	+				−
$d(Charlie)$	+	−	+			−	−
$d(David)$	+	−	+	−		−	−
$d(Edward)$	+	−	+	−	−	−	−

If we read a query $K(X)$ as "X is (certainly) true", and a query $\neg K(X)$ as "X is (certainly) false", then the above table matches our expectations. The cells where neither $K(X)$, nor $\neg K(X)$ are satisfied can be read as "X is possibly true/false, but is not certainly true/false".

However descriptions $d(Alice)$ to $d(David)$ exhibit two problems. Firstly, objects $Alice$ to $David$ satisfy neither $K(rich)$, nor $\neg K(rich)$, what triggers the anomaly 1 (see Section 3) in spite of the presence of the modality K. This implies that possible facts cannot be represented, and so we cannot retrieve persons who "may be rich". Secondly, the query $\neg young$ is possible for $Alice$ whereas $young$ is true: i.e., "Alice is young, but may be old". This is obviously a contradiction. We handle these two problems by a generalization of the logic AIK [Fer01].

5.2 Generalization of AIK

As the first problem is similar to the problem in Section 3, it is tempting to apply the solution in Section 4, i.e., to encapsulate descriptions in the modality O, and queries in the modalities K and $\neg K$. Descriptions are then in the form

$$O(K(d_0) \land (N\neg(d_1) \lor \ldots \lor N\neg(d_n))), \text{ with } n \in \mathbb{N}, \forall i \in 0..n : d_i \in Prop.$$

Unfortunately all formulas in this form have no model, and so are contradictions (see Example 2 of Section 2 in [Ros00]). So, AIK is not a direct answer to our problem, and we propose in the following two successive adaptations of AIK in order to solve it.

Similarly to the formula $O(young \wedge rich)$ that enables us to reason about the set of models of $young \wedge rich$ as if it would be a single model, we would like that the formula $O(K(young) \wedge N\neg(young \wedge rich))$ enables us to reason on the set of models of $K(young) \wedge N\neg(young \wedge rich)$, i.e., on the set of structures (w, R) such that $W_R(young \wedge rich) \subseteq R(w) \subseteq W_R(young)$ (see Lemma 2). To this end, it is necessary that the set $R(w)$ depends on the world w, so as to keep the multiplicity of interpretations of incomplete descriptions. This is why we propose to adapt the logic AIK by removing any condition on the accessiblity relation in Definition 4. The logic AIK can then be seen as an ordinary modal logic, where the modality K is defined on accessible worlds $R(w)$, while the modality N is defined on unaccessible worlds $W \setminus R(w)$, and the modality O is simply a combination of both ($O\phi =_{def} K\phi \wedge N\neg\phi$). Now a family of logics can be derived by applying various conditions on the accessibility relation, as this is already done for modal logics [Bow79]. For example, the usual logic AIK has a transitive and euclidean accessibility relation, and so could be renamed K45-AIK; while our adaptation has an arbitrary relation, and so could be nammed K-AIK.

Definition 5 (logic K-AIK). *The semantics of the logic K-AIK is defined as in Definitions 4, except there is no condition on the accessibility relation.*

In the logic K-AIK, knowledge is stratified because the accessibility relation is neither transitive nor euclidean. The object description $O(d(David)) = O(K(young \wedge \neg happy) \wedge N\neg(young \wedge \neg happy \wedge rich))$ is no more contradictory, and it can be read at three levels of knowledge: a model of

1. $young \wedge \neg happy$ is a world w'' satisfying the proposition $young \wedge \neg happy$;
2. $K(young \wedge \neg happy) \wedge N\neg(young \wedge \neg happy \wedge rich)$ is a world w' such that $R(w')$ is included in the set of models of $young \wedge \neg happy$, and contains all models of $young \wedge \neg happy \wedge rich$;
3. $O(K(young \wedge \neg happy) \wedge N\neg(young \wedge \neg happy \wedge rich))$ is a world w such that $R(w)$ is the set of all models of $K(young \wedge \neg happy) \wedge N\neg(young \wedge \neg happy \wedge rich)$.

So this description expresses the complete knowledge about the incomplete knowledge about the object *David*: "all I know about this person is that he is young and unhappy, and he may be rich as well". The logic K-AIK allows for the following subsumption relation.

Example. $O(d(David)) \sqsubseteq_{K-AIK}$
$\quad K(K(young)) \wedge K(\neg K(\neg young)) \wedge \neg K(K(rich)) \wedge \neg K(\neg K(rich)). \qquad \square$

If the outermost K is read "I know", then we can translate the above entailment as "I know that David is young, and that he is not mal-young, but I do not know whether he is rich or not". So $K(K(q))$ represents (certainly) true facts, and $K(\neg K(q))$ represent (certainly) false facts. Formulas $\neg K(\neg K(q))$, which can be read as "I do not know that not q", represents possibly true facts, and similarly formulas $\neg K(K(q))$ represent possibly false facts. This solves the first problem about the representation of possible facts.

The first problem is now solved, but the second is not because $\neg young$ ("mal-young") is judged as a possible fact for the object *Alice* whereas *young* is judged as true. This means there is a contradiction somewhere. In order to remove this contradiction, we try to exclude models (w', R) at the 2nd level of knowledge such that $R(w') = \emptyset$. This implies that in the description $d(Alice)$, the part $N\neg(0)$ does not mean "at most everything", but rather "at most everything provided it is not contradictory". Technically this is obtained by requiring the accessibility relation to be *serial*[5], which enforces every world to have at least one successor world: the result is the logic KD-AIK.

Definition 6 (logic KD-AIK). *The semantics of the logic KD-AIK is defined like in Definition 4, except the accessibility relation must be serial.*

This time we obtain the expected subsumption for *Alice*, with $\neg young$ being judged as (certainly) false, whereas *rich* remains possibly true/false.

Example. $O(d(Alice)) \sqsubseteq_{KD-AIK}$
    $K(K(young)) \wedge K(\neg K(\neg young)) \wedge \neg K(K(rich)) \wedge \neg K(\neg K(rich))$.   □

In summary we use a version of the logic AIK (KD-AIK) with two levels of modalities in descriptions and queries. In descriptions the modality O is used at the outermost level, while a combination of modalities K and N is used in order to express certain and possible facts about object. In queries the modality K is used at both levels, which implies that negations can occur at three levels. From innermost to outermost levels the logical negation represents respectively *opposition*, *falsity*, and *possibility*. It is noticeable that in order to distinguish these three kinds of negations, there is need to introduce neither new connectors, nor special semantics. A small variation of the logic AIK, which is itself a small variation of well known modal logics, and appropriate combinations of modal operators are sufficient. In the following section we generalize the application of modal operators to other logics than propositional logic, and we introduce a compact and more intuitive syntax in place of combinations of modal operators and negations.

6   The Epistemic Extension of an Arbitrary Logic

Though the logic KD-AIK is suitable to the expression of opposition, negation, and possibility, it has the drawback of being applicable only to contexts whose facts are expressed in the propositional logic. So, we now define the epistemic

---

[5] A relation R is serial iff $\forall w : \exists w' : wRw'$.

extension $\mathcal{L}^2$ (2 levels of modalities) of an arbitrary logic $\mathcal{L} = (L, \sqcap, \sqcup, \neg, \top, \bot, \sqsubseteq)$ (among those applicable in LCA) so as to enable us to distinguish between opposition, falsity, and possibility.

The language $\mathcal{L}^2$ is defined by the following grammar, where L is the language of formulas of $\mathcal{L}$:

$$
\begin{aligned}
L^2 \longrightarrow\ & [L,\ L \mid \ldots \mid L] \\
\mid\ & !L \\
\mid\ & ?L \\
\mid\ & L \text{ and } L \mid L \text{ or } L \mid not\ L \mid true \mid false.
\end{aligned}
$$

The correspondance of the first 3 derivations with AIK modalities is as follows:

- $[d_0,\ d_1 \mid \ldots \mid d_n] \longrightarrow O(K(d_0) \wedge (N\neg(d_0 \sqcap d_1) \vee \ldots \vee N\neg(d_0 \sqcap d_n)))$: d_0 represents all certain facts, while each $d_0 \sqcap d_i$ represents a maximum set of possible facts;
- $!q \longrightarrow K(K(q))$: q is *certainly* true;
- $?q \longrightarrow \neg K(\neg K(q))$: q is *possibly* true.

A complete knowledge where every possible fact is also certain has the form $[d_0, \top]$, which can be shortened as $[d_0]$. At the opposite, an incomplete knowledge where every non-certain fact is possible has the form $[d_0, \bot]$. Boolean operations have an extensional meaning, and enables us to express a *certainly false* fact q as a not possibly true fact, i.e., by the formula *not* $?q$, and a *strictly possibly true* fact q (i.e., possible but not certain) by the formula $?q$ *and not* $!q$. So *not* $!q$ must be read "not certainly q", i.e., "possibly not q". Similarly, the formula *not* $?q$ must be read "not possibly q", and is equivalent to "certainly not q" (i.e., "q is false"). These two equivalences are kinds of Morgan laws like those that exist between conjunction and disjunction, between universal and existential quantifiers, and between modal operators $\square$ and $\lozenge$. Finally, opposition does not appear in the above grammar as it is played by the negation of $\mathcal{L}$ (when defined).

We now give the semantics of the language L^2 as a function of the semantics of L by using the Lemma 2. Moreover, as we use exactly 2 levels of modalities in object descriptions and queries, only the set $\{R(w') \mid w' \in R(w)\}$ semantically matters in a structure (w, R). So interpretations in L^2 are simply sets of sets of interpretations of L.

Definition 7 (semantics). *Let* $S_L = (I, \models)$ *be the semantics of a logic* $\mathcal{L} = (L, \sqcap, \sqcup, \neg, \top, \bot, \sqsubseteq)$. *The semantics* $S_{L^2} = (I^2, \models^2)$ *of the language* L^2 *is defined by (given that for all* $f \in L$, $M_L(f) = \{i \in I \mid i \models f\}$*):*

- $I^2 = 2^{2^I}$;
- *and for all* $i^2 \in I^2$,

$$i^2 \models^2 \begin{cases} [d_0, d_1 \mid \ldots \mid d_n] & \text{iff} & i^2 = \{M \subseteq I \mid M \subseteq M_L(d_0), \\ & & \exists k \in 1..n : M \supseteq M_L(d_0 \sqcap d_k)\} \\ !q & \text{iff} & \forall M \in i^2 : M \subseteq M_L(q) \\ ?q & \text{iff} & \exists M \in i^2 : M \subseteq M_L(q) \\ Q \text{ and } Q' & \text{iff} & i^2 \models^2 Q \text{ and } i^2 \models^2 Q' \\ Q \text{ or } Q' & \text{iff} & i^2 \models^2 Q \text{ or } i^2 \models^2 Q' \\ \text{not } Q & \text{iff} & i^2 \not\models^2 Q \\ \text{true} & \text{iff} & \text{true, i.e., for all } i^2 \\ \text{false} & \text{iff} & \text{false, i.e., for no } i^2 \end{cases}$$

We now have to verify that the anomalies in Section 3 are still solved, and that the semantics of possible facts is like expected. This is demonstrated by the following theorem.

Theorem 2. *Let $K = (\mathcal{O}, \mathcal{L}^2, d)$ be a context, and $d_i, d'_i, q \in \mathcal{L}$ and $Q, Q' \in \mathcal{L}^2$ be formulas. If the subsumption $\sqsubseteq$ of $\mathcal{L}$ is consistent and complete, the following equations are verified:*

1. $ext(!q)$ $= \{o \in \mathcal{O} \mid d(o) = [d_0, d_1 \mid \ldots \mid d_n], d_0 \sqsubseteq q\}$
2. $ext(?q)$ $= \{o \in \mathcal{O} \mid d(o) = [d_0, d_1 \mid \ldots \mid d_n], \exists k \in 1..n : (d_0 \sqcap d_k) \sqsubseteq q\}$
3. $ext(Q \text{ and } Q') = ext(Q) \cap ext(Q')$
4. $ext(Q \text{ or } Q') = ext(Q) \cup ext(Q')$
5. $ext(\text{not } Q) = \mathcal{O} \setminus ext(Q)$
6. $ext(all) = \mathcal{O}$
7. $ext(none) = \emptyset$

Proof: The following proofs use Definition 7 of the semantics of $\mathcal{L}^2$, and use a lot the result that formulas in the form $d = [d_0, d_1 \mid \ldots \mid d_n]$ have a single model

$$m(d) =_{def} \{M \subseteq I \mid M \subseteq M_L(d_0), \exists k \in 1..n : M_L(d_0 \sqcap d_k) \subseteq M\}. \quad (1)$$

For conciseness, we omit to note the context K as indices of ext.

1. $o \in ext(q) \iff d(o) \sqsubseteq^2 q$
   $\iff \{m(d(o))\} \subseteq \{i^2 \in I^2 \mid \forall M \in i^2 : M \subseteq M_L(q)\}$
   $\iff \forall M \subseteq I : M \subseteq M_L(d_0)$
   and $(\exists k \in 1..n : M_L(d_0 \sqcap d_k) \subseteq M)$ implies $M \subseteq M_L(q)$ $\qquad (1)$
   $\iff d_0 \sqsubseteq q$: by contradiction, it is enough to consider the case where $M = M_L(d_0)$ knowing that $d_0 \sqcap d_k \sqsubseteq d_0$ for all k since $\mathcal{L}$ forms a lattice, and its subsumption is consistent and complete
2. $o \in ext(?q) \iff d(o) \sqsubseteq^2 ?q$
   $\iff \{m(d(o))\} \subseteq \{i^2 \in I^2 \mid \exists M \in i^2 : M \subseteq M_L(q)\}$
   $\iff \exists M \subseteq I : M \subseteq M_L(d_0)$
   and $(\exists k \in 1..n : M_L(d_0 \sqcap d_k) \subseteq M)$ et $M \subseteq M_L(q)$ $\qquad (1)$
   $\iff \exists k \in 1..n : M_L(d_0 \sqcap d_k) \subseteq M_L(d_0)$
   $\iff \exists k \in 1..n : d_0 \sqcap d_k \sqsubseteq d_0$. $\qquad$ ($\sqsubseteq$ is consistent and complete)
3. immediate from the Galois connection.
4. $o \in ext(Q \text{ or } Q') \iff d(o) \sqsubseteq^2 Q \text{ or } Q'$
   $\iff \{m(d(o))\} \subseteq M_{L^2}(Q \text{ or } Q') \iff \{m(d(o))\} \subseteq M_{L^2}(Q) \cup M_{L^2}(Q')$
   $\iff \{m(d(o))\} \subseteq M_{L^2}(Q) \text{ or } \{m(d(o))\} \subseteq M_{L^2}(Q')$
   $\iff d(o) \sqsubseteq^2 Q \text{ or } d(o) \sqsubseteq^2 Q' \iff o \in ext(Q) \cup ext(Q')$.

5. $o \in ext(not\ Q) \iff d(o) \sqsubseteq^2 not\ Q$
   $\iff \{m(d(o))\} \subseteq M_{L^2}(not\ Q) \iff \{m(d(o))\} \not\subseteq M_{L^2}(Q)$
   $\iff d(o) \not\sqsubseteq^2 Q \iff o \notin ext(Q).$
6. immediate from the Galois connection.
7. $o \in ext(none) \implies d(o) \sqsubseteq^2 none$
   $\iff \{m(d(o))\} \subseteq \emptyset$: contradiction, hence $ext(none) = \emptyset$.　■

It can be verified that Theorem 2.(4) solves the anomaly 2, and that Theorem 2.(5) solves the anomaly 1. In fact operations in $\mathcal{L}^2$ (*and*, *or*, *not*, *true*, *false*) have a fully extensional interpretation as they match exactly set operations on context extents (intersection, union, complement, the set of all objects, the empty set). This theorem also generalizes Theorem 1 in two ways. Firstly, the logic used for describing objects and elementary queries is arbitrary (among those applicable to LCA), and not necessarily the propositional logic $\mathcal{L}_{Prop}$. Secondly, results apply to arbitrary queries in L^2: variables Q and Q' are not restricted to the language L.

Finally, this theorem gives us a way to compute the answers of queries in $\mathcal{L}^2$, provided we have a decision procedure for the subsumption $\sqsubseteq$ in $\mathcal{L}$. We have thus integrated epistemic knowledge in the querying process, and allowed distinction between opposition, negation, and possibility.

7 Related Work

The issue of representing incomplete knowledge, and distinguishing between certain and possible facts has already been studied in the scope of Formal Concept Analysis [Obi02, BH05]. They use 3-valued contexts which are similar to our distinction between true, possible, and false facts in object descriptions. They also use compound attributes, which are in fact formulas in the propositional logic. A first difference is that instead of having simple attributes we have formulas of an almost arbitrary logic in descriptions and queries, e.g., propositions over valued attributes. On the contrary they have only extensional operations, and no intensional operations like opposition. Another difference is that they have several derivation operators corresponding to the two modalities "certainly" and "possibly", which they respectively denote by $\square$ and $\lozenge$, and which we denote in the epistemic extension (Section 6) by ! and ?. Instead we have only one derivation operator ext that applies to any combination of modalities, and both extensional and intensional operations. A more general difference is that they use contexts as the semantics of Contextual Logic whereas we use logics and their semantics right inside logical contexts.

The *certainly and possibly valid* implications defined by S. Obiedkov can be represented as follows in our framework:

- an implication $\phi \to \psi$ is *certainly valid* in a context K if every object that possibly satisfies ϕ certainly satisfies ψ, i.e., $ext_K(?\phi) \subseteq ext_K(!\psi)$,
- an implication $\phi \to \psi$ is *possibly valid* in a context K if every object that certainly satisfies ϕ possibly satisfies ψ, i.e., $ext_K(!\phi) \subseteq ext_K(?\psi)$.

Opposition has already been introduced as an alternative negation that respects the law of contradiction $(q \wedge \neg q)$, but not the law of excluded middle $(q \vee \neg q)$. Indeed, nobody can be young and old at the same time, but somebody can be neither young, nor old. In the scope of FCA, opposition appears first in the double Boolean algebra [Wil00]. It is also introduced in Description Logics through a new connector $\sim$, and an extended semantics [Kan05]. In both cases a new connector is introduced, whereas we simply use the classical connector in combination with modalities so as to obtain the logical properties of opposition.

8 Conclusion

To the best of our knowledge, our solution is the first that enables us to represent incomplete knowledge, and distinguish negation and opposition in a unique formalism. The logic used to describe objects is let free, and its negation (if defined) plays the role of opposition. Object descriptions contain certainly true facts (opposites or not), and a disjunction of possibly true facts. If a fact does not appear in a description it is considered as certainly false by CWA, which allows for concise descriptions (remind that the language of facts is often infinite). This is more expressive than 3-valued contexts as properties like "possibly either rich or smart" can be represented. Moreover the subsumption of the logic AIK recognizes the hierarchy that exists between opposition, negation, and possibility (from the most specific to the most general).

Surprisingly this result is achieved with no extension of the theory of LCA, and a small variation of the logic AIK that is common place in modal logics. The result comes from the right combination and interpretation of modal operators. The danger of losing the genericity of LCA is escaped by the *epistemic extension* of a logic that enables us to retain, and even extend its genericity. Indeed a detailed study about the combination of logics [FR02] shows that, w.r.t. their application in LCA, less properties are required on logics when this epistemic extension is applied on them. In other words, the epistemic extension ensures that a weak logic (having only consistent and partially complete subsumption) gains all desired properties for LCA.

The epistemic extension is implemented as part of a toolbox of logic components, LOGFUN (see www.irisa.fr/lande/ferre/logfun). It is systematically used in applications of CAMELIS, an implementation of logical information systems [FR04], which can handle efficiently up to several 10,000 objects (see www.irisa.fr/lande/ferre/camelis).

References

[BH05]   P. Burmeister and R. Holzer. Treating incomplete knowledge in formal concept analysis. In B. Ganter, G. Stumme, and R. Wille, editors, *Formal Concept Analysis*, LNCS 3626, pages 114–126. Springer, 2005.

[Bow79] K. A. Bowen. *Model Theory for Modal Logic*. D. Reidel, London, 1979.

[Che94] J. Chen. The logic of only knowing as a unified framework for non-monotonic reasoning. *Fundamenta Informatica*, 21, 1994.

[CM00]   L. Chaudron and N. Maille. Generalized formal concept analysis. In
           G. Mineau and B. Ganter, editors, *Int. Conf. Conceptual Structures*, LNCS
           1867. Springer, 2000.

[DNR97]   F. M. Donini, D. Nardi, and R. Rosati. Autoepistemic description logics. In
           IJCAI, 1997.

[Fer01]   S. Ferré. Complete and incomplete knowledge in logical information sys-
           tems. In S. Benferhat and P. Besnard, editors, *Symbolic and Quantita-
           tive Approaches to Reasoning with Uncertainty*, LNCS 2143, pages 782–791.
           Springer, 2001.

[Fer02]   S. Ferré. *Systèmes d'information logiques : un paradigme logico-contextuel
           pour interroger, naviguer et apprendre.* Thèse d'université, Université de
           Rennes 1, October 2002. Accessible en ligne à l'adresse http://www.irisa.fr/
           bibli/publi/theses/theses02.html.

[FR00]   S. Ferré and O. Ridoux. A logical generalization of formal concept analysis.
           In G. Mineau and B. Ganter, editors, *Int. Conf. Conceptual Structures*,
           LNCS 1867, pages 371–384. Springer, 2000.

[FR02]   S. Ferré and O. Ridoux. A framework for developing embeddable customized
           logics. In A. Pettorossi, editor, *Int. Work. Logic-based Program Synthesis
           and Transformation*, LNCS 2372, pages 191–215. Springer, 2002.

[FR04]   S. Ferré and O. Ridoux. An introduction to logical information systems.
           Information Processing & Management, 40(3):383–419, 2004.

[GK01]   B. Ganter and S. Kuznetsov. Pattern structures and their projections. In
           H. S. Delugach and G. Stumme, editors, *Int. Conf. Conceptual Structures*,
           LNCS 2120, pages 129–142. Springer, 2001.

[JH93]   P. Janton and H.Tonkin. *Esperanto: Language, Literature, and Community.*
           State University of New York Press, 1993.

[Kan05]   K. Kaneiwa. Negations in description logic - contraries, contradictories,
           and subcontraries. In F. Dau, M.-L. Mugnier, and G. Stumme, editors,
           Contributions to ICCS 2005, pages 66–79. Kassel University Press GmbH,
           2005.

[Lev90]   H. Levesque. All I know: a study in autoepistemic logic. *Artificial Intelli-
           gence*, 42(2), March 1990.

[Lif91]   V. Lifschitz. Nonmonotonic databases and epistemic queries. In *12th Inter-
           national Joint Conference on Artificial Intelligence*, pages 381–386, 1991.

[McC86]   J. McCarthy. Applications of circumscription to formalizing common sense
           knowledge. *Artificial Intelligence*, 28(1), 1986.

[Moo85]   R. C. Moore. Semantical considerations on nonmonotonic logic. *Artificial
           Intelligence*, 25(1):75–94, 1985.

[Obi02]   S. A. Obiedkov. Modal logic for evaluating formulas in incomplete contexts.
           In *ICCS*, LNCS 2393, pages 314–325. Springer, 2002.

[Plo70]   G. D. Plotkin. A note on inductive generalization. *Machine Intelligence,
           Edinburgh Univ. Press*, 5:153–163, 1970.

[Pre97]   S. Prediger. Logical scaling in formal concept analysis. *LNCS 1257*, pages
           332–341, 1997.

[Ros00]   R. Rosati. Tableau calculus for only knowing and knowing at most. In Roy
           Dickhoff, editor, *TABLEAUX*, LNCS 1847. Springer, 2000.

[Wil00]   R. Wille. Boolean concept logic. In G. Mineau and B. Ganter, editors, *Int.
           Conf. Conceptual Structures*, LNCS 1867, pages 317–331. Springer, 2000.

A Note on Negation: A PCS-Completion of Semilattices

Léonard Kwuida

Mathematisches Institut, Universität Bern, CH-3012 Bern
kwuida@math-stat.unibe.ch

Abstract. In the paper "Which concept lattices are pseudocomplemented?" ([GK05]) we gave a contextual characterization of pseudocomplemented concept lattices by means of the arrow relations. In this contribution we use this description to embed finite semilattices into pseudocomplemented semilattices. This process can be used to define a negation on concepts.

Keywords: negation, semilattices, pseudocomplement, FCA.

AMS Subject Classification: 06D15, 68T30.

1 Introduction

Boolean algebras arose from the investigations of the laws of thought by George Boole [Bo54]. They are algebras $(L, \wedge, \vee, ', 0, 1)$ of type $(2, 2, 1, 0, 0)$ such that $(L, \wedge, \vee, 0, 1)$ is a bounded distributive lattice and the unary operation is a complementation. The operations $\wedge$, $\vee$ and $'$ were used to model respectively the conjunction, the disjunction and the negation. Since Boole many classes have been introduced to generalize these structures. A frequent approach consists in relaxing some properties or some operations of Boolean algebras. In [KPR04] the authors presented an overview of the classes that mainly focus on generalizations of the unary operation. Among these is the class of pseudocomplemented lattices or *pcl* for short.

The *pseudocomplement* of a lattice element x, if it exists, is the largest element whose meet with x is the zero element of the lattice. In other words x^* is the pseudocomplement of x iff

$$ x \wedge y = 0 \iff y \le x^*. $$

A lattice L with 0 is *pseudocomplemented* iff each $x \in L$ a pseudocomplement; the mapping $x \mapsto x^*$ then defines a unary operation on L, called *pseudocomplementation*. The algebra $(L, \wedge, \vee, {}^*, 0, 1)$ is called a *p-algebra*. In the case of a concept lattice, the pseudocomplement of a concept (A, B) is the most general concept that contradicts (A, B). Such a pseudocomplement may be interpreted as a negation[1] of the concept, and we shall do so in this article.

---

[1] See [Ho89–Ch. 1] or [Wi00].

R. Missaoui and J. Schmid (Eds.): ICFCA 2006, LNAI 3874, pp. 146–160, 2006.
© Springer-Verlag Berlin Heidelberg 2006

The pseudocomplementation is antitone and square extensive, i.e., it satisfies

$$x \leq y \implies y^* \leq x^* \quad \text{and} \quad x \leq x^{**}$$

for all $x, y \in L$. These two properties together imply the *infinite join de Morgan law*

$$\left(\bigvee_{k \in K} x_k \right)^* = \bigwedge_{k \in K} x_k^*.$$

The infinite join de Morgan law is to be understood as follows: if $\bigvee_{k \in K} x_k$ exists and all x_k^* exist then the equality above holds.

Moreover the set of elements satisfying the law of double negation ($x^{**} = x$) with the inherited order, called skeleton, is a Boolean algebra. O. Frink pointed out in [Fr62] that most of the results obtained for p-algebras can be proved even if the operation $\vee$ is omitted, i.e., by considering only the algebra $(L, \wedge, ^*, 0, 1)$. This is called a *pseudocomplemented semilattice* or *pcs* for short. Recall that a *semilattice* is an algebra $(L, \circ)$ such that $\circ$ is a binary operation that is commutative, associative and idempotent. In this contribution we will prove that each finite[2] semilattice can be extended in a natural way to a pcs. This completion process is used to define a negation on concepts.

The rest of the paper is divided as follows: in Section 2 we consider the relationship between lattices and semilattices. Section 3 presents the main result, namely that each closure system can be extended to a pseudocomplemented closure system. Section 4 is devoted to an application: constructing a negation on concepts. The reader is referred to [GW99] for terminologies and further notions on concept lattices. Some results of [GK05] will be mentioned without proofs. For some applications we refer to [So98], [Du97] and [DG00]. An introduction to p-algebras can be found in [Ka80], [CG00] or in [Gr71] and [BD74] for distributive pseudocomplemented lattices.

2   Lattices and Semilattices

From the order theoretical point of view, each lattice is also a (meet-)semilattice. Therefore each pcl is a pcs. Note that each pcs necessary has a top element, the pseudocomplement of 0. Now let $(L, \wedge)$ be a (meet-)semilattice. If a and b are elements of L and have u and v as upper bound, then $u \wedge v$ is also an upper bound of a and b. Therefore if L is a finite semilattice that is not a lattice, this should be because of the lack of upper bounds and not because of too many upper bounds without a least one. Thus there is no difference between the underlying order on nearlattices[3] or chopped lattices and semilattices, provided they are finite. Unfortunately this is no more valid if the poset is infinite. For example if we replace the top element of a 4-element Boolean algebra with an infinite chain without a least element, we obtain a semilattice which is not a nearlattice.

---

[2] The results also apply to complete semilattices.

[3] A nearlattice (see [Ci04]) or chopped lattice (see [GS04]) is a meet semilattice in which any two elements having common upper bounds have a supremum.

However we can obtain a lattice $S \oplus 1$ from a finite semilattice S by adding a new top element. Of course removing the top element of a finite lattice produces a semilattice. Thus there is a one-to-one correspondence between finite lattices and finite (meet)-semilattices. Since each pcs has a top element we can assume w.l.o.g. that our semilattices have a 1. Anyway the process we present later ends up with the same pcs no matter the initial semilattice has a top element or not.

It is well known that complete lattices and closure systems are 1-1 correspondent. What about semilattices? Closure systems are closed under (arbitrary) intersections. They are also called $\bigcap$-closed families. For semilattices closure systems can be replaced by $\cap$-closed families. By a $\cap$-closed family on a set G we mean a family $\mathcal{E}$ of subsets of G such that $\bigcup \mathcal{E} = G$ and $A \cap B \in \mathcal{E}$ for all $A, B \in \mathcal{E}$. Such families are closed under nonempty finite intersections. Closure systems are $\bigcap$-closed families. The converse holds in the finite case. For a concept lattice, being pseudocomplemented is naturally expressed in terms of the closure system of extents. A closure system or a $\cap$-closed family is called pseudocomplemented if each closed set has a pseudocomplement. In the next section we will show how closure systems can be extended to pseudocomplemented closure systems.

3   Embedding Closure Systems into Pseudocomplemented Closure Systems

Let $\mathcal{E}$ be a closure system on a set G, and let $A \mapsto A''$ be the corresponding closure operator. We use the notation $a'' := \{a\}''$ for $a \in G$. For simplicity we assume $\emptyset'' = \emptyset$ and $g'' = h'' \implies g = h$.

A subset T of G is called **transversal** to $\mathcal{E}$ if each nonempty closed set E of $\mathcal{E}$ contains some elements of T. $\mathcal{E}$ is **atomic** iff each element of $\mathcal{E} \setminus \{\emptyset\}$ contains an upper cover of $\emptyset$. The closure system $\mathcal{E}$ is atomic iff

$$G_{\min} := \{g \in G \mid g'' = \{g\}\}$$

is a transversal. A subfamily $\mathcal{U}$ of $\mathcal{E}$ is called **co-initial** if for every $E \in \mathcal{E} \setminus \{\emptyset\}$ there is $U \in \mathcal{U} \setminus \{\emptyset\}$ such that $U \subseteq E$. The closure system $\mathcal{E}$ is atomic iff the subfamily $\{g'' \mid g \in G_{\min}\}$ is co-initial. The following proposition shows the relationship between "transversal" and "co-initial", as well as their importance in checking pseudocomplementation.

Proposition 1. *Let $\mathcal{E}$ be a closure system on G.*

(i) *A subset T of G is transversal to $\mathcal{E}$ iff $\{t'' \mid t \in T\}$ is co-initial.*
(ii) *A subfamily $\mathcal{U}$ of $\mathcal{E}$ is co-initial iff $\bigcup \mathcal{U}$ is transversal to $\mathcal{E}$.*
(iii) *$\mathcal{E}$ is pseudocomplemented iff $\mathcal{E}$ has a co-initial subfamily $\mathcal{U}$ whose elements have pseudocomplements.*

Proof. (i) and (ii) are straightforward. One direction of the equivalence in (iii) is obvious since $\mathcal{U}$ is a subfamily of $\mathcal{E}$. The converse is Proposition 4 of [GK05].

Therefore, an atomic closure system $\mathcal{E}$ will be pseudocomplemented iff all g'', $g \in G_{\min}$ have pseudocomplements. To express the pseudocomplementation we need some operations introduced in [GK05], namely a projection and its inverse image. We consider $\mathcal{E}$ a closure system on a set G with a transversal $G_{\min}$. (See Lemma 1 for the general case.)

The projection on $G_{\min}$ is the map s defined on $\mathcal{P}(G)$ sending any subset A of G to

$$s(A) := \bigcup_{g \in A} g'' \cap G_{\min}.$$

Note that $s(g) := s(\{g\}) = g'' \cap G_{\min}$ and $s(A) = A \cap G_{\min}$ for any $A \in \mathcal{E}$. The inverse image of s denoted by $[\cdot]$ is defined on $\mathcal{P}(G)$ by

$$[A] := \{g \in G \mid s(g) \subseteq s(A)\}.$$

The following proposition gives a characterization of the pseudocomplementation.

Proposition 2. *[GK05–Propn. 6, Propn. 7 & Thm. 1]*

(a) *The operator $[\cdot]$ defines a closure operator on G.*
(b) *An element A of a closure system $\mathcal{E}$ has a pseudocomplement iff $[G_{\min} \setminus A]$ is in $\mathcal{E}$.*
(c) *$\mathcal{E}$ is a pseudocomplemented closure system iff all sets $[G_{\min} \setminus \{a\}]$ with $a \in G_{\min}$ are in $\mathcal{E}$.*

Thus if $\mathcal{E}$ is not pseudocomplemented, then there is an A in $\mathcal{E}$ such that $[A] \notin \mathcal{E}$. The idea is to collect all such $[A]$s and generate a new closure system. It is enough by (c) to generate the new closure system from $\mathcal{E}$ and those $[G_{\min} \setminus \{a\}]$ that are not in $\mathcal{E}$.

Theorem 1. *Let $\mathcal{E}$ be a closure system on G with a transversal $G_{\min}$. We denote by*

$$\tilde{\mathcal{E}} := \langle \mathcal{E} \cup \{[G_{\min} \setminus \{a\}] \mid a \in G_{\min}\} \rangle$$

the closure system generated by $\mathcal{E}$ and $\{[G_{\min} \setminus \{a\}] \mid a \in G_{\min}\}$. The following assertions hold:

(i) *$\tilde{\mathcal{E}}$ is a pseudocomplemented closure system.*
(ii) *The map*

$$i : \mathcal{E} \to \tilde{\mathcal{E}}$$
$$A \mapsto A$$

   is a $\wedge$-embedding, preserving existing pseudocomplements.
(iii) *The closure system $\tilde{\mathcal{E}}$ is the coarsest pseudocomplemented refinement of the system containing $\mathcal{E}$.*
(iv) *If $\varphi: \mathcal{E} \to \mathcal{F}$ is a $\wedge$-embedding of $\mathcal{E}$ into a pseudocomplemented closure system $\mathcal{F}$ that preserves existing pseudocomplements, then there is an order embedding $\psi: \tilde{\mathcal{E}} \to \mathcal{F}$ that preserves pseudocomplements and $\psi \circ i = \varphi$.*

Proof. All $[G_{min} \setminus \{a\}]$ are subsets of G. So $\tilde{\mathcal{E}}$ is a closure system on G. We denote by $''^{\tilde{\mathcal{E}}}$ the corresponding closure operator. Set

$$\widetilde{G_{min}} := \{g \in G \mid g''^{\tilde{\mathcal{E}}} = \{g\}\}.$$

Obviously G_{min} is a subset of $\widetilde{G_{min}}$. We claim that both sets are equal. Otherwise there would be an element $b \in G$ such that $\{b\} \in \tilde{\mathcal{E}}$, which is not in $\mathcal{E}$. Therefore there would be an element $E \in \mathcal{E}$ and a subset $\{a_k \mid k \in K\} \subseteq G_{min}$ such that

$$\{b\} = E \cap \bigcap \{[G_{min} \setminus \{a_k\}] \mid k \in K\}.$$

Note that $b \in [G_{min} \setminus \{a_k\}]$ implies $a_k \notin b''$ for all $k \in K$. If b belongs to $[G_{min} \setminus \{a\}]$ for all $a \in G_{min}$, we will have $s(b) \subseteq G_{min} \setminus \{a\}$ for all $a \in G_{min}$, and by then $a \notin b''$ for all $a \in G_{min}$. This can not happen since G_{min} is transversal. Thus there is at least one $a \in G_{min}$ (let's say a_{i_0}) such that $b \notin [G_{min} \setminus \{a_{i_0}\}]$. The closure b'' contains a_{i_0} with $a_{i_0} \neq b$. Moreover $b \in E$ implies $b'' \subseteq E$; thus a_{i_0} is another element of E (different from b). From the equality

$$\{b\} = E \cap \bigcap \{[G_{min} \setminus \{a_k\}] \mid k \in K\}$$

there should be an element $i_1 \in K$ such that $a_{i_0} \notin [G_{min} \setminus \{a_{i_1}\}]$. Now $a_{i_0} \notin [G_{min} \setminus \{a_{i_1}\}]$ implies $s(a_{i_0}) \not\subseteq G_{min} \setminus \{a_{i_1}\}$. Therefore $a_{i_1} \in s(a_{i_0}) = \{a_{i_0}\}$, since $a_{i_0} \in G_{min}$. Thus $a_{i_0} = a_{i_1}$, which is in contradiction with $a_{i_0} \in b''$ together with $a_{i_1} \notin b''$. This concludes the proof of our claim, and means that there is no new atoms.

We denote by $\tilde{s}$ the projection of $\tilde{\mathcal{E}}$ on $\widetilde{G_{min}}$ and by $[\tilde{\cdot}]$ the corresponding inverse image. We will have

$$\tilde{s}(A) = \bigcup_{g \in A} g''^{\tilde{\mathcal{E}}} \cap \widetilde{G_{min}} = \bigcup_{g \in A} g''^{\tilde{\mathcal{E}}} \cap G_{min} = \bigcup_{g \in A} g'' \cap G_{min} = s(A).$$

Then both projections coincide. It follows that

$$[\tilde{A}] = \{h \in G \mid \tilde{s}(h) \subseteq \tilde{s}(A)\} = \{h \in G \mid s(h) \subseteq s(A)\} = [A].$$

Thus for all $a \in \widetilde{G_{min}}$ the set $[\widetilde{G_{min} \setminus \{a\}}]$ belongs to $\tilde{\mathcal{E}}$. By (c) of Proposition 2 we can conclude that $\tilde{\mathcal{E}}$ is a pseudocomplemented closure system and (i) is proved.

For (ii) it is evident that the map i preserves the meet operation since the meet in closure systems is the intersection. Let $a \in G_{min}$ such that $\{a\}$ has a pseudocomplement. Then $[G_{min} \setminus \{a\}] \in \mathcal{E}$ is its pseudocomplement. In fact if $E \in \mathcal{E}$ and $\{a\} \cap E = \emptyset$ then $s(h) \subseteq G_{min} \setminus \{a\}$ for all $h \in E$. Thus E is a subset of $[G_{min} \setminus \{a\}]$. Of course $[G_{min} \setminus \{a\}] \cap \{a\} = \emptyset$. Since the two projections coincide it follows that $i(\{a\}^*) = (i(\{a\}))^* = \{a\}^*$ because

$$(\dagger) \qquad i(\{a\}^*) = i([G_{min} \setminus \{a\}]) = [G_{min} \setminus \{a\}] \text{ and}$$

$$(\ddagger) \qquad (i(\{a\}))^* = [\widetilde{G_{min} \setminus i(\{a\})}] = [G_{min} \setminus \{a\}] = \{a\}^*,$$

where $*$ in (†) is the pseudocomplementation in $\mathcal{E}$ and in (‡) the pseudocomplementation in the newly generated closure system $\tilde{\mathcal{E}}$. Note that if $A \in \mathcal{E}$ has a pseudocomplement then

$$A^* = \bigcap_{a \in s(A)} [G_{\min} \setminus \{a\}] = \bigcap_{a \in \tilde{s}(A)} \widetilde{[G_{\min} \setminus \{a\}]} = A^{\tilde{*}},$$

where $*$ denotes the (partial operation of) pseudocomplementation on $\mathcal{E}$ and $\tilde{*}$ the (total operation of) pseudocomplementation on $\tilde{\mathcal{E}}$. From here on, we will no make this distinction again, since they coincide wherever they exist.

For (iii) observe that $\tilde{\mathcal{E}}$, as closure system generated by $\mathcal{E}$ and $\{[G_{\min} \setminus \{a\}] \mid a \in G_{\min}\}$, is the smallest closure system containing these two sets. Its elements are intersections of elements of $\mathcal{E}$ and subsets of the form $[G_{\min} \setminus \{a\}]$ with $a \in G_{\min}$. The later are exactly pseudocomplements of atoms of $\mathcal{E}$ in $\tilde{\mathcal{E}}$. Then if a closure system is peudocomplemented and contains $\mathcal{E}$ then it contains also pseudocomplements of all elements of $\mathcal{E}$ and by then contains $\tilde{\mathcal{E}}$.

For (iv) note that for each element $X \in \tilde{\mathcal{E}}$ there is $E \in \mathcal{E}$ and $A \subseteq G_{\min}$ such that

$$X = E \cap \bigcap \{[G_{\min} \setminus \{a\}] \mid a \in A\}.$$

We set the following notations:

$$\hat{X} := \bigcap \{E \in \mathcal{E} \mid X \subseteq E\}$$

and

$$\check{X} := \{a \in G_{\min} \mid X \subseteq [G_{\min} \setminus \{a\}]\}.$$

We define ψ as follows: If $X \in \mathcal{E}$ then we set $\psi(X) := \varphi(X)$. If $X = [G_{\min} \setminus \{a\}] \notin \mathcal{E}$ then we set $\psi(X) := (\varphi(\{a\}))^*$. Elsewhere we set

$$\psi(X) := \varphi(\hat{X}) \cap \bigcap \{(\varphi(\{a\}))^* \mid a \in \check{X}\}.$$

ψ defines an order embedding of $\tilde{\mathcal{E}}$ into $\mathcal{F}$ that preserves existing pseudocomplements and $\psi \circ i = \varphi$. ∎

Remark 1. Not all closure systems have a transversal of the form $G_{\min}$. An example is the closure system of a complete atom-free Boolean algebra. The existence of a transversal of the form $G_{\min}$ does not only keep the size smaller, but also make the proof easier. However, Theorem 1 can be reformulated for any transversal T of $\mathcal{E}$. By [GK05–Theorem 1] the closure system

$$\tilde{\mathcal{E}} := \langle \mathcal{E} \cup \{[G_{\min} \setminus \{a\}] \mid a \in G_{\min}\}\rangle$$

will be replaced by

$$\tilde{\mathcal{E}} := \langle \mathcal{E} \cup \{[T \setminus t''] \mid t \in T\}\rangle.$$

In the proof of Theorem 1 the crucial step was to prove that $\tilde{\mathcal{E}}$ generates no new atoms. Now we should prove that all $t''^{\tilde{\mathcal{E}}} \mid t \in T$, are in $\mathcal{E}$. This is the result of the following lemma. The projection s is now on T. i.e., for $A \subseteq G$ and $g \in G$,

$$s(A) := \bigcup_{g \in A} g'' \cap T \text{ and } s(g) := s(\{g\}) = g'' \cap T.$$

Lemma 1. *Let $\mathcal{E}$ be a closure system on G with a transversal T. Let*

$$\tilde{\mathcal{E}} := \langle \mathcal{E} \cup \{[T \setminus t''] \mid t \in T\}\rangle.$$

be the closure system generated by $\mathcal{E}$ and $\{[T \setminus t''] \mid t \in T\}$. Then for all $t \in T$, $t''^{\tilde{\mathcal{E}}} = t''^{\mathcal{E}}$.

Proof. For each $t \in T$ we have $t''^{\tilde{\mathcal{E}}} \subseteq t''^{\mathcal{E}}$. Now, if $t''^{\tilde{\mathcal{E}}} \subsetneq t''^{\mathcal{E}}$ then there is an index set K and an element $E \in \mathcal{E}$ such that

$$t''^{\tilde{\mathcal{E}}} = E \cap \bigcap_{k \in K} [T \setminus t_k''^{\mathcal{E}}]$$

and $t''^{\mathcal{E}} \setminus t''^{\tilde{\mathcal{E}}}$ contains an element a. On one hand we have

$$
\begin{aligned}
t \in t''^{\tilde{\mathcal{E}}} &\implies t \in E \text{ and } t \in [T \setminus t_k''^{\mathcal{E}}] \text{ for all } k \in K \\
&\implies t''^{\mathcal{E}} \subseteq E \text{ and } s(t) \subseteq [T \setminus t_k''^{\mathcal{E}}] \text{ for all } k \in K \\
&\implies t''^{\mathcal{E}} \subseteq E \text{ and } t''^{\mathcal{E}} \cap T \subseteq T \setminus t_k''^{\mathcal{E}} \text{ for all } k \in K \\
&\implies t''^{\mathcal{E}} \subseteq E \text{ and } t''^{\mathcal{E}} \cap T \cap t_k''^{\mathcal{E}} = \varnothing \text{ for all } k \in K \\
&\implies t''^{\mathcal{E}} \subseteq E \text{ and } s(t_k) \subseteq [T \setminus t''^{\mathcal{E}}] \text{ for all } k \in K,
\end{aligned}
$$

and on the other hand

$$
\begin{aligned}
a \in t''^{\mathcal{E}} \setminus t''^{\tilde{\mathcal{E}}} &\implies a''^{\mathcal{E}} \subseteq t''^{\mathcal{E}} \text{ and } \exists k_0 \in K \text{ such that } a \notin [T \setminus t_{k_0}''^{\mathcal{E}}] \\
&\implies s(a) \subseteq s(t) \text{ and } s(a) \not\subseteq T \setminus t_{k_0}''^{\mathcal{E}} \text{ for some } k_0 \in K \\
&\implies s(a) \subseteq s(t) \text{ and } \exists k_1 \in K \text{ such that } t_{k_1} \in s(a) \setminus (T \setminus t_{k_0}''^{\mathcal{E}}) \\
&\implies s(a) \subseteq s(t) \text{ and } t_{k_1} \in a''^{\mathcal{E}} \cap t_{k_0}''^{\mathcal{E}} \\
&\implies s(a) \subseteq s(t) \text{ and } s(t_{k_1}) \subseteq s(a) \cap s(t_{k_0}) \\
&\implies s(t_{k_1}) \subseteq s(a) \subseteq s(t).
\end{aligned}
$$

As k_1 belongs to K, it follows that $s(t_{k_1}) \subseteq [T \setminus t''^{\mathcal{E}}]$. This leads to a contradiction since

$$s(t_{k_1}) \subseteq [T \setminus t''^{\mathcal{E}}] \text{ and } s(t_{k_1}) \subseteq s(t) \implies t_{k_1} \in \varnothing = [T \setminus t''^{\mathcal{E}}] \cap t''^{\mathcal{E}},$$

and completes the proof of the lemma. $\square$

Since each closure system has a transversal we obtain the general version of Theorem 1.

Theorem 2. *Each closure system has a coarsest pseudocomplemented refinement.*

Proof. By Lemma 1, the projections are equal. The rest of the proof is similar to that of Theorem 1 [(i)-(iii)], replacing $G_{\min}$ by T and $\{a\}$ by a''. $\square$

From the one-to-one correspondence between closure systems and concept lattices we can rewrite Theorem 2 together with (iv) of Theorem 1 as follows:

Theorem 3. *Each concept lattice L can be $\bigwedge$-embedded into a smallest pseudo-complemented concept lattice $\tilde{L}$.*

In Theorem 3, $\tilde{L}$ is smallest in the sense that each $\bigwedge$-embedding of L into a pseudocomplemented concept lattice L_1 that preserves existing pseudocomplements can be extended to an order embedding of $\tilde{L}$ into L_1 that preserves pseudocomplements. An immediate consequence is the following corollary for finite semilattices with 1 (see Section 2). We say that **a pcs S_1 contains a semilattice S** if S is a subsemilattice of S_1 and the elements of S having pseudocomplements in S keep their pseudocomplements in S_1. We will call a map between pcs that preserves pseudocomplements a **∗-map**.

Corollary 1. *For each finite semilattice S with 1, there is a pcs $\tilde{S}$ containing S such that $\tilde{S}$ ∗-order-embeds into every pcs containing S.*

The pcs $\tilde{S}$ from Corollary 1 is, up to pcs-isomorphism, unique. In term of cardinality, it is also the smallest pcs containing S. We call it the pcs-completion of S.

Definition 1. *By pcs-completion of a semilattice S, we mean a pcs $\tilde{S}$ containing S that ∗-order-embeds into any pcs containing S.*

From Section 2 we can always replace a finite semilattice S without 1 by $S \oplus 1$. Whether each semilattice has a pcs-completion is still open. Theorem 2 says that each finite or complete semilattice has a pcs-completion. (See also Remark 3.)

4   An *Extensional* Negation on Concept Lattices

4.1   Construction

We now apply the results to construct an *extensional* negation on concept lattices. This applies to atomic concept lattices (i.e., when the closure system of extents is atomic). In order to have the arrow-relations[4] at hand we shall also assume that the concept lattice is doubly founded. These conditions include all finite lattices. Moreover we suppose w.l.o.g. that the formal context (G, M, I) is clarified with $M' = \emptyset$.

Let $\mathfrak{B}(G, M, I)$ be a concept lattice and $\mathcal{E}$ the closure system of extents. We proved in [GK05–Prop. 10] that if $[G_{\min} \setminus \{a\}]$ is an extent and $a \in G_{\min}$, then there is an attribute $m_a \in M$ such that $m'_a = [G_{\min} \setminus \{a\}]$. Therefore generating

---

[4] In a context (G, M, I), for $g, h \in G$ and $m, n \in M$ the arrow-relations are defined by:

$$g \swarrow m : \Longleftrightarrow m \notin g' \text{ and } g' \subsetneq h' \text{ implies } m \in h',$$
$$g \nearrow m : \Longleftrightarrow g \notin m' \text{ and } m' \subsetneq n' \text{ implies } g \in n',$$
$$g \swarrownearrow m : \Longleftrightarrow g \swarrow m \text{ and } g \nearrow m.$$

a new closure system with $[G_{\min} \setminus \{a\}]$, $a \in G_{\min}$ is equivalent to adding new attributes m_a in the context whenever $[G_{\min} \setminus \{a\}]$ is not an extent. These attributes have exactly $[G_{\min} \setminus \{a\}]$ as extent. Theorem 2 says that the so obtained lattice is pseudocomplemented and has $(\mathfrak{B}(G, M, I), \wedge)$ as subsemilattice. The following result is the arrow characterization of the pseudocomplementation.

Theorem 4 ([GK05]). *Let $\mathfrak{B}(G, M, I)$ be atomic and doubly founded. Then $\mathfrak{B}(G, M, I)$ is pseudocomplemented if and only if the following condition holds for all $g \in G$:*

If $g \swarrow n$ for all $n \notin g'$ and $g \nearrow m$ then

- *if $h \swarrow m$ then $g' = h'$, and*
- *if $g \nearrow n$ then $n' = m'$.*

The arrow configuration described in Theorem 4 is displayed in Figure 1. The subcontext $(G_{\min}, \{m_a \mid a \in G_{\min}\})$ is a contranominal context, with exactly one double arrow in each row and column and crosses elsewhere. The rows of the atoms $G_{\min}$ contain no empty cells (arrowless non-incidences) and no upward arrows except for the double arrows mentioned. The columns corresponding to the attributes $\{m_a \mid a \in G_{\min}\}$ have no other downward arrows. What Theorem 4 expresses is that the configuration displayed in Figure 1 is characteristic for atomic doubly founded p-algebras.

Fig. 1. Arrow configuration in the context of an atomic pseudocomplemented concept lattice

In practice what one has to do is to first enter the arrow relations and check if one can obtain the configuration of Figure 1. If this is not the case one should add new attributes m_a for the atoms a whose inverse images are not extents and compute the new concept lattice.

4.2   Some Examples

Example 1. This example demonstrates this construction in details for the context of Figure 2. This context does not have the configuration presented in

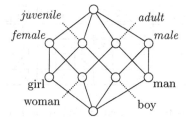

Fig. 2. A context and its concept lattice

Figure 1. Since each concept (apart from top and bottom) of this context has exactly one label we will use its label to name it. The atoms are the concepts *girl, woman, boy* and *man*. None of them has a pseudocomplement. Therefore we should add 4 attributes $m_{girl}, m_{woman}, m_{boy}$ and m_{man}. Their extents are

$$m'_{girl} = \{woman, boy, man\},$$
$$m'_{woman} = \{girl, boy, man\},$$
$$m'_{boy} = \{girl, woman, man\}, \text{ and}$$
$$m'_{man} = \{girl, woman, boy\}.$$

The newly generated context and its concept lattice is presented in Figure 3. It has the configuration presented in Figure 1. Its concept lattice is pseudo-complemented. The attribute m_a can be interpreted as $not - a$ for $a \in \{girl, woman, boy, man\}$.

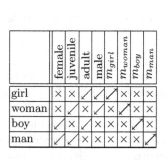
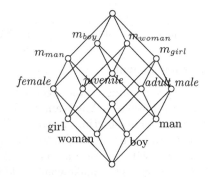

Fig. 3. The context generated from Figure 2 and its concept lattice

In addition to the attribute concepts generated as pseudocomplements of atoms we get two other concepts. The structure obtained is a Boolean algebra. The embedding i in this case is only $\wedge$-preserving. The lattice M_4 generates the same algebra.

Example 2. The second example presents the pcs-completion of the reduced context of the interordinal scale I_4.

I_4-reduced	≤3	≥2	≤1	≤2	≥3	≥4
4	↗	×	↙	↙	×	×
1	×	↗	×	×	↙	↙
2	×	×	↙	×	↗	↙
3	×	×	↙	↙	×	↗

I_4-pcs	≤3	≥2	m_2	m_3	≤1	≤2	≥3	≥4
4	↗	×	×	×	↙	↙	×	×
1	×	↗	×	×	×	×	↙	↙
2	×	×	↙	×	↙	×	↙	↙
3	×	×	×	↙	↙	↙	×	↙

Fig. 4. The reduced context of the interordinal scale I_4 and its pcs-completion context

The atoms are the extents $\{1\}, \{4\}, \{2\}$ and $\{3\}$. The first two have pseudo-complements. Then we need two new attributes for the remaining ones: m_2 and m_3. Their extents are

$$m_2' = \{1, 3, 4\} \quad \text{and} \quad m_3' = \{1, 2, 4\}.$$

The context and the concept lattice we obtained are in Figure 5.

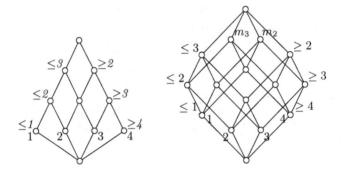

Fig. 5. The concept lattice of the reduced context of I_4 and its pcs-completion

4.3   Lessons Learnt

Imagine you are working within a context $\mathbb{K}$. This context can be understood as the universe of discourse. You want the negation of a concept $\mathfrak{X}$ to be the most general concept contradicting $\mathfrak{X}$. Some concepts might have a negation within this context, and other concepts might not. If you want all of them to have a negation, you will generate a context $\tilde{\mathbb{K}}$ associated to the closure system $\tilde{\mathcal{E}}$. Informally you decide to "make your world beautiful" by enlarging[5] the universe of discourse, as you realize that it was not large enough to discuss negation inside. Something important to be pointed out here, is that this generation is a step by step process. In the case only few concepts are of interest, then you can generate a subcontext of $\tilde{\mathbb{K}}$ instead of $\tilde{\mathbb{K}}$. For example if you want a pseudocomplement of a special concept $\mathfrak{X}$ (which does not have a pseudocomplement in the initial context), all you have to do is to generate a new closure system from the closure system of extents with the subsets $[G_{\min} \setminus \{a\}]$, $a \in G_{\min}$ and $a \in \text{ext}(\mathfrak{X})$.

---

[5] The idea of enlarging the context was also discussed in [GW99a], [KTS04] and [Kw04].

Remark 2. The process described in this paper can be applied to get dual pcs from finite ∨-semilattices. Then if you want the negation of a concept $\mathfrak{X}$ to be the least general concept satisfying together with $\mathfrak{X}$ the *principle of excluded middle* you can perform the same construction on the closure system of intents. In this case you enlarge the set of objects. We can iterate to construct pcs, then dual pcs, and pcs again, and so on. Does this process converge? If it does, then you can encode a negation with a double pseudocomplementation, a pair $(^*,^+)$ such that * is a pseudocomplementation and $^+$ a dual pseudocomplementation. The conjecture is that

this iteration process converges to a double p-algebra.

Note that if the atoms are exactly the join irreducible elements, we need perform only the first step since it generates a Boolean algebra. This was the case for examples 1 and 2 above. In general the price to pay would be the destruction of the join and the meet operations. However you would get an order embedding preserving existing pseudo- and dual pseudocomplements. For this purpose, it would be nice to get an arrow-like characterization of a double pseudocomplementation.

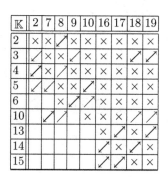
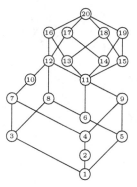

$\mathbb{K}$	2	7	8	9	10	16	17	18	19
2	×	×	↗	×	×	×	×	×	×
3	↗	×	×	↙	×	×	×	↗	↗
4	↗	×	↗	×	×	×	×	×	×
5	↗	↗	×	×	↗	×	×	×	×
6		×	↙	↗	↗	×	×	×	×
10	↙	↗			×	×	×	↗	↗
13						×	↗	×	↗
14						↗	×	↗	×
15						↗	↗	×	×

Fig. 6. A lattice and its reduced context

As an illustration consider the lattice in Figure 6. The atoms 2 and 5 have pseudocomplements, but 3 not. Therefore we need add one attribute m_3 with extent

$$m_3' := [G_{\min} \setminus \{3\}] = \{4, 6, 13, 14, 15\}.$$

The context $\mathbb{K}$-pcs in Figure 7 is obtained as the pcs-completion of $\mathbb{K}$ (Figure 6). Its concept lattice is obtained by adding one coatom (called m_3) as the join of the concepts 18 and 19. It is not dual pseudocomplemented. The other coatoms (the concepts 16 and 17) have no dual pseudocomplements. Now working on the closure system of extents we need add two objects g_{16} and g_{17} with

$$g_{16}' := \{m_3, 17, 18, 19\} \quad \text{and} \quad g_{17}' := \{m_3, 16, 18, 19\}.$$

K-pcs	8	10	m3	2	7	9	16	17	18	19
2	↗	×	×	×	×	×	×	×	×	×
5	×	↗	×	↗	↗	×	×	×	×	×
3	×	×	↗	↗	×	↗	×	×	↗	↗
4	↗	×	×	↗	×	×	×	×	×	×
6	×	↗	×			↗	×	×	×	×
10	↗	×	↗		↗		×	×		
13			×				×	↗	×	↗
14			×				↗	×	↗	×
15			×				↗	↗	×	×

K-pcs-dpcs	8	10	m3	2	7	9	16	17	18	19
2	↗	×	×	×	×	×	×	×	×	×
5	×	↗	×	↗	↗	×	×	×	×	×
3	×	×	↗	↗	×	↗	×	×	↗	↗
4	↗	×	×	↗	×	×	×	×	×	×
6	×	↗	×			↗	×	×	×	×
10	↗	×	↗		↗		×	×		
13			×				×	↗	×	↗
14			×				↗	×	↗	×
15			×				↗	↗	×	×
g_{16}			×				↗	×	×	×
g_{17}			×				×	↗	×	×

Fig. 7. Context of the pcs-completion of the context in Figure 6 (left), and its dual pcs-completion (right)

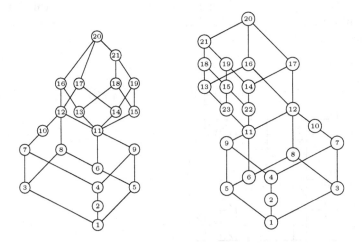

Fig. 8. The pcs-completion (left) of the lattice in Figure 6 and its dual pcs-completion (right)

The context $\mathbb{K}$-pcs-dpcs (Figure 7) is obtained from $\mathbb{K}$-pcs as dual pcs-completion. Its concept lattice (Figure 8) is pseudocomplemented and dual pseudocomplemented.

Remark 3. The main result describes the pcs-completion of concept (semi-) lattices. To get the arrow characterization in [GK05] we needed the atomicity and the double foundedness. Theorem 2 holds for all concept lattices. If we omit the finiteness/completeness condition in Corollary 1, there is no guaranty that the so constructed pcs is the smallest pcs containing the initial semilattice. Is it possible to carry such constructions from semilattices to pcs without requiring the completeness? Consider as illustration the semilattice $L := S \oplus \mathbb{Q}$, where S is the 4 element $\wedge$-semilattice with 3 atoms and $\mathbb{Q}$ the poset of rational numbers. Note that L is neither a pcs nor a nearlattice. The pcs-completion of S is the Boolean algebra B_3 (with 3 atoms). Denote by $\bar{B}_3$ the semilattice obtained from B_3 by removing the top element. The semilattice $\tilde{L} := \bar{B}_3 \oplus \mathbb{Q} \oplus 1$ is the pcs-completion of L. It is not a complete lattice and not even doubly founded.

5   Conclusion

As mentioned in [GK05] the arrow configuration of Figure 1 is a useful visual tool for checking whether an atomic doubly founded concept lattice is pseudocomplemented. This could be used not only to construct more examples of pseudocomplemented lattices, but to give a natural construction extending a finite semilattice to a pseudocomplemented one.

References

[BD74]   R. Balbes & P. Dwinger. *Distributive lattices*. University of Missouri Press. (1974).

[Bo54]   G. Boole. *An investigation of the laws of thought on which are founded the mathematical theories of logic and probabilities*. Macmillan 1854. Reprinted by Dover Publ. New York (1958).

[Ci04]   J. Cīrulis. *Knowledge representation systems and skew nearlattices*. Contributions to General Algebra **14**, 43-52 (2004).

[CG00]   I. Chajda & K. Glazek. *A basic course on general algebra*. Zielona Góra: Technical University Press. (2000).

[CM93]   C. Chameni Nembua & B. Monjardet. *Finite pseudocomplemented lattices and "permutoedre"*. Discrete Math. **111**, No.1-3, 105-112 (1993).

[Du97]   I. Düntsch. *A logic for rough sets.* Theor. Comput. Sci. **179**, No.1-2, 427-436 (1997).

[DG00]   I. Düntsch & G. Gediga. *Rough set data analysis*. In Encyclopedia of Computer Science and Technology vol. **43**, Marcel Dekker, 281-301 (2000).

[Fr62]   O. Frink. *Pseudo-complements in semi-lattices*. Duke Math. J. **29**, 505-514 (1962).

[GK05]   B. Ganter & L. Kwuida. *Which concept lattices are pseudocomplemented?* LNAI **3403**, 408-416 (2005).

[GW99]   B. Ganter & R. Wille. *Formal Concept Analysis – Mathematical Foundations*. Springer (1999).

[GW99a]   B. Ganter & R. Wille. *Contextual Attribute Logic*. Springer LNAI **1640**, 377-388 (1999).

[Gr71]   G. Grätzer.   *Lattice Theory. First concepts and distributive lattices*. W. H. Freeman and Company (1971).

[GS04]   G. Grätzer & E. T. Schmidt. *Finite lattices and congruences. A survey*. Algebra Universalis **52**, 241-278 (2004).

[Ho89]   L. R. Horn.   *A natural history of negation*. University of Chicago Press. (1989).

[Ka80]   T. Katrinak. *P-algebras. Contributions to lattice theory, Szeged/Hung. 1980*, Colloq. Math. Soc. Janos Bolyai **33**, 549-573 (1983).

[Kw04]   L. Kwuida. *Dicomplemented lattices. A contextual generalization of Boolean algebras*. Dissertation, TU Dresden. Shaker Verlag. (2004).

[KTS04]   L. Kwuida & A. Tepavčević & B. Šešelja.   *Negation in Contextual Logic.* Springer LNAI **3127**, 227-241 (2004).

[KPR04]   L. Kwuida & C. Pech & H. Reppe. *Generalizations of Boolean algebras. An attribute exploration.* Preprint MATH-AL-02-2004, TU Dresden (2004).

[So98]   V. Sofronie-Stokkermans. *Representation theorems and automated theorem proving in certain classes of non-classical logics.* Proceedings of the ECAI-98 workshop on many-valued logic for AI applications.

[Wi00]   R. Wille. *Boolean Concept Logic.* Springer LNAI **1867**, 317-331 (2000).

Towards a Generalisation of Formal Concept Analysis for Data Mining Purposes

Francisco J. Valverde-Albacete and Carmen Peláez-Moreno*

Dpto. de Teoría de la Señal y de las Comunicaciones,
Universidad Carlos III de Madrid,
Avda. de la Universidad, 30. Leganés 28911, Spain
{fva, carmen}@tsc.uc3m.es

Abstract. In this paper we justify the need for a generalisation of Formal Concept Analysis for the purpose of data mining and begin the synthesis of such theory. For that purpose, we first review semirings and semimodules over semirings as the appropriate objects to use in abstracting the Boolean algebra and the notion of extents and intents, respectively. We later bring to bear powerful theorems developed in the field of linear algebra over idempotent semimodules to try to build a Fundamental Theorem for $\mathcal{K}$-Formal Concept Analysis, where $\mathcal{K}$ is a type of idempotent semiring. Finally, we try to put Formal Concept Analysis in new perspective by considering it as a concrete instance of the theory developed.

1 Introduction and Motivation

When using Formal Concept Analysis for data mining purposes on non-binary data one is always forced to perform scaling procedures [9, 16] which carry a heuristic component sometimes difficult to justify in terms of the original data and requiring, in any case, a good deal of experience from the knowledge engineer.

From the point of view of the data it would be interesting to have alternative domains over which formal contexts could be defined and their lattices later built. There exist at least one such domains, the fuzzy domain tackled in [4, 2]. Unfortunately, this domain presents operative problems when trying to build the concept lattices associated to them, mainly the fact that it is unclear whether the intuitions and tools developed in the standard case [9] can be translated to such undertaking. In particular, lattice building algorithms become much more demanding computationally speaking.

It would be much more interesting to develop an abstract theory of concept lattices sharing as many mathematical and algorithmic results and intuitions as

---

* This work has been partially supported by two grants for "Estancias de Tecnólogos Españoles en el International Computer Science Institute" of the Spanish Ministry of Industry and a Spanish Government-Comisión Interministerial de Ciencia y Tecnología project TEC2005-04264/TCM.

R. Missaoui and J. Schmid (Eds.): ICFCA 2006, LNAI 3874, pp. 161–176, 2006.

possible with the concrete instance of Formal Concept Analysis, but somehow parameterized in the basic domain over which incidences could be defined.

For a proper generalisation, such an abstract theory should cater for the needs of Galois connection-induced concept lattices as well as the less understood adjunction-induced neighbourhood lattices [12]. Of course, such a theory should also encompass Boolean-defined incidences and their induced lattices, as we know them, as special cases, perhaps with outstanding or representative properties.

On the one hand, for the general enterprise of data mining any advance in this direction would be enlightening for a number of problems nowadays tackled with tools from conventional algebra. On the other hand, for the enterprise of coordinatising logical thought on the basis of an "algebra of concepts" as sketched in ([16], cfr. 4) this would broaden the range of tools at our disposal.

In a flight of fancy, let us suppose that we could have a "linear algebra" over extents and intents, that is, a conceptual "geometry". Then we could translate a wealth of methods and intuitions from "vector spaces" into Formal Concept Analysis with the appropriate *caveats*. For instance, if the polars could actually be represented by linear operators as matrices I, I^t it would be feasible to solve problems like the following:

- *Problem 1.* find all the sets of objects closing to a particular intent, $B \in 2^{p \times 1}$", that is, solve $I^t \cdot x = B$ for $x \in 2^{n \times 1}$.
  Note that standard Formal Concept Analysis asserts that $I \cdot B$ is one such set, specifically its extent, but we might also be interested in describing the variety without resorting to the enumeration of all candidates, for example to apply the alternative recipe to reduce contexts of [9]([§1.5.1.2).
- *Problem 2.* given a set of objects $A_1 \in 2^{n \times 1}$ find all other sets $A_2 \in 2^{n \times 1}$ that map to its intent, that is, those such that $I^t \cdot A_1 = I^t \cdot A_2$.
  Note that this problem amounts to finding the quotient class of A_1 under I^t. In data mining terms this amounts to finding those input patterns that map to the same output pattern, i.e. classes of output-undistinguishable input patterns.
- *Problem 3.* find the quotient space of input patterns, $2^{n \times 1}/I^t$
  This amounts to solving exhaustively problem 2 without finding first "representatives" like A_1 to guide it.

We could also think of dual problems involving the space of output patterns, produced by the transpose of the initial incidence.

The idea suggests itself that in this case the "geometric" properties may depend only secondarily on the Boolean domain and primarily on the "vector space" character of such sets. In which case, changing the domain in which sets and incidences take value would produce a flavour of "geometry", each different from the vanilla, standard Boolean flavour. Hence we put forward the following:

Hypothesis 1. *Standard Formal Concept Analysis is a particular instance of $\mathcal{K}$-valued Formal Concept Analysis in which incidences and sets of objects and attributes take values in a suitable algebraic structure, $\mathcal{K}$.*

Consequently, in the rest of the paper we first introduce the mathematical notions leading to a suitable algebraic structure $\mathcal{K}$ and "modules over $\mathcal{K}$" that can replace "modules over fields" (also "vector spaces") paying special attention to adjunctions and Galois connections between ordered sets. Later in section 3 we demonstrate a basic theorem for $\mathcal{K}$-valued Formal Concept Analysis and consider standard Formal Concept Analysis in as an instance of such construction. We conclude with a summary of contributions.

2 Mathematical Preliminaries

2.1 Residuation Theory, Adjunctions and Galois Connections

Lower semicontinuous functions are (isotone) maps commuting with joins in partial orders, and *upper semicontinuous* functions are (isotone) maps commuting with meets in partial orders [1]. Given two partial orders $\langle P, \leq \rangle$ and $\langle Q, \leq \rangle$, we have:

- A map $f : P \to Q$ is *residuated* if inverse images of principal (order) ideals of Q under f are again principal ideals. Its *residual map* or simply *residual*, $f^{\#} : Q \to P$ is:

$$f^{\#}(q) = max\{\, p \in P \mid f(p) \leq q \,\}$$

- A map $g : Q \to P$ is *dually residuated* if the inverse images of principal dual (order) ideals under g are again dual ideals. Its *dual residual map* or simply *dual residual*, $g^{\flat} : P \to Q$ is:

$$g^{\flat}(p) = min\{\, q \in Q \mid p \leq g(q) \,\}$$

Residuated maps are lower semicontinuous, while dually residuated maps are upper semicontinuous ([1], Th. 4. 50).

This abundance of concepts is fortunately simplified by a well-known theorem stating that residual maps are dually residuated, while dual residual maps are residuated, hence we may maintain only the two notions of residuated maps and their residuals. In fact, the two notions are so intertwined that we give a name to them: An *adjoint pair of maps* (λ, ρ) is a pair $(\lambda : P \to Q, \rho : Q \to P)$ between two quasi ordered sets so that $\forall p \in P, q \in Q$,

$$p \leq \rho(q) \iff \lambda(p) \leq q \quad \text{equivalently} \quad p \leq \rho(\lambda(p)) \quad \& \quad \lambda(\rho(q)) \leq q$$

If the order relation is partial the *lower or left adjoint*, λ is uniquely determined by its *right or upper adjoint*, ρ, and conversely ([8], §1.1). The characterization theorem for adjoint maps ([8], p. 7) states that (λ, ρ) are adjoint if and only if, λ is residuated with residual ρ, or equivalently, ρ is dually residuated with λ its dual residual. Hence adjunctions admit the following forms, using the following notation $(\lambda, \rho) : P \leftrightharpoons Q$ to make the sets evident:

$$(\lambda, \rho) : P \leftrightharpoons Q \quad \iff \quad (\lambda, \lambda^{\#}) : P \leftrightharpoons Q \quad \iff \quad (\rho^{\flat}, \rho) : P \leftrightharpoons Q$$

For Formal Concept Analysis, the more interesting notion of *Galois connection*, a contravariant pair of maps between the orders P and Q, reads:

$$p \leq \rho(q) \iff q \leq \lambda(p) \quad \text{equivalently} \quad p \leq \rho(\lambda(p)) \quad \& \quad q \leq \lambda(\rho(q))$$

A Galois connection can be equivalently described as an adjunction with the second order dualised:

$$(\lambda, \rho) : P{\multimap}Q \overset{\Delta}{=} (\lambda, \rho) : P{\leftrightharpoons}Q^d$$

We introduce the diagram to the left of figure 1 as the pattern that carries the structures described in ([8], §1.2):

- A closure system, $Q\rho$, the closure range of the right adjoint (see below).
- An interior system, $P\lambda$, the kernel range of the left adjoint (see below).
- A closure function (also "closure operator", [12, 6]) $\gamma = \rho \circ \lambda$, from P to the closure range $Q\rho$, with adjoint inclusion map $\hookrightarrow$.
- A kernel function (also "interior operator", [12]) $\kappa = \lambda \circ \rho$, from Q to the kernel range $P\lambda$, with adjoint inclusion map $\hookrightarrow$.

However, due to the dualisation of the second set in Galois connections ranges are closures systems and both compositions closure operators (we write κ^* for the new closure operator), resulting in the well-known dual isomorphism between closure ranges, illustrated to the right of figure 1. Recall that a *perfect adjunction* is an adjunction where the diagram collapses to an order isomorphism between P and Q, or equivalently, to an isomorphism between $\overline{P}$ and $\underline{Q}$. Dually, for a *perfect Galois connection*, the diagram collapses to a single dual (antitone) order isomorphism between $\overline{P}$ and $\overline{Q}$. Prerequisites for this to happen are that the closure maps be identities, $\gamma(p) = p$, $\kappa^*(q) = q$ ([8], p. 12).

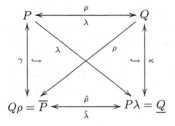
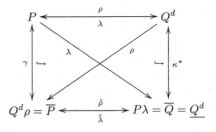

Fig. 1. Diagrams visually depicting the maps and structures involved in the adjunction $(\lambda, \rho) : P{\leftrightharpoons}Q$ (left) and Galois connection $(\lambda, \rho)P{\multimap}Q$ (right) between two partially ordered sets (adapted from [8])

2.2 Idempotent Semirings

This section aims at presenting the algebra that abstracts the features of the Boolean algebra which are adequate in our belief to generalise Formal Concept Analysis.

A *semiring* $\mathcal{K} = \langle K, \oplus, \otimes, \epsilon, e \rangle$ is a structure where the additive structure, $\langle K, \oplus, \epsilon \rangle$, is a commutative monoid and the multiplicative one, $\langle K \backslash \{\epsilon\}, \otimes, e \rangle$, a monoid whose multiplication distributes over addition from right and left:

$$\lambda \otimes (\mu \oplus \nu) = \lambda \otimes \mu \oplus \lambda \otimes \nu \qquad (\mu \oplus \nu) \otimes \lambda = (\mu \otimes \lambda) \oplus (\nu \otimes \lambda)$$

and whose neutral element is absorbing for $\otimes$, $\epsilon \otimes x = \epsilon, \forall x \in K$. On any semiring $\mathcal{K}$ left and right multiplications can be defined:

$$
\begin{aligned}
&L_a : K \to K && R_a : K \to K \qquad\qquad (1)\\
&b \mapsto L_a(b) = ab && b \mapsto R_a(b) = ba
\end{aligned}
$$

Hence a *commutative semiring* is a semiring whose multiplicative structure is commutative, and a *semifield* one whose multiplicative structure is a group. Thus, compared to a ring, a semiring crucially lacks additive inverses.

An *idempotent semiring* $\mathcal{K}$ is a semiring whose addition is idempotent:

$$\forall a \in K, a \oplus a = a$$

All idempotent commutative monoids $(K, \oplus, \epsilon)$ are endowed with a *natural order* $\forall a, b \in K, a \leq b \iff a \oplus b = b$, which turns them into join-semilattices with least upper bound defined as $a \vee b = a \oplus b$. Moreover, for the additive structure of semiring $\mathcal{K}$ the neutral element is the infimum for this natural order, $\epsilon_\mathcal{K} = \bot$.

An idempotent semiring $\mathcal{K}$ is *complete*, if it is complete as a naturally ordered set and left (L_a) and right (R_a) multiplications are lower semicontinuous, that is, residuated.

Therefore, complete idempotent semirings, as sup-semilattices with infimum are automatically complete lattices [6] with join ($\vee$, max or sup) and meet ($\wedge$, min or inf) ruled by the equations:

$$\forall a, b \in K, a \leq b \iff a \vee b = b \iff a \wedge b = a \qquad (2)$$

Example 2. *1. The Boolean semiring $\mathcal{B} = \langle \mathbb{B}, \vee, \wedge, 0, 1 \rangle$, with $\mathbb{B} = \{0, 1\}$, is complete, idempotent and commutative.*

2. The completed Maxplus semiring $\overline{R}_{max,+} = \langle \mathbb{R} \cup \{\pm\infty\}, \max, +, -\infty, 0 \rangle$, is a complete, idempotent semifield when defining $-\infty + \infty = -\infty$, that $\epsilon_\mathcal{K} \otimes \top_\mathcal{K} = \epsilon_\mathcal{K}$ for $\mathcal{K} \equiv \overline{R}_{max,+}$

3. The completed Minplus semiring $\overline{R}_{min,+} = \langle \mathbb{R} \cup \{\pm\infty\}, \min, +, \infty, 0 \rangle$ is a complete, idempotent semifield with a similar completion to that of ex. 2 with $\infty + (-\infty) = \infty$, that is $\epsilon_\mathcal{K} \otimes \top_\mathcal{K} = \epsilon_\mathcal{K}$ for $\mathcal{K} \equiv \overline{R}_{min,+}$.

The "values" populating a semiring are essentially positive or zero, hence we cannot expect to find "additive inverses" for them. The situation is less radical with multiplications in case the semiring exhibits the adequate order properties, as in idempotent semirings, because we may resort to residuation theory [3, 1] to try and invert such operations. But in that case we have the additional complexity of tracking the side of the multiplication, which applies particularly in case we want to "invert" the left and right multiplications of eqs. 1, in which case the residuals are:

$$L_a^\# : K \to K \qquad\qquad\qquad R_b^\# : K \to K \qquad\qquad (3)$$

$$L_a^\#(c) = \vee\{\lambda \in K \mid ab \le c\} = a\backslash c \quad R_b^\#(c) = \vee\{\lambda \in K \mid ab \le c\} = c/b$$

where the notation $a\backslash b$ reads "a under b" and b/a reads "b over a", with:

$$ab \le c \iff a \le c/b \iff b \le a\backslash c$$

Finally, note that if K is commutative, then $a\backslash b = b/a$.

Example 3.   *1. For the boolean semiring, $a\backslash b = b/a = a \to b$, where $\to$ is the logical conditional, $a \mapsto 0/a$ is the negation and $b \mapsto b/1$ the identity.*

2. For generic semiring $K = \langle K, \oplus, \otimes, \epsilon, e \rangle$, the expression of a vector $x = [x_j] \in K^{p\times 1}$ multiplying a matrix $R = [r_{ij}] \in K^{n\times p}$ is, $(R \cdot x)_i = \oplus_j(r_{ij} \otimes x_j)$. Similarly, let $A, D \in K^{m\times n}$, $B \in K^{m\times p}$ and $C \in K^{n\times p}$, then the residuals for vectors and matrices over an idempotent semimodule may be obtained as ([1], p. 199):

$$C = A\backslash B \qquad\qquad\qquad D = B/C \qquad\qquad (4)$$

$$c_{ij} = \bigwedge_{k=1}^{m} a_{ki}\backslash b_{kj} \qquad\qquad d_{ij} = \bigwedge_{k=1}^{p} b_{ik}/c_{jk}$$

2.3   Semimodules over Idempotent Semirings

A semimodule over a semiring is defined in a similar way to a module over a ring, but allowances have to be made as to the sides of the multiplication[1].

A *left K-semimodule*[14, 13], $\mathcal{X} = \langle X, \oplus, \epsilon_X \rangle$, is an additive commutative monoid endowed with a map $(\lambda, x) \mapsto \lambda \cdot x$ such that for all $\lambda, \mu \in K$, $x, z \in X$, and following the convention of dropping the symbol for the scalar action and multiplication for the semiring:

$$(\lambda\mu)x = \lambda(\mu x) \qquad\qquad \epsilon_K x = \epsilon_X \qquad\qquad (5)$$

$$\lambda(x \oplus z) = \lambda x \oplus \lambda z \qquad\qquad e_K x = x$$

The definition of a *right K-semimodule*, $\mathcal{Y}$, follows the same pattern with the help of a *right action*, $(\lambda, y) \mapsto y \cdot \lambda$ and similar axioms to those of 5. A (K, S)-*semimodule* is a set M endowed with left K-semimodule and a right S-semimodule structures, and a (K, S)-*bisemimodule* a (K, S)-semimodule such that the left and right multiplications commute. For a left K-semimodule, $\mathcal{X}$, the left and right multiplications are defined as:

$$L_\lambda^K : X \to X \qquad\qquad\qquad R_v^{\mathcal{X}} : K \to X \qquad\qquad (6)$$

$$x \mapsto L_\lambda^K(x) = \lambda x \qquad\qquad \lambda \mapsto R_x^{\mathcal{X}}(\lambda) = \lambda x$$

And similarly, for a right K-semimodule. If $\mathcal{X}$, $\mathcal{Z}$ are left semimodules a *morphism of left semimodules or left linear map $F : \mathcal{X} \to \mathcal{Z}$* is a map that preserves finite sums and commutes with the action: $F(\lambda v \oplus \mu w) = \lambda F(v) \oplus \mu F(w)$, and similarly, *mutatis mutandis* for right morphisms of right semimodules.

---

[1] We are following essentially the notation of [5].

Idempotency and natural order in Semimodules. A $\mathcal{K}$-semimodule $\mathcal{M}$ over an idempotent semiring $\mathcal{K}$ inherits the idempotent law, $v \oplus v = v, \forall v \in M$, which induces a *natural order* on the semimodule by $v \leq w \iff v \oplus w = w, \forall v, w \in M$ whereby it becomes a sup-semilattice, with $\epsilon_{\mathcal{M}}$ the minimum.

When $\mathcal{K}$ is a complete idempotent semiring, a left $\mathcal{K}$-semimodule, $\mathcal{M}$ is *complete* if it is complete as a naturally ordered set and its left and right multiplications are (lower semi)continuous. Therefore, if $\mathcal{M}$ is complete for the natural order, it is also a complete lattice, with join and meet operations given by:

$$v \leq w \iff v \vee w = w \iff v \wedge w = v$$

All these definitions can be extended naturally to bisemimodules.

Example 4.   *1. Each semiring, $\mathcal{K}$, is a left (right) semimodule over itself, with left (right) action the semiring product. Therefore, it is a $(\mathcal{K}, \mathcal{K})$-bisemimodule over itself, because both actions commute by associativity. Such is the case for the Boolean $(\mathcal{B}, \mathcal{B})$-bisemimodule, the Maxplus and the Minplus bisemimodules. These are all complete and idempotent.*

  2. The set of matrices $K^{n \times p}$ for finite n and p is a $(K^{n \times n}, K^{p \times p})$-bisemimodule with matrix multiplication-like left and right actions and componentwise addition. The set of column vectors $K^{p \times 1}$ for finite p is a $(K^{p \times p}, K)$-bisemimodule and the set of row vectors $K^{1 \times n}$ for finite n a $(K, K^{n \times n})$-bisemimodule with similarly defined operations. If K is idempotent (resp. complete), then all are idempotent (resp. complete) with the componentwise partial order their natural order.

Like in the semiring case, because of the natural order structure, the actions of idempotent semimodules also admit residuation: given a complete, idempotent left $\mathcal{K}$-semimodule, $\mathcal{X}$, we define for all $x, z \in X, \lambda \in K$:

$$\left(L_\lambda^{\mathcal{K}}\right)^{\#} : X \to X \qquad \left(L_\lambda^{\mathcal{K}}\right)^{\#}(z) = \bigvee \{\, x \in X \mid \lambda x \leq z \,\} = \lambda \backslash z \qquad (7)$$

$$\left(R_x^{\mathcal{X}}\right)^{\#} : X \to K \qquad \left(R_x^{\mathcal{X}}\right)^{\#}(z) = \bigvee \{\, \lambda \in K \mid \lambda x \leq z \,\} = z / x$$

and likewise for a right semimodule, $\mathcal{Y}$.

There is a remarkable operation that changes the character of a semimodule while at the same time reversing its order by means of residuation: let $\mathcal{Y}$ be a complete, idempotent right $\mathcal{K}$-semimodule, its *opposite semimodule* is the complete *left* $\mathcal{K}$-semimodule $\mathcal{Y}^{op} = \langle Y, \overset{op}{\oplus}, \overset{op}{\to} \rangle$ with the same underlying set Y, addition defined by $(x, y) \mapsto x \overset{op}{\oplus} y = x \wedge y$ where the infimum is for the natural order of $\mathcal{Y}$, and *left* action:

$$K \times Y \to Y \quad (\lambda, y) \mapsto \lambda \overset{op}{\to} y = y / \lambda$$

Consequently, the order of the opposite is the *dual* of the original order. For the opposite semimodule the residual definitions are:

$$\lambda \overset{op}{\diagdown} x = \left(L_\lambda^{\mathcal{Y}^{op}} \right)^{\#} (x) = \bigwedge \{ y \in X \mid x \le y/\lambda \} = x \cdot \lambda \qquad (8)$$

$$x \overset{op}{\diagup} y = \left(R_y^{\mathcal{Y}^{op}} \right)^{\#} (x) = \bigvee \{ \lambda \in K \mid x \le y/\lambda \} = x \backslash y$$

Note that we can define *mutatis mutandis* the opposite semimodule of a left $\mathcal{K}$-semimodule, $\mathcal{X}$, with right action $x \overset{op}{\twoheadleftarrow} \lambda = \lambda \backslash x$. Also, noticing that the first residual in eq. 8 is in fact an involution we may conclude that the operation of finding the opposite of a complete (left, right) $\mathcal{K}$-semimodule is an involution: $(\mathcal{M}^{op})^{op} = \mathcal{M}$.

Example 5 (Opposite Boolean semimodule). *All semirings, $\mathcal{K}$, taken as $(\mathcal{K}, \mathcal{K})$-bisemimodules accept an opposite semiring, $\mathcal{K}^{op}$. In particular, the opposite of the boolean bisemimodule of ex. 1 $B^{op} = \langle \mathbb{B}, \overset{op}{\oplus}, \overset{op}{\odot}, \mathbf{1}, \mathbf{0} \rangle$ is also a complete bisemimodule where addition is the min operation, notated by the meet $v \overset{op}{\oplus} w = v \wedge w$. Consequently, its natural order is the inverse of the usual order for the lattice $\mathbf{2}$, the additively neutral element is $\epsilon_{B^{op}} = \mathbf{1}$, which is the bottom for the opposite natural order, the unit is $e_{B^{op}} = \mathbf{0}$ and the action is the residual of the original action, $\lambda \overset{op}{\cdot} x = \lambda \backslash x = x / \lambda$. In fact, the truth table for this connective is that of the logical conditional $\lambda \overset{op}{\to} x = x \overset{op}{\twoheadleftarrow} \lambda = \lambda \to x$.*

Semimodules as vector spaces [2]. The elements of a semimodule are *vectors*. Given a semiring $\mathcal{K}$ and a left $\mathcal{K}$-semimodule $\mathcal{X}$, for each finite, nonvoid set $W \subseteq X$, there exists an homomorphism $\alpha : K^W \to X$, $f \mapsto \bigoplus_{w \in W} f(w)w$. Moreover, α induces a congruence of semimodules $\equiv_\alpha$ on K^W, by $f \equiv_\alpha g \iff \alpha(f) = \alpha(g)$. Then W is a *set of generators* or a *generating family* precisely when α is surjective, in which case any element $x \in X$ can be written as $x = \bigoplus_{w \in W} \lambda_w w$, and we will write $\mathcal{X} = \langle W \rangle$, that is, $\mathcal{X}$ *is the span of W*. A semimodule is *finitely generated* if it has a finite set of generators. For individual vectors, we say that $x \in W$ is *dependent (in W)* if $x = \bigoplus_{w \in W \backslash \{x\}} \lambda_w w$ otherwise, we say that it is *free (in W)*. The set W is *linearly independent* if and only if $\equiv_\alpha$ is the trivial congruence, that is, when $\bigoplus_{w \in W} f(w)w = \bigoplus_{w \in W} h(w)w \iff f = h$, otherwise, W is *linearly dependent*. Let $\ker \alpha = \{ f \in K^W \mid \alpha(f) = 0 \}$, W is *weakly linearly independent* if and only if $\ker \alpha = \{0\}$, otherwise it is *weakly linearly dependent*. A *basis for $\mathcal{X}$ (over $\mathcal{K}$)* is a linearly-independent set of generators, and a semimodule generated by a basis is *free*. By definition, in a free semimodule $\mathcal{X}$ with with basis $\{ x_i \}_{i \in I}$ each element $x \in X$ can be uniquely written as $x = \bigoplus_{i \in I} \alpha_i x_i$, with $[a_i]_{i \in I}$ the *coordinates* of x with respect to the basis. A weakly linearly-independent set of generators for $\mathcal{X}$ is a *weak basis for $\mathcal{X}$ (over $\mathcal{K}$)*. The cardinality of a (weak) basis is the *(weak) rank* of the semimodule.

In such framework, notions in usual vector spaces have to be imported with care. For instance, the *image* of a linear map $F : \mathcal{X} \to \mathcal{Z}$ is simply the semimodule $\text{Im} F = \{ F(x) \mid x \in X \}$, but it is in general not free. Similarly, the following variant definition makes more sense: the *(bi)kernel* of the linear map $F : \mathcal{X} \to \mathcal{Z}$, is the congruence of semimodules $\text{Ker} F = \{ (x, x') \in X^2 \mid F(x) = F(x') \}$.

---

[2] Most of the material in this section is from [14], §17, and [10, 11, 15].

Given a free semimodule $\mathcal{X}$ with basis $\{\, x_i \,\}_{i \in I}$, for each family $\{\, y_i \,\}_{i \in I}$ of elements of an arbitrary semimodule $\mathcal{Y}$ there is a unique morphism of semimodules $F : \mathcal{X} \to \mathcal{Y}$ such that $F(x_i) = y_i, \forall i \in I$, namely $F\left(\bigoplus_{i \in I} \lambda_i x_i \right) = \bigoplus_{i \in I} \lambda_i y_i$ and all the linear maps $Lin(\mathcal{X}, \mathcal{Y})$ are obtained in this way([10], prop. §73; [14], prop. §17.12). That is, linear maps from free semimodules are characterized by the images of the elements of a basis.

On the other hand, a semiring $\mathcal{K}$ has the *linear extension property* if for all free, finitely generated $\mathcal{K}$-semimodules $\mathcal{X}, \mathcal{Y}$, for all finitely generated subsemimodules $\mathcal{Z} \subset X$ and for all $F \in Lin(\mathcal{Z}, \mathcal{Y})$, there exists $H \in Lin(\mathcal{X}, \mathcal{Y})$ such that $\forall x \in X, H(x) = F(x)$. The importance of this property derives from the fact that when the linear extension property holds, each linear map between finitely generated subsemimodules of free semimodules is represented by a matrix. In particular, when it holds for free, finitely generated (left) semimodules, $\mathcal{X}$ and $\mathcal{Y}$ with bases $\{x_i\}_{i \in I}$ and $\{y_j\}_{j \in J}$, each linear map is characterized by the $n \times p$-matrix $I = (F(x_i)_j)$, which sends vector x with coordinates $x \simeq (\alpha_i)_{1 \le i \le n}$ to the vector $F(x) \simeq ((xI)_1, \ldots, (xI)_p)$.

Idempotent vector spaces. Idempotent semimodules have additional properties which make them easier to work with as spaces. Therefore, when $\mathcal{K}$ is an idempotent semiring if a $\mathcal{K}$-semimodule has a (weak) basis, it is unique up to a rescaling map $y_i = \lambda x_i$([15], th. §3.1); and every finitely generated $\mathcal{K}$-semimodule has a weak basis ([15], cor. §3.6).

Importantly, the linear property holds in every idempotent semiring which is a distributive lattice for the natural order ([10], th. §83). This is the case for $\mathcal{K} = \mathbb{B}$, the boolean semiring and $\mathcal{K} = \overline{R}_{\max}$. Therefore, in such semimodules, modulo a choice of bases for $\mathcal{X}$ and $\mathcal{Y}$, we may identify $\mathcal{X} \cong \mathcal{K}^{1 \times n}$ and $\mathcal{Y} \cong \mathcal{K}^{1 \times p}$, and linear maps to matrix transformations $\mathrm{Lin}(\mathcal{X}, \mathcal{Y}) \cong \mathcal{K}^{n \times p}$, $I : \mathcal{K}^{1 \times n} \to \mathcal{K}^{1 \times p}, x \mapsto xI$. When passing from left to right semimodules this should read $\mathcal{K}^{p \times 1} \to \mathcal{K}^{n \times 1}, y \mapsto Iy$.

Constructing Galois connections in semimodules. Given a complete idempotent semiring $\mathcal{K}$, we call *predual pair* a complete left $\mathcal{K}$-semimodule $\mathcal{X}$ together with a complete right $\mathcal{K}$-semimodule $\mathcal{Y}$ equipped with a bracket $\langle \cdot \mid \cdot \rangle : X \times Y \to Z$ to a complete $\mathcal{K}$-bisemimodule $\mathcal{Z}$, such that, for all $x \in X, y \in Y$ the maps:

$$R_x : Y \to Z \qquad\qquad L_y : X \to Z \qquad\qquad (9)$$
$$y \mapsto \langle x \mid y \rangle \qquad\qquad x \mapsto \langle x \mid y \rangle$$

are respectively left and right linear, and continuous. The most usual choice of bisemimodule $\mathcal{K}$ gives the bilinear forms, but one may also choose $\mathcal{K}^{op}$.

The following construction is due to Cohen et al. [5]: for a bracket $\langle \cdot \mid \cdot \rangle : X \times Y \to Z$ and an arbitrary element $\varphi \in Z$, which we call the *pivot*, define the maps:

$$i_l : X \to Y \qquad\quad i_l(x) = L_x^{\#}(\varphi) = \bigvee \{\, y \in Y \mid \langle x \mid y \rangle \le \varphi \,\} \qquad (10)$$
$$i_r : Y \to X \qquad\quad i_r(y) = R_y^{\#}(\varphi) = \bigvee \{\, x \in X \mid \langle x \mid y \rangle \le \varphi \,\}$$

which may be shortened to: $i_l(x) = x^-$ and $i_r(y) = {}^-y$. We have $\langle x \mid y \rangle \leq \varphi \iff y \leq x^- \iff x \leq {}^-y$, whence $i_l : (X, \leq) \to (Y, \leq^d)$ is residuated with residual $i_l^{\#} = i_r$ ([5], proof of prop. 24) hence, (i_l, i_r) is an adjunction between $\mathcal{X}$ and $\mathcal{Y}^{op}$ or equivalently, a Galois connection between $\mathcal{X}$ and $\mathcal{Y}$. Figure 2 depicts the morphisms and structures induced by such Galois connection. Note that the closure lattices $\overline{X} = \mathcal{Y}i_r$ and $\underline{Y} = \mathcal{X}i_l$ do not agree with their ambient vector spaces [3] in their joins, but only in their meets!

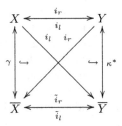

Fig. 2. The Galois connection $(i_l, i_r) : \mathcal{X} {\multimap} \mathcal{Y}$ of the maps induced by $(\langle \cdot \mid \cdot \rangle, \varphi)$

This construction is affected crucially by the choice of a suitable pivot φ: in the operations of residuation only those pairs $(x, y) \in X \times Y$ are considered whose degree amounts to *at most* φ. Therefore we can think of the pivot as a *maximum degree of existence* allowed for the pairs.

Example 6.   *1.* **(Involutions).** *The above construction can be used to obtain a family of different Galois connections between* $\mathcal{X}_m \simeq \mathcal{K}^{m \times 1}$ *and* $\mathcal{Y}_m \simeq \mathcal{K}^{1 \times m}$: *define* $\langle x \mid y \rangle = \bigoplus_i x(i) \otimes y(i)$, *which is a predual pair for* $\mathcal{Z} = \mathcal{K}$ ([5], *ex.* §21), *then* $\psi_m \overset{def}{=} (i_l, i_r)_m : \mathcal{X}_m {\multimap} \mathcal{Y}_m$, *as above, is a Galois Connection for each finite* m.

2. **(Galois connection between free row and column semimodules,** [5] §4.5, *adapted)* *Given a matrix[4]* $R \in \mathcal{K}^{n \times p}$, *the free complete semimodules* $\mathcal{X} = \mathcal{K}^{1 \times n}$ *and* $\mathcal{Y} = \mathcal{K}^{p \times 1}$ *form a predual pair for the bracket* $\langle x \mid y \rangle_R = xRy$. *For such construction, define for a specific* $\varphi \in K$:

$$x_\varphi^- = \bigvee \{ y \mid xRy \leq \varphi \} = (xR)\backslash \varphi \tag{11}$$

$$_\varphi^- y = \bigvee \{ x \mid xRy \leq \varphi \} = \varphi/(Ry)$$

hence we have: $\overline{Y} = \{ (xR)\backslash\varphi \mid x \in \mathcal{K}^{n \times 1} \}$ *and* $\overline{X} = \{ \varphi/(Ry) \mid y \in \mathcal{K}^{1 \times p} \}$ *whence* $\psi_R \overset{def}{=} (\cdot_\varphi^-, {}_\varphi^-\cdot) : \mathcal{K}^{1 \times n} {\multimap} \mathcal{K}^{p \times 1}$ *is a Galois connection.*

Furthermore, the notion of a *left (resp. right) reflexive*, $(\mathcal{K}, \varphi)$, semiring is introduced in [5] as a complete idempotent semiring such that $((\langle \cdot \mid \cdot \rangle : K \times K \to K, \varphi)$

---

[3] Recall $\mathcal{X}$ and $\mathcal{Y}$ are both complete lattices as well as free vector spaces.
[4] Note that we are avoiding here giving using a generic I for a relation because that name traditionally denotes unitary matrices.

with $\langle \lambda \mid \mu \rangle = \lambda\mu$ induces a perfect Galois connection under the above-mentioned construction, that is for all $\lambda \in K$, $^-(\lambda^-) = \lambda$ (resp. $(^-\lambda)^- = \lambda$). For the Boolean semiring we must choose $\varphi = 0_\mathcal{B}$, the bottom in the order. For other semirings any invertible element may be chosen, e.g. $\varphi = e_\mathcal{K}$.

Note that φ need not be unique: if $(\mathcal{K}, \varphi)$ is right (or left) reflexive, for any $\lambda \in K$ invertible, $(\mathcal{K}, \varphi\lambda)$ is left reflexive (and $(\mathcal{K}, \lambda\varphi)$ is right reflexive.) Finally, Cohen et al. [5] prove that idempotent semifields are left and right reflexive.

3 $\mathcal{K}$-Formal Concept Analysis

We model ($\mathcal{K}$-valued) sets of objects, $x \in \mathcal{X} \cong \mathcal{K}^{1\times n}$, with row vectors in a left $\mathcal{K}$-semimodule and sets of attributes, $y \in \mathcal{Y} \cong \mathcal{K}^{p\times 1}$, with column vectors in a right $\mathcal{K}$-semimodule, as generalisations of characteristic functions in the powersets $\mathbf{2}^G, \mathbf{2}^M$, respectively[5].

Definition 7 ($\mathcal{K}$-valued formal context). *Given two finite set of objects G and attributes M, where $|G| = n$ and $|M| = p$, an idempotent semiring, $\mathcal{K}$, and a $\mathcal{K}$-valued incidence between them, $R \in \mathcal{K}^{n\times p}$, where $R(g,m) = \lambda$ reads as "object g has attribute m in degree λ" and dually "attribute m is manifested in object g to degree λ", the triple $(G, M, R)_\mathcal{K}$ is called a $\mathcal{K}$-valued formal context.*

Clearly the context can be represented as a $\mathcal{K}$-valued table with the value $R(g,m) = \lambda$ in the crossing of row g with column m. Also, we are forced to admit that objects are isomorphic to elements of the space $\mathcal{K}^{1\times p}$, that is *rows of R* or *object descriptions*, vectors of as many values as attributes and attributes are isomorphic to elements of the space $\mathcal{K}^{n\times 1}$, *columns of R* or *attribute descriptions*.

Proposition 1. *Let $(\mathcal{K}, \varphi)$ be a reflexive, idempotent semiring. For a $\mathcal{K}$-valued formal context $(G, M, R)_\mathcal{K}$, with finite $|G| = n$ and $|M| = p$, there is at least one Galois connection between the power-sets of $\mathcal{K}$-valued sets of objects $\mathcal{K}^{1\times n}$ and attributes $\mathcal{K}^{p\times 1}$.*

Proof. Recall that $\mathcal{X} = \mathcal{K}^{1\times n}$ is a left semimodule and $\mathcal{Y} = \mathcal{K}^{p\times 1}$ a right semimodule, whence $\mathcal{X}^{op}$ and $\mathcal{Y}^{op}$ are right and left semimodules, respectively, whose multiplications are $R \overset{op}{\cdot} x = x^t \backslash R$ and $y \overset{op}{\cdot} R = R/y^t$. We build a new bracket over the opposite semiring $\mathcal{K}^{op}$ as given by $\langle y \mid x \rangle_R = y \overset{op}{\cdot} R \overset{op}{\cdot} x = y^t \backslash R / x^t$ which is formally identical to the bracket over a relation in example 2. Therefore, by the construction of section 2.3[6] the maps:

$$y_\varphi^- = \overset{op}{\bigvee} \{ x \in X \mid \langle y \mid x \rangle_R \overset{op}{\leq} \varphi \} \qquad _\varphi^- x = \overset{op}{\bigvee} \{ y \in Y \mid \langle y \mid x \rangle_R \overset{op}{\leq} \varphi \}$$

form a Galois connection $(\cdot_\varphi^-, {}_\varphi^-\cdot) : \mathcal{Y}^{op} \multimap \mathcal{X}^{op}$. As requested in that section 2.3, for $\mathcal{B}$ we must choose $\varphi = \epsilon_\mathcal{B} = \mathbf{0}$, but we are operating now in the opposite

---

[5] This section follows in the tracks of §1.1 of [9].

[6] And the demonstration of proposition §24 of [5].

semiring $\mathcal{B}^{op}$, hence we choose the bottom thereof, $\varphi = \epsilon_{\mathcal{B}^{op}} = \mathbf{1}$. For any other semiring we may choose $\varphi = e_{\mathcal{K}}$.   □

Note that in an idempotent semifield we are guaranteed enough φ to build as many connections: choose any invertible $\lambda \in K$, so that $\varphi = \lambda \otimes e_{\mathcal{K}}$.

Definition 8 (φ-polars). *Given a reflexive, idempotent semiring $(\mathcal{K}, \varphi)$ and a $\mathcal{K}$-valued formal context $(G, M, R)_{\mathcal{K}}$ satisfying the conditions of proposition 1, we call φ-polars the dually adjoint maps of the corresponding Galois connection*

$$y_\varphi^- = \overset{op}{\bigvee} \{ x \in X \mid \langle y \mid x \rangle_R \overset{op}{\leq} \varphi \} = \bigwedge \{ x \in X \mid \langle y \mid x \rangle_R \geq \varphi \} = \left(y \overset{op}{\to} R \right) \overset{op}{\overset{\backslash}{}} \varphi \tag{12}$$

$$_\varphi^- x = \overset{op}{\bigvee} \{ y \in Y \mid \langle y \mid x \rangle_R \overset{op}{\leq} \varphi \} = \bigwedge \{ y \in Y \mid \langle y \mid x \rangle_R \geq \varphi \} = \varphi \overset{op}{/} \left(R \overset{op}{\leftarrow} x \right)$$

However, in this dualised construction the pivot describes a *minimum degree of existence* required for pairs $(x, y) \in X \times Y$ to be considered for operation.

Definition 9 (Formal φ-Concepts and φ-concept lattices). *Given a reflexive, idempotent semiring $(\mathcal{K}, \varphi)$, a $\mathcal{K}$-valued formal context $(G, M, R)_{\mathcal{K}}$ with finite $|G| = n$ and $|M| = p$ and $\mathcal{K}$-valued vector spaces of rows $\mathcal{X} \cong \mathcal{K}^{1 \times n}$ and columns $\mathcal{Y} \cong \mathcal{K}^{p \times 1}$*

1. *A (formal) φ-concept of the formal context $(G, M, R)_{\mathcal{K}}$ is a pair $(a, b) \in \mathcal{X} \times \mathcal{Y}$ such that $_\varphi^- a = b$ and $b_\varphi^- = a$. We call a the* extent *and b the* intent *of the concept (a, b), and φ its (minimum) degree of existence.*
2. *If (a_1, b_1) (a_2, b_2) are φ-concepts of a context, they are ordered by the relation*
$$(a_1, b_1) \leq (a_2, b_2) \iff a_1 \leq a_2 \iff b_1 \overset{op}{\leq} b_2, \text{ called the hierarchical order.}$$
The set of all concepts ordered in this way is called the φ-concept lattice, $\mathfrak{B}_\varphi(G, M, R)_{\mathcal{K}}$, of the $\mathcal{K}$-valued context $(G, M, R)_{\mathcal{K}}$

Of course, the structure for the latter definition is proved next.

Theorem 2 (Fundamental theorem of $\mathcal{K}$-valued Formal Concept Analysis, finite version, 1st half). *Given a reflexive, idempotent semiring $(\mathcal{K}, \varphi)$, the φ-concept lattice $\mathfrak{B}_\varphi(G, M, R)_{\mathcal{K}}$ of a $\mathcal{K}$-valued formal context $(G, M, R)_{\mathcal{K}}$ with finite $|G| = n$ and $|M| = p$ is a (finite, complete) lattice in which infimum and supremum are given by:*

$$\bigwedge_{t \in T} (a_t, b_t) = \left(\overset{op}{\bigoplus_{t \in T}} a_t, \overset{-}{\underset{\varphi}{\overset{op}{\bigoplus_{t \in T}}}} a_t \right) \qquad \bigvee_{t \in T} (a_t, b_t) = \left(\left[\overset{op}{\bigoplus_{t \in T}} b_t \right]_\varphi^-, \overset{op}{\bigoplus_{t \in T}} b_t \right) \tag{13}$$

Proof. Recall that $a_t \in \mathcal{X}$ and $b_t \in \mathcal{Y}$ with $\mathcal{X} \cong \mathcal{K}^{1 \times n}$ and $\mathcal{Y} \cong \mathcal{K}^{p \times 1}$. The two dually isomorphic lattices $\overline{Y}$ and $\overline{X}$ are join semilattices of their ambient spaces, $\overline{Y} \subseteq \mathcal{X}^{op}$ and $\overline{X} \subseteq \mathcal{Y}^{op}$.

$$\overline{Y} = \left\{ \varphi \overset{op}{/} \left(R \overset{op}{\leftarrow} x \right) \mid x \in X \right\} \qquad \overline{X} = \left\{ \left(y \overset{op}{\to} R \right) \overset{op}{\overset{\backslash}{}} \varphi \mid y \in Y \right\} \tag{14}$$

Therefore by the inversion of the orders in opposite semimodules they are meet semilattices of $\mathcal{X}$ and $\mathcal{Y}$ respectively, hence the meets for $a_t \in \mathcal{X}$ and $b_t \in \mathcal{Y}$, and their φ-polars obtain the missing part of the concept.   □

Standard Formal Concept Analysis: an Example. At the end of section 1 we proposed a hypothesis about the origin of standard Formal Concept Analysis, for which we now provide the following corollary and informal proof. Of course, the wealth of results of Standard Formal Concept Analysis will not be available, specially those involving the second half of the Main Theorem, here missing.

Corollary 3. *Standard Formal Concept Analysis is the concrete case of $\mathcal{K}$-valued FCA where $\mathcal{K}$ is the Boolean semiring.*

Proof (Informal). Recall that for the construction of proposition 1 the recommendation was to choose in the dualised semiring $\varphi = \epsilon_{\mathbb{B}^{op}} = \mathbf{1}$ as pivot. In such case, we obtain $\mathfrak{B}(G, M, I) = \mathfrak{B}_1(G, M, R)_{\mathbb{B}}$ and most of the basic results in Formal Concept Analysis follow from definitions 7–9 and theorem 2.   □

We now turn to an example about calculating extents, intents and concepts of a $\mathfrak{B}_1(G, M, I)_{\mathbb{B}}$ lattice. For this purpose, we will use the context of [9], fig. 1.5 and associated concept lattice reproduced in figure 3.

 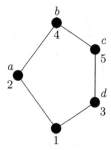

Fig. 3. The context and its concept lattice in [9] figs. §1.5 and 1.6

We represent the context in figure 4 as a matrix having $\mathbf{1}$ in the places occupied by crosses and $\mathbf{0}$ in the rest. In the same figure, we list the object and attribute concepts of the context adapted to our notation. Note that we introduce the singletons generating the concepts as row and column vectors, that is for objects, $\tilde{\gamma}(5) = \tilde{\gamma}([00001])$, and for attributes $\tilde{\mu}(c) = \tilde{\mu}([0010]^t)$.

$$I = \begin{bmatrix} 1 & 1 & 1 & 1 \\ 1 & 1 & 0 & 0 \\ 0 & 1 & 1 & 1 \\ 0 & 1 & 0 & 0 \\ 0 & 1 & 1 & 0 \end{bmatrix}$$

$$\begin{aligned}
\tilde{\gamma}([10000]) &= &= ([10000], [1111]^t) \\
\tilde{\gamma}([01000]) &= \tilde{\mu}([1000]^t) = ([11000], [1100]^t) \\
\tilde{\gamma}([00100]) &= \tilde{\mu}([0001]^t) = ([10100], [0111]^t) \\
\tilde{\gamma}([00010]) &= \tilde{\mu}([0100]^t) = ([11111], [0100]^t) \\
\tilde{\gamma}([00001]) &= \tilde{\mu}([0010]^t) = ([10101], [0110]^t)
\end{aligned}$$

Fig. 4. The context as a Boolean matrix and its object and attribute concepts

To illustrate the calculations involved in the concept of object #5 in the right hand side of the lattice with sets of objects and sets of attributes, we refer to table 1. In the first three columns of the table, we show, respectively, the extent of each concept, its left product by the matrix, the result of the whole bracket with the intent y_5 and whether this product complies with the restriction $\langle y_5 \mid x_i \rangle \geq 1$, for the actual intent y_5. In the next four columns, the same operations are done based in the intents for the extent x_5.

Table 1. Table showing the calculations described in the text

x_i	$R \overset{op}{\cdot} x_i$	$\langle y_5\|x_i\rangle$	$\langle y_5\|x_i\rangle \overset{op}{\leq} \varphi$?	y_i	$y_i \overset{op}{\cdot} I$	$\langle y_i\|x_5\rangle$	$\langle y_i\|x_5\rangle \overset{op}{\leq} \varphi$?
[10000]	[1111]	1	Yes	$[1111]^t$	$[10000]^t$	0	No
[11000]	[1100]	0	No	$[1100]^t$	$[11000]^t$	0	No
[10100]	[0111]	1	Yes	$[0111]^t$	$[10100]^t$	0	No
[11111]	[0100]	0	No	$[0100]^t$	$[11111]^t$	1	Yes
[10101]	[0110]	1	Yes	$[0110]^t$	$[10101]^t$	1	Yes

Considering the lattice of extents, we see that the extents of the concepts that comply with the restriction $\langle y_5|x_i\rangle \geq 1$ are $\text{ext}(\tilde{\gamma})(1) = [10000]$, $\text{ext}(\tilde{\gamma}(3)) = [10100]$ and $\text{ext}(\tilde{\gamma}(5)) = [10101]$. Of these, $\text{ext}(\tilde{\gamma}(5))$ is the minimum (*in the opposite order.*) Likewise, for intents complying with the restriction $\langle y_i|x_5\rangle \geq 1$, that is $\text{int}(\tilde{\gamma}(4)) = [0100]$ and $\text{int}(\tilde{\gamma}(5)) = [0110]$, the latter is the minimum (*in the opposite order*).

4   Discussion and Conclusions

At the beginning of this paper, we started with a number of constraints and requirements for our endeavour. Where do they stand now?

Linear algebra and $\mathcal{K}$-Formal Concept Analysis. In section 1 we introduced a number of problems of interest in data mining. Specifically, recall problem 2 (adapted): for a set of objects a, find all other sets a' such that $R \overset{op}{\leftarrow} a = R \overset{op}{\leftarrow} a'$. After the results in section 3, this amounts to finding the quotient class of the set of objects a by the polar, $a \pmod{\overset{-}{\varphi} \cdot}$; and similarly for the set of attributes b finding the class, $b \pmod{\cdot \overset{-}{\varphi}}$. Note that problem 3 is essentially finding the quotient spaces without finding representatives for the classes.

As to the importance of such procedures for standard Formal Concept Analysis, a related procedure involving the closure maps is invoked in [9], §1.5 as an alternative to the standard reduction procedure based in arrow relations: find the kernels of the polars and form the quotient sets on objects and attributes modulo these kernels; the reduced incidence is actually the incidence between the corresponding classes in the quotient sets. We are confident that our results will help develop this alternative to context reduction.

Conclusions. We have tried to introduce in this paper a linear-algebraic perspective into Formal Concept Analysis whereby contexts may actually be represented as matrices, and the basic operations as multiplications in adequate algebras. And this with the twofold intention of bringing some light into the relation of logical operators and Formal Concept Analysis, and making the latter better adapted to deal with a broader class of quantitative problems.

These algebras happen to be a special kind of semirings, reflexive idempotent ones, and we have provided a construction for $\mathcal{K}$-Formal Concept Analysis generalising the standard analysis to allow for semiring-valued incidences. These results are not really surprising, since semirings seem to be closely related to Baer semigroups [3]. We still wonder whether idempotent semirings, will not actually be a sufficiently rich algebra to allow the kind of processing we put forward here.

One instance of the above structure is the Boolean semiring and its opposite semimodule. We provide and example how standard Formal Concept Analysis seems to be the particularisation of our technique for these semirings. Of course, the demonstration of the technique for actual data mining is still missing and will be the object of future papers.

References

1. F. Baccelli, G. Cohen, G. Olsder, and J. Quadrat. *Synchronization and Linearity.* Wiley, 1992.
2. R. Belohlávek. Lattices generated by binary fuzzy relations. In *Int. Conf. on Fuzzy Set Theory and Applications*, Slovakia, 1998.
3. T. Blyth and M. Janowitz. *Residuation Theory.* Pergamon press, 1972.
4. A. Burusco and R. Fuentes-González. The study of the l-fuzzy concept lattice. *Mathware and Soft Computing*, 1(3):209–218, 1994.
5. G. Cohen, S. Gaubert, and J.-P. Quadrat. Duality and separation theorems in idempotent semimodules. *Linear Algebra and Its Applications*, 379:395–422, 2004.
6. B. Davey and H. Priestley. *Introduction to lattices and order.* Cambridge University Press, Cambridge, UK, 2nd edition, 2002.
7. K. Denecke, M. Erné, and S. Wismath, editors. *Galois Connections and Applications.* Number 565 in Mathematics and Its Applications. Kluwer Academic, Dordrecht, Boston and London, 2004.
8. M. Erné. *Adjunctions and Galois connections: Origins, History and Development,* pages 1–138. In Denecke et al. [7], 2004.
9. B. Ganter and R. Wille. *Formal Concept Analysis: Mathematical Foundations.* Springer, Berlin, Heidelberg, 1999.
10. S. Gaubert. Two lectures on max-plus algebra. Support de cours de la 26–iéme École de Printemps d'Informatique Théorique, May 1998. http://amadeus.inria.fr/gaubert/papers.html.
11. S. Gaubert and the Maxplus Group. Methods and applications of $(max, +)$ linear algebra. Technical Report 3088, INRIA –, 1997.
12. G. Gediga and I. Dütsch. Approximation operators in qualitative data analysis. Technical Report CS-03-01, Department of Computer Science, Brock University, St. Catharines, Ontario, Canada, May 2003.

13. J. S. Golan. *Power Algebras over Semirings. With Applications in Mathematics and Computer Science*, volume 488 of *Mathematics and its applications*. Kluwer Academic, Dordrecht, Boston, London, 1999.
14. J. S. Golan. *Semirings and Their Applications*. Kluwer Academic, 1999.
15. E. Wagneur. Moduloïds and pseudomodules 1. dimension theory. *Discrete Mathematics*, 98:57–73, 1991.
16. R. Wille. *Dyadic Mathematics – Abstractions from Logical Thought*, pages 453–498. In Denecke et al. [7], 2004.

Interactive Association Rules Discovery

Raoul Medina, Lhouari Nourine, and Olivier Raynaud

L.I.M.O.S. Université Blaise Pascal,
Campus des Cézeaux, Clermont-Ferrand, France
{Medina, Nourine, Raynaud}@isima.fr

Abstract. An interactive discovery method for finding association rules
is presented. It consists in a user-guided search using reduction operators
on a rule. Rules are generated on-demand according to the navigation
made by the user. Main interest of this approach is that, at each step,
the user has only a linear number of new rules to analyze and that all
computations are done in polynomial time. Several reduction operators
are presented. We also show that the search space can be reduced when
clone items are present.

1 Introduction

Originally introduced in [1], association rules mining has been a major topic in
data mining research during the last decade. An association rule is an expression $X \to Y$, where X and Y are sets of items. Meaning is quite intuitive: if a
transaction contains X then it probably contains Y also. Some quality measures
are attached to association rules. The *support* of a rule expresses the fraction
of transactions in the database containing both antecedents and consequent of
a rule. The *confidence* of a rule expresses the conditional probability $p(X \mid Y)$.
Some other quality measures have been defined [3, 10]. All those measures are
(more or less) relevant for analysis purposes, but they are mainly used to narrow the number of generated rules [2, 9, 12, 16]. Indeed, the main problem when
mining association rules is that their number exponentially grows with the number of items. A solution to reduce the number of generated rules is to generate
only a subset of rules from which all other rules can be derived [8, 17]. Such a
set is called an implication base [5, 13]. Minimum bases are implication bases
with minimal number of rules. Unfortunately, such bases might still contain an
exponential number of rules [14] and their computation is still an open problem.

Current methods for association rules mining are highly iterative and require
interaction with the end-user. Basic steps of the methods are:

- Generate a (potentially exponential) set of rules,
- Browse rules to find *interesting* association rules.
- Use discovered knowledge to refine the analysis and restart the process.

Those methods are *downstream* navigation processes: navigation is done *after*
the rules have been generated. Refinement of the analysis is done by applying
some user-defined constraints. For instance, the search space can be pruned

R. Missaoui and J. Schmid (Eds.): ICFCA 2006, LNAI 3874, pp. 177–190, 2006.

using a minimal threshold for quality measures on the rules [2, 9, 12]. We focus on rules belonging to implication bases, and especially *left-reduced rules* (i.e. with minimal antecedents). Another constraint can be the generation of rules where an item (or set of items) appears either in the antecedent or the consequent.

Defining which constraint(s) to apply is the only interaction between the end-user and the knowledge discovery process. It is also the most crucial phase in the analysis process. Knowledge discovery process is highly human-centered and thus relies on the user (the "expert") skills and intuitions. Response times of the computational phases should not curb the user creativity. Performance of the process is either improved during the generation or the navigation [11] step depending on when the constraints are applied. Unfortunately, despite highly optimized association mining algorithms, response times hardly allow real time interactivity during the whole process. A constant "association rule" verified by every current method seems to be *"whenever an association rule mining process is launched, the coffee machine of the company works full-time"*. The objective of our paper is a first attempt in keeping the end-user behind his desk during the whole process.

Our method is an *upstream navigation* process: rules are generated depending on the navigation process. We first propose a set of *general rules* and at each step the end-user can refine the rules by applying a *reduction operator* on the antecedent. At each step the number of new rules to analyze is linear in the number of items, and their computation is done in polynomial time with a single scan of the database.

As a first attempt, the following restrictions apply to our method.

- We consider rules whose confidence is equal to 1. In other words, we focus only on *exact* rules, or implications.
- We focus on rules of the form $I \rightarrow x$ (with $x \notin I$);
- Quality measures are not taken into account. They can however be computed for each rule.

2   Notations and Definitions

Throughout this paper, the following notations are used: Let $\mathcal{F}$ be a closure system over X. Elements of X are then called items. The closure operator of $\mathcal{F}$ is denoted by $\overline{}$. We say that I is a closed set if and only if $I = \overline{I}$. Meet-irreducible (resp. join-irreducible) elements are elements of $\mathcal{F}$ which are not intersections (resp. unions) of other elements of $\mathcal{F}$. We denote the set of meet-irreducible elements of $\mathcal{F}$ by $\mathcal{M}$ and the set of join-irreducible elements are denoted by $\mathcal{J}$.

Usually a sub-collection $\mathcal{R}$ of $\mathcal{F}$ is given rather than the whole closure system $\mathcal{F}$ and its closure operator $\overline{}$. An equivalent closure operator is easily deduced from $\mathcal{R}$ since meet-irreducible elements of $\mathcal{F}$ are present in $\mathcal{F}$. Note that the sub-collection $\mathcal{R}$ is often given as a *context*. A context is a binary relation between a set of objects and a set of attributes. In this case, the closure operator is $''$ as defined in [6].

We denote by $\mathcal{I}$ the lattice of ideals of the poset $(\mathcal{J}, \subseteq)$. The principal ideal of x is denoted by $\downarrow x$ and the principal filter of x is denoted by $\uparrow x$. An implication rule $I \rightarrow x$ means that $x \in \overline{I}$. In this paper we focus on rules such that x belongs to $\overline{I} \setminus I$. We denote by Σ an implication base of $\mathcal{F}$.

The Guigues-Duquenne base of $\mathcal{F}$ is a minimum implication base of $\mathcal{F}$ such that the antecedents are pseudo-closed sets.

Definition 1. *Quasi-closed and Pseudo-closed set*
Let $\mathcal{F}$ be a closure system and $^-$ the closure operator associated to $\mathcal{F}$. Let $P \in 2^X$ and $P \notin \mathcal{F}$.

- *P is a Quasi-closed set iff for all $Q \subset P$, $\overline{Q} \subset P$ or $\overline{Q} = \overline{P}$.*
- *A quasi-closed set P is a pseudo-closed set if there is no quasi-closed set $Q \subset P$ with $\overline{Q} = \overline{P}$.*

Finally small caps $(a, b, ...)$ represent items, capitals $(A, B, ...)$ represent sets and calligraphic capitals $(\mathcal{A}, \mathcal{B}, ...)$ represent collections of sets.

3   Interactive Navigation Principle

We assume that the end user is interested in rules where an item x is implied. There might exist many rules implying x. Rather than generating all possible rules implying x, we give to the user some of those rules, let him pick one rule and then check if the antecedent can be reduced. New rules will be generated according to the user's choice.

Here we are only interested in association rules of the form $I \rightarrow x$ such that $x \in \overline{I} \setminus I$. The set of all possible antecedents implying an item x is defined by $\mathcal{E}_x$.

Definition 2. *Let $x \in X$, we define the set $\mathcal{E}_x = \{I \mid x \in \overline{I} \setminus I\}$.*

The set $\mathcal{E}_x$ is the set of all antecedents that imply item x (see Figure 2). The user can navigate in this family and search for interesting rules for him. Our navigation principle needs a well-defined starting point : the rule implying x with the largest antecedent. Next proposition shows that whenever there exists at least one rule implying x, this largest rule always exists.

Proposition 1. *Let x be in X, if $\mathcal{E}_x \neq \emptyset$ then $X \setminus x = Max(\mathcal{E}_x)$.*

Proof. Let us first show that any element of $\mathcal{E}_x$ is a subset of $X \setminus x$. Let I be in $\mathcal{E}_x$ then I does not contain x by definition of $\mathcal{E}_x$ and so we have $I \subseteq X \setminus x$. Let us show now that $X \setminus x$ belongs to $\mathcal{E}_x$, in other words that x belongs to $\overline{(X \setminus x)} \setminus (X \setminus x)$ by definition of $\mathcal{E}_x$. This is equivalent to show that $X \setminus x \neq \overline{X \setminus x}$. Let I be in $\mathcal{E}_x$ and suppose the set $X \setminus x$ is a closed set. Then we know that $I \subseteq (X \setminus x)$, by monotony we obtain $\overline{I} \subseteq \overline{X \setminus x}$. By hypothesis we have $\overline{I} \subseteq X \setminus x$ and so x does not belong to $\overline{I}$, which contradicts the fact that I is in $\mathcal{E}_x$. And so $X \setminus x$ is not a closed set. □

A corollary to the previous proposition is that there exists at least one rule implying x if and only if $X \setminus x$ is not a closed set.

Corollary 1. *Let x be in X, the following assertions are equivalent :*

- *There exists $I \to x$ with $x \in \overline{I} \setminus I$;*
- *$\mathcal{E}_x \neq \emptyset$;*
- *$X \setminus x \neq \overline{X \setminus x}$;*

Proof. Follows directly from the previous proposition.   □

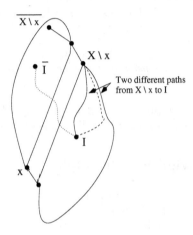

Fig. 1. From $X \setminus x$ to I many different paths exist. And all along this path, non closed sets imply item x.

Now we show that all other rules implying x are reachable from this maximal rule using a path in $\mathcal{E}_x$. In this path, a rule can be generated from its predecessor by the removal of an item in the antecedent.

Proposition 2. *Let x in X, $\forall I \in \mathcal{E}_x$ and $\forall A \subseteq ((X \setminus x) \setminus I)$ we have $I \cup A \in \mathcal{E}_x$.*

Proof. Let us show that $\forall A \subseteq (X \setminus x) \setminus I$ we have $I \cup A \in \mathcal{E}_x$. By hypothesis we have $I \subseteq I \cup A \subseteq X \setminus x$ and by monotony we obtain $\overline{I} \subseteq \overline{I \cup A} \subseteq \overline{X \setminus x}$. Suppose that $I \cup A$ does not belongs to $\mathcal{E}_x$, in other words that x is not in $\overline{I \cup A}$. Then x could not be in $\overline{I}$. This is a contradiction with hypothesis of the proposition.   □

The previous property states that each element of the set $\mathcal{E}_x$ is reachable by successive deletions of single elements from the set $X \setminus x$. Moreover, each intermediary obtained set belongs to $\mathcal{E}_x$ and thus implies x. Thus there exist different navigation paths from the top $(X \setminus x)$ to each element of $\mathcal{E}_x$, depending on the order of deletion (see figure 1). Principle of our navigation scheme is to let the user choose the wanted direction at each step.

Initially all the valid rules of the form $X \setminus x \to x$ are proposed to the user. He picks one rule among them corresponding to the element x he is interested in. From this rule, we then compute all the valid rules of the form $I \to x$ such that I is obtained by removing a single element from the left part of the chosen rule. At this step, the user chooses the rule having the most interesting left part for him. Next step will be to reduce this left part by iteratively applying the selection and rules generation process. When no new valid rules can be obtained, the set of rules is empty and thus the navigation is done for this path. Of course, at any stage, the user can go backwards and choose another direction if he desires.

Note that when the rule $I \rightarrow x$ is chosen, the new set of rules which is generated is exactly equal to $\{I \setminus y \rightarrow x \mid I \setminus y \in \mathcal{E}_x\}$. Thus the number of rules at each step is linear in the size of X.

The navigation process can be summarized by the following algorithm.

Algorithm 1. Navigation process

   Data   : A closure system $\mathcal{F}$ defined over X
   Result : A set of rules
   begin
      $Rules = \{X \setminus x \rightarrow x \mid X \setminus x \notin \mathcal{F}\}$;
      repeat
         Let the user choose $I \rightarrow x \in Rules$;
         $NewRules = Reduction(I)$;
         if $NewRules = \emptyset$ **then**
            $Display$ "$I \rightarrow x$ cannot be reduced";
         $Rules = Rules \cup NewRules$;

      until *End of work*;
   end

In order to gain display clarity the whole set of rules might be stored in a tree where each node is a rule. The end-user can then browse the tree, store it on disk for future work, etc. We call this tree the navigation tree. Note that a rule might appear in different nodes of the tree since there are different paths leading from the maximal rule to this rule.

Given a rule $I \rightarrow x$, the *reduction(I)* function simply returns the set $\{I \setminus y \mid I \setminus y \in \mathcal{E}_x\}$. But one could imagine some others reduction operators, for instance by adding conditions on $I \setminus y$. For example the user may want to display only rules which have element a in the left-part. In the next section we propose some other reduction operators.

4   Reduction Operators

In this section we discuss in detail different reduction operators. First, we show how to implement the reduction operator presented in the previous section. Then we present two other reduction operators. The first one speeds up the navigation process by removing more than one element (when possible). The second one allows the user to navigate only on ideals and thus reduce the number of displayed rules.

4.1 Testing $I \in \mathcal{E}_x$

The reduction operator presented in the previous section is as follows:

$$Reduction(I) = \{I \setminus y \to x \mid I \in \mathcal{E}_x\}$$

To test if I belongs to $\mathcal{E}_x$, one has to test if $x \notin I$ and $x \in \bar{I}$. The first test is not necessary in our navigation scheme since we start it from $X \setminus x$ and then navigate by removing one of its element. Thus x cannot belong to I. Testing $x \in \bar{I}$ usually requires to compute the closure of I. Another way could be to test that $x \notin \bar{I}$ is false. Indeed, from the monotony property of closure operators we know that $x \notin \bar{I}$ if there exists a closed set $M \subseteq X \setminus x$ such that $I \subseteq M$. Thus it is sufficient to test if I is a subset of at least one of the maximal closed sets which does not contain x (see Figure 2).

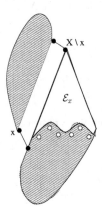

Fig. 2. White nodes represent element of $\mathcal{M}_x$. Any subset below one of these nodes cannot imply x.

Definition 3. *We note by $\mathcal{M}_x$ the set of maximal closed sets which do not contain x, i.e. $\mathcal{M}_x = Max(\{M \subseteq X \mid M = \overline{M}, M \subseteq X \setminus x\})$.*

Note that sets in $\mathcal{M}_x$ are meet-irreducible elements of the lattice of closed sets. And thus they must be present in the context and can be trivially found. The reduction operator can then be rewritten as follows :

$$Reduction(I) = \{I \setminus y \to x \mid \not\exists M \in \mathcal{M}_x \text{ such that } I \setminus y \subseteq M\}$$

When comparing the two approaches (computing the closure or testing the inclusion in a maximal closed set) one has to notice that the theoretical complexity remains unchanged. However, since $|\mathcal{M}_x| \leq |\mathcal{M}|$ and since only inclusion tests are done, the practical efficiency should be improved when using the second approach.

4.2 The Jump Reduction Operator

Usually, the end user wants to obtain the rules with the most reduced left-part. Those rules are exactly the rules whose antecedents are in $Min(\mathcal{E}_x)$ (see

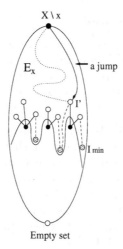

Fig. 3. Black nodes represent elements of $\mathcal{M}_x$. The minimal elements of $\mathcal{E}_x$ (dashed nodes) are reachable from immediate successors of elements of $\mathcal{M}_x$ (the white nodes).

Figure 3). Reaching those rules might require several steps for the user. The idea of the jump reduction operator is to reduce this number of steps. It relies on the following proposition.

Proposition 3. *Let x be in X, for any I in $Min(\mathcal{E}_x)$ there exists M in $\mathcal{M}_x$ and $y \notin M$ such that $I \subseteq M \cup \{y\}$.*

Proof. Let I in $Min(\mathcal{E}_x)$ then $\forall y \in I$, $I \setminus y \notin \mathcal{E}_x$. Thus x does not belong to $\overline{I \setminus y}$ and thus there exists a closed set M in $\mathcal{M}_x$ such that $I \setminus y \subseteq M$. As a consequence, $I \subseteq M \cup \{y\}$. □

Note that if $\mathcal{E}_x \neq \emptyset$ then, for any M in $\mathcal{M}_x$ and $y \notin M$, we have $M \cup \{y\}$ in $\mathcal{E}_x$. Thus, according to Propositions 2 and 3, any I in $Min(\mathcal{E}_x)$ can be reached from those sets $M \cup \{y\}$ (see Figure 3). The idea is thus to propose those rules as the starting points for the user. The jump reduction operator can be written as follows :

$$Jump - reduction(I) = \{M \cup \{y\} \rightarrow x \mid M \in \mathcal{M}_x\}$$

Those rules can be efficiently computed since $\mathcal{M}_x$ can be computed in polynomial time by a single scan of the context. Moreover, the number of rules is bounded by $|X| \times |\mathcal{M}_x|$, which is still a reasonable size for the end user to search in.

One could be tempted to reiterate such jump reduction from any rule $I \rightarrow x$ chosen by the user. It suffices to consider maximal closed subsets of I rather than $\mathcal{M}_x$. Unfortunately, those maximal closed sets are not necessary meet-irreducible elements of the lattice of closed sets and thus might not be present in the context. Computing such maximal closed sets would then require to compute all the closed sets included in I, which is prohibitive in our goal to achieve interactivity.

4.3   Ideal Refinement Operator

In this section, we show, using an example, how to navigate while maintaining a specific property for the set I. In this example we choose to maintain the property "I is an ideal of $(\mathcal{J}, \subseteq)$". The consequence is that the reduction operator now has to preserve this property. It thus has to be rewritten:

$$Ideal - reduction(I) = \{I \setminus y \to x \mid y \in Max(I) \text{ and } I \setminus y \in \mathcal{E}_x\}$$

If I is an ideal, choosing y in $Max(I)$ guarantees that $I \setminus y$ is also an ideal.

The modification of the operator does not change the overall complexity of the process. Moreover, by doing this, we navigate in the ideal lattice rather than the power set. The search space is thus reduced. The number of new rules is bounded by the width of the poset $(\mathcal{J}, \subseteq)$. Another remark that can be done is that the starting point of the process can be refined. Indeed, $X \setminus x$ is not necessarily an ideal. We thus have to start from the biggest ideal that does not contain x, i.e. the set $X \setminus \uparrow x$ (cf. figure 4).

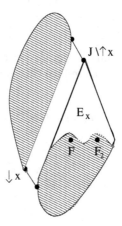

Fig. 4. The set $\mathcal{E}_x$ with elements F and F_2 of $\mathcal{M}_x$

There is a link between quasi-closed sets (and thus pseudo-closed sets) and ideals. This relation is interesting since pseudo-closed sets are elements defining a minimal basis (in the number of rules) of a context: the Guigues-Duquenne base.

Proposition 4. *Let Q be a quasi-closed set of $\mathcal{F}$. If $|Q| > 1$ then Q is an ideal of $(\mathcal{J}, \subseteq)$.*

We can thus decompose the sets of pseudo-closed sets of the context in two categories : the simple pseudo-closed sets (those containing exactly one element) and pseudo-closed sets which are ideals. Note that simple pseudo-closed sets can be trivially computed from $(\mathcal{J}, \subseteq)$. Indeed, the rule $a \to \bar{a}$ is in the Guigues-Duquenne basis if and only if there exists an element b such that $b < a$ in $(\mathcal{J}, \subseteq)$ (see section 6.1 for an example). The Guigues-Duquenne basis Σ can be rewritten as follows : $\Sigma = \Sigma_{\mathcal{J}} \cup \Sigma_{\downarrow}$, where $\Sigma_{\mathcal{J}}$ are the rules which left-part is a simple pseudo-closed set, and $\Sigma_{\downarrow}$ are the rules which antecedent is an ideal.

Authors in [4] have shown that closed-sets can be characterized by intervals pruning of the power set 2^X :

Proposition 5. *[4] Let Σ be a base of $\mathcal{F}$ a closure system on X:*

$$\mathcal{F} = 2^X \setminus \bigcup_{A \to B \in \Sigma} \bigcup_{x \in B} [A, X \setminus x] \tag{1}$$

Note that our basic navigation schema navigates in those intervals. But now we want to navigate over ideals which are non closed. Sets in intervals which bottom set is a simple pseudo-closed set are not ideals. We will not navigate in those intervals. We navigate only in intervals containing ideals. Thus, their bottom element is an ideal as well as their top element (which is of the form $X \setminus \uparrow x$). This allows us to rephrase the previous characterization :

$$\mathcal{F} = \mathcal{I} \setminus \bigcup_{A \to B \in \Sigma \downarrow} \bigcup_{x \in B} [A, X \setminus \uparrow x] \tag{2}$$

The only rules that holds in I are the rules of $\Sigma_{\mathcal{J}}$, i.e. the simple rules. Thus it is interesting to note that a refinement of [4] is to prune the closure system defined by $\Sigma_{\mathcal{J}}$ with the intervals defined by $\Sigma_{\downarrow}$!

From all the above, navigating in the lattice of ideals seems to make sense. But unfortunately, for the moment there is no known polynomial algorithm to test if an ideal is a pseudo-closed set.

5 Reducing the Navigation Space Using Clone Items

In [7], authors had shown that symmetries between items could be hidden in large sets of rules. If for all rules containing a in the antecedent there exists the same rule where a is replaced by b, they say that items a and b are clone items. Those symmetries could be seen as redundant information. Thus, we could ignore all rules containing b but no a without loss of information.

In this section we propose to detect those kinds of symmetries in order to reduce the search space. The user chooses to keep one major item a by clone classes. Rules that contain other clone items but not the major item a are redundant information and thus are simply forgotten. Of course, those rules can be computed and displayed on user's demand.

To define clone items, authors in [7] use the following swapping function :

Let x be a set of items $\{x_1, ..., x_{|X|}\}$ and $\mathcal{C}$ be a collection of sets on X. We denote by $\varphi_{a,b} : 2^X \to 2^X$ the mapping which associates to any subset of X its image by swapping items a and b. More formally :

$$\varphi_{a,b}(M) = \begin{cases} (M \setminus \{a\}) \cup \{b\} & \text{if } b \notin M \text{ and } a \in M \\ (M \setminus \{b\}) \cup \{a\} & \text{if } a \notin M \text{ and } b \in M \\ M & \text{otherwise} \end{cases}$$

The definition of clone items is then :

Definition 4. *Let $\mathcal{C}$ be a collection of sets defined on X. We say that items a and b are clone items in $\mathcal{C}$ **if and only if** for any M in $\mathcal{C}$, $\varphi_{a,b}(M)$ also belongs to $\mathcal{C}$.*

In our case, desired symmetries concern the collection of the left-parts of the rules of the form $I \rightarrow x$ with $x \in \overline{I} \setminus I$. The collection is then constituted with non closed sets. One can remark that if two items a and b are clone items on non closed sets, they remain clone items in the corresponding closure system. Lastly in [7] authors showed that clone items of a closure system correspond to clone items of the corresponding meet-irreducible elements.

To summarize we have the following equivalent assertions :

- a and b are clone items with respect to the collection of non closed sets;
- a and b are clone items in the closure system;
- a and b are clone items in the collection $\mathcal{M}$ of meet-irreducible elements.

In [7] authors propose clone classes detection algorithm as well as a clone reduction algorithm which removes the clone classes in a context. They also show how to rebuild the removed rules from a rule of the reduced context. In [15] an improvement of the detection algorithm is proposed. All these algorithms run in polynomial time. Thus, clone items might be interesting to reduce the search space used by our navigation principle.

This reasoning can be refined when the user picks a rule $X \setminus x \rightarrow x$. Now that x has been fixed, we are interested in clone items present in $\mathcal{E}_x$. Indeed, some items might be clone in $\mathcal{E}_x$ and not clone in the closed-sets lattice. We will say that they are locally clone with respect to x. Detecting clone items in $\mathcal{E}_x$ is equivalent to the detection of clone items on the closure system defined by :

$$\mathcal{F}_x = \{X \setminus x\} \cup \{F \mid \exists M \in \mathcal{M}_x \text{ such that } F \subseteq M\}$$

In other words, we consider that all subsets of maximal closed sets included in $X \setminus x$ are closed sets. In order to detect the clone items of the lattice $\mathcal{F}_x$, we need to compute all its meet-irreducible elements. To stay the more general possible, we consider that the navigation is done using the ideals reduction operator. Indeed, the ideal lattice is a distributive lattice and the power set is just a particular case of distributive lattice.

Thus, we navigate in the ideal lattice. As a consequence, the meet-irreducible elements of the ideal lattice which are closed set are also meet-irreducible elements of $\mathcal{F}_x$. The sets in $\mathcal{M}_x$ are also trivially meet-irreducible elements of $\mathcal{F}_x$. And, finally, some immediate predecessors of sets in $\mathcal{M}_x$ can be meet-irreducible elements too. Indeed, given $M \in \mathcal{M}_x$, its immediate predecessors are co-atoms (and thus meet-irreducible elements) of the sub-lattice $[\emptyset, M]$. And if there are not predecessors of another $M' \in \mathcal{M}_x$, they will be meet-irreducible elements of $\mathcal{F}_x$.

More formally, the meet-irreducible elements of $\mathcal{F}_x$ are :

$$\mathcal{M}_{\mathcal{F}_x} = \overset{\mathcal{M}_x}{\cup} \{(X \setminus x) \setminus \uparrow y \mid y \in (X \setminus x) \text{ and } \exists M \in \mathcal{M}_x \text{ with}(X \setminus x) \setminus \uparrow y \subseteq M\}$$
$$\cup \{M \setminus y \mid M \in \mathcal{M}_x, y \in Max(M) \text{ and } \nexists M' \in \mathcal{M}_x \text{ with } M \setminus y \subset M'\}$$

To summarize we have the following equivalent assertions :

- a and b are clone items in $\mathcal{E}_x$;
- a and b are clone items in $\mathcal{M}_{\mathcal{F}_x}$.

The collection $\mathcal{M}_{\mathcal{F}_x}$ can be computed in polynomial time. Recall that the clone detection as well as the clone reduction have to be done on $\mathcal{M}_{\mathcal{F}_x}$. Thus, reducing the search space by removing clone items has initially an extra cost. However, this extra cost is balanced by less rules to compute and display in the remaining of the navigation. The user gains in readability.

6   Application to Databases

6.1   Context

In this section we present the extraction context of association rules in databases. Let X be a set of items, and $\mathcal{T}$ be a set of transaction identifiers or *tids*. An input database is a binary relation $\mathcal{R} = X \times \mathcal{T}$.

Example 1. Example of extraction context.

R1	a	b	c	d	e
i_1	1	0	0	0	0
i_2	1	1	0	0	0
i_3	0	0	1	0	0
i_4	0	0	1	1	0
i_5	0	0	1	1	1
i_6	1	1	1	0	0
i_7	1	1	1	1	1

The simple rules in $\Sigma_\mathcal{J}$ (and as a consequence, the poset $(J, \subseteq)$) are obtained by computing the inclusion order of items in the binary relation (i.e. the inclusion of columns). The binary relation might be previously reduced by merging identical items (i.e. with same columns) and by removing items which are intersections of other items. In other words, we only keep the join-irreducible elements.

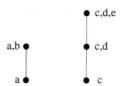

Fig. 5. The poset $(\mathcal{J}, \subseteq)$ for the binary relation R1

6.2   Navigation Tree

First step of our process consists in the computation of the simple rules and of the maximal rules (i.e. of the form $X \setminus \uparrow x \rightarrow x$). All those rules can be computed in polynomial time using a single scan of the database. Rules obtained at this step will be the first nodes of our tree (Cf. figure 6 a)). Each rule might be reducible or not. In this later case, the rule is a leaf of the tree. Note that simple rules are not reducible.

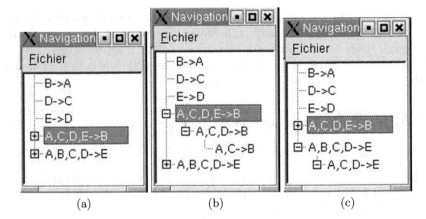

Fig. 6. The navigation tree : (a) The first nodes of the tree; (b) First branch is developed; (c) Second branch developed. For each rule, only one item is displayed in the consequent: the item defining the top of the interval.

Second step of our process is to let the end-user choose a single rule he is interested in, among the set of reducible rules in the tree. The chosen rule is then reduced and the obtained rules become new nodes of the tree (their father being the chosen rule). Again, among those rules, some might be reducible. The end-user can repeat the second step of our process as many times as he wishes, or until he has developed all branches of the tree. It is important to notice that, at each step, computation is done in polynomial time and a single scan of the database is required.

The end-user can define constraints to guide the navigation process. For instance, if the user is searching for left-reduced rules, nodes which correspond to minimal generators could be highlighted and the corresponding left-reduced rule be printed afterward. Another constraint could be the freezing of an item (or a set of items) in the antecedent. In this way, a rule will be considered reducible as long as one of its reductions still contains the frozen item(s).

7   Conclusion

Current methods for mining association rules suffer from several practical difficulties:

- The number of generated rules is potentially exponential in the size of the context;
- The computation of the rules is often prohibitive in time.

The consequence of those practical difficulties is that often few items are considered in order to reduce the combinatorial explosion and only a subset of the rules is generated (according to some quality measures). Thus, usually the user is not really in the center of the mining process. He interacts only to define the constraints to apply on the rules and waits for the next set of rules to be generated.

The aim of our paper is to recenter the mining process on the user. This is still a theoretical study which has to be refined and validated by experimentations. A prototype is under development.

Our first objective was to achieve interactivity : the user should stay behind his desk during the whole process. To achieve this we propose a downstream approach: few rules are proposed to the user at each step and the user picks the direction he wants to go. Only user required rules are generated. To ease his work, only a reasonable number of rules is proposed at each step. Searching the rules should thus gain in clarity and readability.

The navigation is done using reduction operators on the antecedent of a chosen rule. We proposed several reduction operators which all run in polynomial time and do not require a scan of the context (only a subset of the context has to be considered).

Using this approach, it seems that more items could be considered by the user in its mining process. This has however to be validated through experimentation. But when considering lot of items, we then have a representation problem for displaying the rules. Clone items could be used to reduce the antecedents of the rules.

But our theoretical approach suffers from some drawbacks too. The quality measures are not taken into account and navigation is done from the rule with lowest support to the rules with highest support. On the other hand, with this approach, the user can search in rules that usually would have not been considered by other methods. Concerning the quality measures, they can be computed on demand at the cost of a scan of the context.

The next step in our study is to achieve the prototype and do some experimentations on real cases. Those experimentations will then confirm or not if this approach has practical interests.

References

1. R. Agrawal, T. Imielinski, and A. Swami. Mining association rules between sets of items in large databases. In *ACM SIGMOD'93*. Washington, USA, 1993.
2. R. Agrawal and R. Srikant. Fast algorithm for mining association rules. In *20th International Conference of Very Large DataBasis (VLDB)*, pages 487–499. Santiago, Chile, September, 1994.
3. S. Brin, R. Motwani, J.D. Ullman, and S. Tsur. Dynamic itemset counting and implication rules for market basket data. In *SIGMOD'97*, 1997.
4. J. Demetrovics, L. Libkin, and I.B. Muchnik. Functional dependencies in relational databases: A lattice point of view. *Discrete Applied Mathematics*, 40(2):155–185, 1992.
5. V. Duquenne and J-L. Guigues. Famille minimale d'implications informatives résultant d'un tableau de données binaires. *Mathématiques Sciences Humaines*, 24, 1986.
6. B. Ganter and R. Wille. Formal concept analysis. In *Mathematical foundation*. Berlin-Heidelberg-NewYork:Springer, 1999.
7. A. Gely, R. Medina, L. Nourine, and Y. Renaud. Uncovering and reducing hidden combinatorics in guigues-duquenne covers. In *ICFCA'05*, 2005.

8. R. Godin and R. Missaoui. An incremental concept formation approach for learning from databases. *Theoritical Computer Science: Special issue on formal methods in databases and softwear e ngineering*, 133(2):387–419, October, 1994.

9. J. Han, J. Pei, and Y. Yin. Mining frequent patterns without candidate generation. In *in Proceedings of the 2000 ACM-SIGMOD International Conference on Management of Data. Dallas, Texas, USA*, 2000.

10. R. J. Hilderman and H.J. Hamilton. Knowledge discovery and interestingness measures : a survey. *Technical report CS99-04, Departement of computer science, University of Regina*, 1999.

11. J. Hipp and U. Guentzer. Is pushing constraints deeply into the mining algorithms really what we want?. *SIGKDD Explorations*, pages 4(1)50–55, 2002.

12. J. Hipp, U. Guentzer, and G. Nakhaeizadeh. Algorithms for association rules mining - a general survey and comparison. *SIGKDD Exploration*, 2(1):58–64, 2000.

13. D. Maier. *The theory of relational data bases*. Computer Science Press Rockville, 1983.

14. H. Manilla and K.J. Räihä. On the complexity of inferring functionnal dependencies. *Discret Applied Mathematics*, 40(2):237–243, 1992.

15. R. Medina, C. Noyer, and O. Raynaud. Efficient algorithms for clone items detection. In *CLA'05*, pages 70–81, 2005.

16. N. Pasquier, Y. Bastide, R. Taouil, and L. Lakhal. Efficient mining of association rules using closed itemset lattices. *Information Systems*, 24, 1:P. 25–46, 1999.

17. N. Pasquier, Y. Bastides, R. Taouil, and L. Lakhal. Closed set based discovery of small covers for association rules. In *Proceeding in BDA conf.*, pages 361–381, October 1999.

About the Family of Closure Systems Preserving Non-unit Implications in the Guigues-Duquenne Base

Alain Gély and Lhouari Nourine

Université Clermont II Blaise Pascal,
LIMOS - Campus des Cézeaux,
63170 Aubière Cedex - France
{gely, nourine}@isima.fr

Abstract. Consider a Guigues-Duquenne base $\Sigma^{\mathcal{F}} = \Sigma_J^{\mathcal{F}} \cup \Sigma_{\downarrow}^{\mathcal{F}}$ of a closure system $\mathcal{F}$, where Σ_J the set of implications $P \to P^{\Sigma^{\mathcal{F}}}$ with $|P| = 1$, and $\Sigma_{\downarrow}^{\mathcal{F}}$ the set of implications $P \to P^{\Sigma^{\mathcal{F}}}$ with $|P| > 1$. Implications in $\Sigma_J^{\mathcal{F}}$ can be computed efficiently from the set of meet-irreducible $\mathcal{M}(\mathcal{F})$; but the problem is open for $\Sigma_{\downarrow}^{\mathcal{F}}$. Many existing algorithms build $\mathcal{F}$ as an intermediate step.

In this paper, we characterize the cover relation in the family $\mathcal{C}_{\downarrow}(\mathcal{F})$ with the same $\Sigma_{\downarrow}$, when ordered under set-inclusion. We also show that $\mathcal{M}(\mathcal{F}_{\perp})$ the set of meet-irreducible elements of a minimal closure system in $\mathcal{C}_{\downarrow}(\mathcal{F})$ can be computed from $\mathcal{M}(\mathcal{F})$ in polynomial time for any $\mathcal{F}$ in $\mathcal{C}_{\downarrow}(\mathcal{F})$. Moreover, the size of $\mathcal{M}(\mathcal{F}_{\perp})$ is less or equal to the size of $\mathcal{M}(\mathcal{F})$.

Keywords: Closure system, Guigues-Duquenne base, Equivalence relation.

1 Introduction

Computing a minimum implication base from a context is a challenging problem for which many researches are ongoing. This problem deals with information retrieval in Formal Concept Analysis, and have a lot of applications in other fields of computer science, as database, graph theory, Artificial intelligence, lattice algorithmic and so on (see [1–4]).

This problem remains open, even for particular cases as finding all keys of a multi-valued context, enumerating minimal transversal of an hypergraph or for the problem of dualization of monotone boolean functions. The best known result for these particular cases is the one of Fredman and Kachiyan [5], and it is a quasi-polynomial algorithm.

Let $\mathcal{F}$ be a closure system on a finite set J, and $\Sigma^{\mathcal{F}}$ a minimum implicational base for $\mathcal{F}$. $\Sigma^{\mathcal{F}}$ is supposed to be in the Guigues-Duquenne form, i.e. premises of implications are pseudo-closed sets and conclusions are their closure [6]. This base may be partitioned in three parts: $\Sigma^{\mathcal{F}} = \Sigma_J^{\mathcal{F}} \cup \Sigma_{\downarrow}^{\mathcal{F}} \cup \Sigma_{\emptyset}^{\mathcal{F}}$, where

R. Missaoui and J. Schmid (Eds.): ICFCA 2006, LNAI 3874, pp. 191–204, 2006.
© Springer-Verlag Berlin Heidelberg 2006

- $\Sigma_{j}^{\mathcal{F}}$ is the set of implications $X \to Y$ such that $|X| = 1$ (unit implications).
- $\Sigma_{\uparrow}^{\mathcal{F}}$ is the set of implications $X \to Y$ such that $|X| > 1$ (non unit implications).
- $\Sigma_{\emptyset}^{\mathcal{F}}$ is the set of implications $X \to Y$ such that $X = \{\emptyset\}$.

In this paper, without loss of generality, we partition $\Sigma^{\mathcal{F}}$ into the two implicational families $\Sigma_{j}^{\mathcal{F}}$ and $\Sigma_{\uparrow}^{\mathcal{F}}$. i.e. we consider there is no implication with $\emptyset$ as premise.

The set of closure systems having the same set of unit implications has been studied in [7, 8]. Here, we study the set of all closure systems $\mathcal{C}_{\downarrow}(\mathcal{F})$ that have the same set of non unit implications in their Guigues-Duquenne base as $\mathcal{F}$. To keep $\Sigma_{\uparrow}^{\mathcal{F}}$ unchanged, we must add or delete only unit implications. First, we give necessary and sufficient conditions to add a unit implication, and then we characterize the cover relation in $\mathcal{C}_{\downarrow}(\mathcal{F})$. Secondly, we give a polynomial time algorithm to compute the meet-irreducible elements of a minimal closure system in $\mathcal{C}_{\downarrow}(\mathcal{F})$ from $\mathcal{M}(\mathcal{F})$. This leads us to say that computing the minimal cover is better on a minimal closure system (or context), which may improve the efficiency of existing algorithms as Ganter's one[9], since they need to compute the set of closed sets.

The rest of this paper is organized as follows:

Section 2 recall some notations and definitions on closure systems. Section 3 characterizes the cover relation in $\mathcal{C}_{\downarrow}(\mathcal{F})$. Section 4 gives the main theorem. Section 5 illustrates the main theorem. Finally, section 6 gives a polynomial time algorithm to compute the meet-irreducible elements of the smallest closure system preserving non unit implications.

2   Definitions

We briefly recall some definitions and notations:

1. Let J be a finite set. Small letters $(a, b, \dots)$ denote elements of J. capital letters $(A, B, \dots)$ sets, and cursive letters $(\mathcal{A}, \mathcal{B}, \dots)$ family of sets. Where no ambiguity is possible, we abusively denote the set $X = \{x, y, z\}$ by xyz.
2. $\mathcal{F} \subseteq 2^{J}$ is a closure system iff $J \in \mathcal{F}$ and $F_1 \in \mathcal{F}$, $F_2 \in \mathcal{F}$ implie $F_1 \cap F_2 \in \mathcal{F}$. Sets in $\mathcal{F}$ are called closed sets.
3. A pair (X, Y) denoted by $X \to Y$ is an implication on J with X as premise and Y as conclusion. A family of implications is denoted by Σ.
4. The set A is said Σ-closed if $X \subseteq A$ implies $Y \subseteq A$ for each implication $X \to Y \in \Sigma$. There exists a linear time algorithm to compute X^{Σ}, the smallest Σ-closed set which contains X [10]. The family $\mathcal{F}(\Sigma) = \{S \in 2^{J} \mid \text{ is } \Sigma\text{-closed}\}$ is a closure system on J.
5. A family of implications Σ is an implicational base for the closure system $\mathcal{F}$, denoted by $\Sigma^{\mathcal{F}}$, iff $\mathcal{F} = \mathcal{F}(\Sigma)$.
6. $X \to Y$ is derived from Σ, noted by $\Sigma \vdash X \to Y$, if $Y \subseteq X^{\Sigma}$.
7. Let $\mathcal{F}$ be a closure system and $P \notin \mathcal{F}$. Then P is

- a quasi-closed set iff for all $Q \subset P$, $Q^{\Sigma^{\mathcal{F}}} \subset P$ or $Q^{\Sigma^{\mathcal{F}}} = P^{\Sigma^{\mathcal{F}}}$.
- a pseudo-closed set if P is a quasi-closed set and there does not exist a quasi closed set $Q \subset P$ with $Q^{\Sigma^{\mathcal{F}}} = P^{\Sigma^{\mathcal{F}}}$.

A pseudo-closed set P is closed under $\Sigma^{\mathcal{F}} \setminus \{P \to P^{\Sigma^{\mathcal{F}}}\}$.

In this paper, we use Guigues-Duquenne base (or GD-base for short), since it is uniquely defined and there is a bijection between GD-bases and closure systems on J.

Theorem 1. *Duquenne-Guigues base [6]*
Let $\mathcal{F}$ be a closure system. $\Sigma^{\mathcal{F}} = \{P \to P^{\Sigma^{\mathcal{F}}} \mid P$ is a pseudo-closed set $\}$ is an implicational base of $\mathcal{F}$ and has a minimum number of implications.

Notice that if we remove an implication from the GD-base of a closure system $\mathcal{F}$, we obtain a new GD-base corresponding to a closure system $\mathcal{F}'$, where $\mathcal{F} \subset \mathcal{F}'$. Now, which implications can be added in a GD-base such that $\Sigma_\downarrow^{\mathcal{F}}$ is preserved? Based on the partition of GD-base, we will give an equivalence relation on the set of closure systems on J. Let $\mathcal{F}$ and $\mathcal{F}'$ be two closure systems on a finite set J. We say that $\mathcal{F}$ is $\Sigma_\downarrow$-equivalent to $\mathcal{F}'$ if $\Sigma_\downarrow^{\mathcal{F}} = \Sigma_\downarrow^{\mathcal{F}'}$.

This equivalence relation induces a partition of the set $\mathcal{C}(J)$ of all closure systems on J into equivalence classes. We denote by $\mathcal{C}_\downarrow(\mathcal{F}) = \{\mathcal{F}' \mid \Sigma_\downarrow^{\mathcal{F}} = \Sigma_\downarrow^{\mathcal{F}'}\}$ the equivalence class containing $\mathcal{F}$.

For example, the following three closure systems are $\Sigma_\downarrow^{\mathcal{F}}$-equivalent:

$$\mathcal{F} = \{\emptyset, a, b, c, ac, bc, bd, bcd, abcd\}$$
$$\mathcal{F}' = \{\emptyset, a, b, bd, ac, abcd\}$$
$$\mathcal{F}'' = \{\emptyset, a, b, abcd\}$$
with
$$\Sigma_\downarrow^{\mathcal{F}} = \{ab \to abcd, d \to bd\}$$
$$\Sigma_\downarrow^{\mathcal{F}'} = \{ab \to abcd, d \to bd, c \to ac\}$$
$$\Sigma_\downarrow^{\mathcal{F}''} = \{ab \to abcd, c \to abcd, d \to abcd\}$$

The GD-base $\Sigma_\downarrow^{\mathcal{F}'}$ is obtained from the GD-base $\Sigma_\downarrow^{\mathcal{F}}$ by adding the implication $c \to a$, and $\Sigma_\downarrow^{\mathcal{F}''}$ from $\Sigma_\downarrow^{\mathcal{F}'}$ by adding $c \to d$ and $d \to c$.

3 Properties of the Family $\mathcal{C}_\downarrow(\mathcal{F})$

Consider a closure system $\mathcal{F}$ and $\mathcal{C}_\downarrow(\mathcal{F})$ the family of all closure systems that have the same set of non unit implications in their GD-base as $\mathcal{F}$. Since a GD-base remains a GD-base if we remove an implication $P \to P^{\Sigma^{\mathcal{F}}}$, then there is a maximal closure system $\mathcal{F}^\top$ $\Sigma_\downarrow$-equivalent to $\mathcal{F}$, with $\Sigma^{\mathcal{F}^\top} = \Sigma_\downarrow^{\mathcal{F}}$. $\mathcal{F}^\top$ corresponds to the top element of $\mathcal{C}_\downarrow(\mathcal{F})$ which is atomistic. The set $\mathcal{C}_\downarrow(\mathcal{F})$ is not a closure system. We denote by $\mathcal{F}_\perp$ a minimal closure system having the same non unit implications as $\mathcal{F}$.

Note that two closure systems which are $\Sigma_\downarrow$-equivalent differ by implications in Σ_J. The family of closure systems which differ by implications in $\Sigma_\downarrow$ and which are Σ_J-equivalent is a closure system [7, 8]. This is not the case for $\mathcal{C}_\downarrow(\mathcal{F})$.

Before giving the first proposition of this paper, we recall a result in [11] which is first appeared (in variant form) in [6] and [12]. This result is useful to compute a minimum base from any base.

Theorem 2. *[11]*

- *Let Σ be a non redundant base of a closure system $\mathcal{F}$. If all implications of Σ are of the form $X \rightarrow X^{\Sigma^{\mathcal{F}}}$ then Σ is minimum.*
- *$\Sigma^{\mathcal{F}} = \{P \rightarrow P^{\Sigma^{\mathcal{F}}} \mid P$ is pseudoclosed$\}$ is a canonical minimum base of $\mathcal{F}$ (Duquenne-Guigues base). Each minimum base Σ of $\mathcal{F}$ consists of implications $X \rightarrow Y$ where $P \rightarrow P^{\Sigma^{\mathcal{F}}} \in \Sigma^{\mathcal{F}}$ and $X \subseteq P$, $X^{\Sigma^{\mathcal{F}}} = P^{\Sigma^{\mathcal{F}}}$.*

The following corollary follows from theorem 2 and the definition of pseudo-closed sets.

Corollary 1. *If a non redundant implicational base of a closure system $\mathcal{F}$ is such that*

1. *all the conclusions of implications are closed sets*
2. *all the premises of implications are quasi-closed sets,*

then it is the GD-base of $\mathcal{F}$.

The following proposition gives necessary and sufficient conditions to add a unit implication without changing non unit implications in the GD-base.

Proposition 1. *Let $\Sigma^{\mathcal{F}} = \Sigma_\downarrow^{\mathcal{F}} \cup \Sigma_J^{\mathcal{F}}$ be a GD-base of a closure system $\mathcal{F}$, and $a, b \in J$ such that $\Sigma_J^{\mathcal{F}} \not\vdash a \rightarrow b$. Then $\Sigma_\downarrow^{\mathcal{F}} = \Sigma_\downarrow^{\mathcal{F}'}$ with $\Sigma^{\mathcal{F}'}$ the GD-base corresponding to $\Sigma^{\mathcal{F}} \cup \{a \rightarrow b\}$ iff for all $P \rightarrow P^{\Sigma^{\mathcal{F}}} \in \Sigma_\downarrow^{\mathcal{F}}$*

$$
\begin{aligned}
&(i) \; a \in P & &\Rightarrow b \in P \\
&(ii) \; a \in P^{\Sigma^{\mathcal{F}}} & &\Rightarrow b \in P^{\Sigma^{\mathcal{F}}} \\
&(iii) \; j \in P \;, a \in P \cap j^{\Sigma^{\mathcal{F}}} \Rightarrow (jb)^{\Sigma^{\mathcal{F}}} \neq P^{\Sigma^{\mathcal{F}}}
\end{aligned}
$$

Proof. Let $\Sigma^{\mathcal{F}} = \Sigma_\downarrow^{\mathcal{F}} \cup \Sigma_J^{\mathcal{F}}$ be a GD-base of a closure system $\mathcal{F}$, $a, b \in J$ such that $\Sigma_J^{\mathcal{F}} \not\vdash a \rightarrow b$ and $\Sigma_\downarrow^{\mathcal{F}} = \Sigma_\downarrow^{\mathcal{F}'}$ with $\Sigma^{\mathcal{F}'}$ the GD-base corresponding to $\Sigma^{\mathcal{F}} \cup \{a \rightarrow b\}$. Suppose there is a condition not satisfied. We have 3 cases:

(i) $a \in P$ and $b \notin P$, then P is not a pseudo-closed set in $\mathcal{F}'$ since P is not closed under $a \rightarrow b$, and then $\Sigma_\downarrow^{\mathcal{F}} \neq \Sigma_\downarrow^{\mathcal{F}'}$.

(ii) $a \in P^{\Sigma^{\mathcal{F}}}$ and $b \notin P^{\Sigma^{\mathcal{F}}}$. Then $P^{\Sigma^{\mathcal{F}}} \neq P^{\Sigma^{\mathcal{F}'}}$, which implies $\Sigma_\downarrow^{\mathcal{F}} \neq \Sigma_\downarrow^{\mathcal{F}'}$.

(iii) $a \in P \cap j^{\Sigma^{\mathcal{F}}}$ and $(jb)^{\Sigma^{\mathcal{F}}} = P^{\Sigma^{\mathcal{F}}}$. This implies that P is not a minimal quasi-closed set in $\mathcal{F}'$. Indeed, if $a \in j^{\Sigma^{\mathcal{F}}}$, we have $j^{\Sigma^{\mathcal{F}'}} = (jb)^{\Sigma^{\mathcal{F}}} = P^{\Sigma^{\mathcal{F}}}$. Thus, $\Sigma_\downarrow^{\mathcal{F}} \neq \Sigma_\downarrow^{\mathcal{F}'}$.

Conversely, suppose that conditions (i), (ii) and (iii) are satisfied. Let $\Sigma^{\mathcal{F}}$ be the GD-base of $\mathcal{F}$ and $\Sigma^{\mathcal{F}'}$ the GD-base of the closure system $\mathcal{F}'$, which corresponds to the implicational base $\Sigma^{\mathcal{F}} \cup \{a \to b\}$. We have to show that $\Sigma_{\downarrow}^{\mathcal{F}} = \Sigma_{\downarrow}^{\mathcal{F}'}$. To do so, we compute $\Sigma_{\mathcal{J}}^{\mathcal{F}'}$ the GD-base for $\Sigma_{\mathcal{J}}^{\mathcal{F}} \cup \{a \to b\}$, and then show that $\Sigma_{\downarrow}^{\mathcal{F}} \cup \Sigma_{\mathcal{J}}^{\mathcal{F}'}$ satisfies conditions of corollary 1.

- By definition, $(\Sigma_{\downarrow}^{\mathcal{F}} \cup \Sigma_{\mathcal{J}}^{\mathcal{F}} \cup \{a \to b\})$ is an implicational base of $\mathcal{F}'$. Let

$$
\begin{aligned}
\Sigma_{\mathcal{J}}^{\mathcal{F}'} = \ & \{j \to j^{\Sigma^{\mathcal{F}}} \mid j \to j^{\Sigma^{\mathcal{F}}} \in \Sigma_{\mathcal{J}}^{\mathcal{F}}, \ a \notin j^{\Sigma^{\mathcal{F}}}\} \\
& \cup \{j \to (jb)^{\Sigma^{\mathcal{F}}} \mid j \to j^{\Sigma^{\mathcal{F}}}, \ a \in j^{\Sigma^{\mathcal{F}}}\} \\
& \cup \{a \to (ab)^{\Sigma^{\mathcal{F}}} \mid \forall \, j \to j^{\Sigma^{\mathcal{F}}} \in \Sigma_{\mathcal{J}}^{\mathcal{F}}, \ j \neq a\}.
\end{aligned}
$$

Then $\Sigma_{\mathcal{J}}^{\mathcal{F}'}$ is the GD-base of $\Sigma_{\mathcal{J}}^{\mathcal{F}} \cup \{a \to b\}$. In effect, for $j \to j^{\Sigma^{\mathcal{F}}}$, if $a \notin j^{\Sigma^{\mathcal{F}}}$ then $j^{\Sigma^{\mathcal{F}}} = j^{\Sigma^{\mathcal{F}'}}$. If $a \in j^{\Sigma^{\mathcal{F}}}$, then $j^{\Sigma^{\mathcal{F}'}}$ is the smallest closed set in $\mathcal{F}$ which contain j and b (since $\Sigma^{\mathcal{F}'} \vdash a \to b$), so $j^{\Sigma^{\mathcal{F}'}} = (jb)^{\Sigma^{\mathcal{F}}}$. If $a = a^{\Sigma^{\mathcal{F}}}$, we must add the implication $a \to (ab)^{\Sigma^{\mathcal{F}}}$.
- Now, let us show that $\Sigma_{\downarrow}^{\mathcal{F}} \cup \Sigma_{\mathcal{J}}^{\mathcal{F}'}$ satisfies conditions of corollary 1.

  - Let $P \to P^{\Sigma^{\mathcal{F}}}$ in $\Sigma_{\downarrow}^{\mathcal{F}}$. We show that $P^{\Sigma^{\mathcal{F}}} = P^{\Sigma^{\mathcal{F}'}}$.

    By hypothesis, if $a \in P^{\Sigma^{\mathcal{F}}}$ then $b \in P^{\Sigma^{\mathcal{F}}}$ (condition (ii)). So $P^{\Sigma^{\mathcal{F}}}$ is closed under $\Sigma^{\mathcal{F}} \cup \{a \to b\}$ and then belongs to $\mathcal{F}'$. We conclude that $P^{\Sigma^{\mathcal{F}}} = P^{\Sigma^{\mathcal{F}'}}$.
  - We show that $\Sigma_{\downarrow}^{\mathcal{F}} \cup \Sigma_{\mathcal{J}}^{\mathcal{F}'}$ is non redundant.

    It is clear that each implication in $\Sigma_{\mathcal{J}}^{\mathcal{F}'}$ is non redundant, since by construction, for two implications $j_1 \to j_1^{\Sigma^{\mathcal{F}'}}$ and $j_2 \to j_2^{\Sigma^{\mathcal{F}'}}$ in $\Sigma_{\mathcal{J}}^{\mathcal{F}'}$ we have $j_1 \neq j_2$.

    Let show that $P \to P^{\Sigma^{\mathcal{F}}} \in \Sigma_{\downarrow}^{\mathcal{F}}$ is non redundant in $\Sigma_{\downarrow}^{\mathcal{F}} \cup \Sigma_{\mathcal{J}}^{\mathcal{F}'}$. To do that, we will show that for all implications $Q \to Q^{\Sigma^{\mathcal{F}'}}$ in $\Sigma_{\downarrow}^{\mathcal{F}} \cup \Sigma_{\mathcal{J}}^{\mathcal{F}'} \backslash \{P \to P^{\Sigma^{\mathcal{F}}}\}$, we have $Q \subset P$ implies $Q^{\Sigma^{\mathcal{F}'}} \subset P$.

    Since for any implication $Q \to Q^{\Sigma^{\mathcal{F}}} \in \Sigma_{\downarrow}^{\mathcal{F}}$, Q is pseudo-closed set in $\mathcal{F}$ and $Q^{\Sigma^{\mathcal{F}}} = Q^{\Sigma^{\mathcal{F}'}}$, we have: for all $Q \to Q^{\Sigma^{\mathcal{F}}} \in \Sigma_{\downarrow}^{\mathcal{F}} \backslash \{P \to P^{\Sigma^{\mathcal{F}}}\}$, $Q^{\Sigma^{\mathcal{F}'}} \subset P$.

    We must show that, for all $j \to j^{\Sigma^{\mathcal{F}'}} \in \Sigma_{\mathcal{J}}^{\mathcal{F}'}$, we have $j^{\Sigma^{\mathcal{F}'}} \subset P$. We distinguish two cases:
    * if $a \notin j^{\Sigma^{\mathcal{F}}}$ or $b \in j^{\Sigma^{\mathcal{F}}}$, then $j^{\Sigma^{\mathcal{F}}} = j^{\Sigma^{\mathcal{F}'}}$. Since P is a pseudo-closed set in $\mathcal{F}$, we deduce $j^{\Sigma^{\mathcal{F}'}} \subset P$.
    * if $a \in j^{\Sigma^{\mathcal{F}}}$ and $b \notin j^{\Sigma^{\{}}$, then $j^{\Sigma^{\mathcal{F}'}} = (jb)^{\Sigma^{\mathcal{F}}} \neq j^{\Sigma^{\mathcal{F}}}$. Notice that since $j \in P$, we have $j^{\Sigma^{\mathcal{F}}} \subset P$ (P is a pseudo-closed set in $\mathcal{F}$) and therefore $a \in P$. Condition (i) implies that $b \in P$. Thus, since $jb \subset P$ we have $(jb)^{\Sigma^{\mathcal{F}}} \subset P$ or $(jb)^{\Sigma^{\mathcal{F}}} = P^{\Sigma^{\mathcal{F}}}$. By condition (iii), we have $(jb)^{\Sigma^{\mathcal{F}}} \neq P^{\Sigma^{\mathcal{F}}}$. We conclude that $j^{\Sigma^{\mathcal{F}'}} \subset P$.

Note that P is a quasi-closed set, since its closed under $\Sigma^{\mathcal{F}'} \backslash \{P \to P^{\Sigma^{\mathcal{F}}}\}$.

We conclude from corollary 1 that $\Sigma_{\downarrow}^{\mathcal{F}} \cup \Sigma_{J}^{\mathcal{F}'}$ is a GD-base. Indeed, $\Sigma_{\downarrow}^{\mathcal{F}} \cup \Sigma_{J}^{\mathcal{F}'}$ is a non redundant implicational base such that conclusions of implications are closed sets and premises are quasi-closed sets. □

The family $\mathcal{C}_{\downarrow}(\mathcal{F})$ of closure systems preserving non unit implications has a top element $\mathcal{F}^{\top}$ corresponding to the closure system with $\Sigma^{\mathcal{F}^{\top}} = \Sigma_{\downarrow}^{\mathcal{F}}$. The following example shows that the family $\mathcal{C}_{\downarrow}(\mathcal{F})$ is not a closure system.

Example 1. Let $J = \{a, b, c, d\}$. Consider the closure system with $\mathcal{F}$ with $\Sigma^{\mathcal{F}} = \{c \rightarrow ac, b \rightarrow ab, d \rightarrow abd, abc \rightarrow abcd\}$. The closure system $\mathcal{F}'$ with $\Sigma^{\mathcal{F}'} = \{c \rightarrow bc, a \rightarrow ab, d \rightarrow abd, abc \rightarrow abcd\}$ belongs to $\mathcal{C}_{\downarrow}(\mathcal{F})$. But the closure system $\mathcal{F}'' = \mathcal{F} \cap \mathcal{F}'$ does not belong to $\mathcal{C}_{\downarrow}(\mathcal{F})$, since $\Sigma^{\mathcal{F}''} = \{c \rightarrow abcd, a \rightarrow ab, b \rightarrow ab, d \rightarrow abd\}$.

The following theorem characterizes the cover relation in $\mathcal{C}_{\downarrow}(\mathcal{F})$ when ordered under set-inclusion.

Theorem 3. *Let $\mathcal{F}$ and $\mathcal{F}'$ be two closure systems in $\mathcal{C}_{\downarrow}(\mathcal{F})$ such that $\Sigma^{\mathcal{F}} \nvdash a \rightarrow b$, $\Sigma^{\mathcal{F}'} \vdash a \rightarrow b$. Then $\mathcal{F}' \prec \mathcal{F}$ iff,*

$$(ii') \quad \text{for all } P \rightarrow P^{\Sigma^{\mathcal{F}}} \in \Sigma^{\mathcal{F}}, P^{\Sigma^{\mathcal{F}}} \neq a^{\Sigma^{\mathcal{F}}} \quad a \in P^{\Sigma^{\mathcal{F}}} \Rightarrow b \in P^{\Sigma^{\mathcal{F}}}$$

Proof. Let $\mathcal{F}$ and $\mathcal{F}'$ be two closure systems in $\mathcal{C}_{\downarrow}(\mathcal{F})$ such that $\Sigma^{\mathcal{F}} \nvdash a \rightarrow b$ and $\Sigma^{\mathcal{F}'} \vdash a \rightarrow b$. Suppose $\mathcal{F}' \prec \mathcal{F}$ and there exists $P \rightarrow P^{\Sigma^{\mathcal{F}}} \in \Sigma^{\mathcal{F}}$ with $P^{\Sigma^{\mathcal{F}}} \neq a^{\Sigma^{\mathcal{F}}}$ such that $a \in P^{\Sigma^{\mathcal{F}}}$ and $b \notin P^{\Sigma^{\mathcal{F}}}$. According to proposition 1, we have $P \rightarrow P^{\Sigma^{\mathcal{F}}} \in \Sigma_{J}^{\mathcal{F}}$, otherwise condition (ii) is not satisfied. Thus it remains to prove it for $\Sigma_{J}^{\mathcal{F}}$.

Let $j \rightarrow j^{\Sigma^{\mathcal{F}}} \in \Sigma_{J}^{\mathcal{F}}$. Then $\Sigma_{J}^{\mathcal{F}} \vdash j \rightarrow a$ and $\Sigma_{J}^{\mathcal{F}} \nvdash j \rightarrow b$. Consider $\mathcal{F}''$ be the closure system corresponding to $\Sigma^{\mathcal{F}''} = \Sigma^{\mathcal{F}} \cup \{j \rightarrow b\}$. We show that $\Sigma_{\downarrow}^{\mathcal{F}''} = \Sigma_{\downarrow}^{\mathcal{F}}$ and $\mathcal{F}' \subset \mathcal{F}'' \subset \mathcal{F}$, which contradicts $\mathcal{F}' \prec \mathcal{F}$.

Since conditions of proposition 1 are satisfied for $\mathcal{F}'$ and $a, b \in J$, then the same conditions are satisfied for $\mathcal{F}''$ and $j, b \in J$. Thus, $\Sigma_{\downarrow}^{\mathcal{F}''} = \Sigma_{\downarrow}^{\mathcal{F}}$.

Now let us show that $\mathcal{F}' \subset \mathcal{F}'' \subset \mathcal{F}$. Clearly, we have $\mathcal{F}'' \subset \mathcal{F}$ since $\Sigma_{J}^{\mathcal{F}} \vdash j \rightarrow b$ and $\Sigma_{J}^{\mathcal{F}} \nvdash j \rightarrow b$. Moreover $\mathcal{F}' \subseteq \mathcal{F}''$. Indeed, $\Sigma^{\mathcal{F}'} \vdash j \rightarrow b$, since $\Sigma^{\mathcal{F}'} \vdash j \rightarrow a$ and $\Sigma^{\mathcal{F}'} \vdash a \rightarrow b$. Suppose that $\mathcal{F}'' = \mathcal{F}'$. Then $\Sigma^{\mathcal{F}'} \vdash a \rightarrow j$, which is impossible since $j^{\Sigma^{\mathcal{F}}} \neq a^{\Sigma^{\mathcal{F}}}$.

Conversely, suppose $\mathcal{F}' \not\prec \mathcal{F}$. Then there exists $\mathcal{F}''$ such that $\mathcal{F}' \subset \mathcal{F}'' \subset \mathcal{F}$ with $\Sigma^{\mathcal{F}'} \vdash \Sigma^{\mathcal{F}''} \vdash \Sigma^{\mathcal{F}}$. Thus there exist $j, y \in J$ such that $\Sigma^{\mathcal{F}''} \vdash j \rightarrow y$ and $\Sigma^{\mathcal{F}} \nvdash j \rightarrow y$. This implies that $y = b$, since $\Sigma^{\mathcal{F}} \cup \{a \rightarrow b\} \vdash j \rightarrow y$ but $\Sigma^{\mathcal{F}} \nvdash j \rightarrow y$, i.e. any new implication must contains b in its conclusion. Since $\Sigma^{\mathcal{F}} \nvdash j \rightarrow b$ and $\Sigma^{\mathcal{F}} \cup \{a \rightarrow b\} \vdash j \rightarrow b$ then $\Sigma^{\mathcal{F}} \vdash j \rightarrow a$, otherwise $\Sigma^{\mathcal{F}} \cup \{a \rightarrow b\} \nvdash j \rightarrow b$, which implies that $\mathcal{F}' \not\subset \mathcal{F}''$. Moreover, $j^{\Sigma^{\mathcal{F}}} \neq a^{\Sigma^{\mathcal{F}}}$ since $\mathcal{F}'' \neq \mathcal{F}'$. This implies that (ii') is false, since $a \in j^{\Sigma^{\mathcal{F}}}$ and $b \notin j^{\Sigma^{\mathcal{F}}}$. □

Notice that condition (ii') is stronger than condition (ii) of proposition 1. Indeed, condition (ii') must be satisfied for all $P \rightarrow P^{\Sigma^{\mathcal{F}}}$ in $\Sigma^{\mathcal{F}}$ such that $(a)^{\Sigma^{\mathcal{F}}} \neq P^{\Sigma^{\mathcal{F}}}$, but in proposition 1, it must be satisfied only for implications in $\Sigma_{\downarrow}^{\mathcal{F}}$.

Property 1. Let $\mathcal{F}$ be a closure system, $\Sigma^{\mathcal{F}}$ its GD-base and $\Sigma^{\mathcal{F}} \nvdash a \to b$. If $\mathcal{F}$ satisfies conditions (i) and (ii') then the following conditions are equivalents :

- (iii) For all $P \to P^{\Sigma^{\mathcal{F}}} \in \Sigma^{\mathcal{F}}_{\downarrow}$, $j \in P$ and $a \in P \cap j^{\Sigma^{\mathcal{F}}}$ implies $(jb)^{\Sigma^{\mathcal{F}}} \neq P^{\Sigma^{\mathcal{F}}}$
- (iii') For all $P \to P^{\Sigma^{\mathcal{F}}} \in \Sigma^{\mathcal{F}}_{\downarrow}$, $a \in P$ implies $(ab)^{\Sigma^{\mathcal{F}}} \neq P^{\Sigma^{\mathcal{F}}}$

Proof. Let $\mathcal{F}$ be a closure system satisfying conditions (i) and (ii'). Suppose condition (iii) is satisfied. Then it is satisfied when $j = a$, and thus (iii') is satisfied.

Now suppose (iii') is satisfied. Let $P \to P^{\Sigma^{\mathcal{F}}} \in \Sigma^{\mathcal{F}}_{\downarrow}$ such that $a \in P \cap j^{\Sigma^{\mathcal{F}}}$ with $j \in P$. We have two cases :

- If $j^{\Sigma^{\mathcal{F}}} = a^{\Sigma^{\mathcal{F}}}$, then $(ab)^{\Sigma^{\mathcal{F}}} = (jb)^{\Sigma^{\mathcal{F}}} \neq P^{\Sigma^{\mathcal{F}}}$.
- If $j^{\Sigma^{\mathcal{F}}} \neq a^{\Sigma^{\mathcal{F}}}$, then according to condition (ii') we have $a \in j^{\Sigma^{\mathcal{F}}}$ implies $b \in j^{\Sigma^{\mathcal{F}}}$. Thus $(jb)^{\Sigma^{\mathcal{F}}} = j^{\Sigma^{\mathcal{F}}} \neq P^{\Sigma^{\mathcal{F}}}$.

$\square$

Thus, given $\Sigma^{\mathcal{F}}$, we can compute in polynomial time $\Sigma^{\mathcal{F}_\perp}$ where $\mathcal{F}_\perp$ is a minimal closure system with $\Sigma^{\mathcal{F}_\perp}_{\downarrow} = \Sigma^{\mathcal{F}}_{\downarrow}$. Since the input is usually a context, proposition 1 cannot be applied. The next section translates the three properties of proposition 1 on closure systems and section 6 on contexts.

4   Characterization of $\Sigma_{\downarrow}$-Equivalence Relation Using Closure Systems

To avoid dispatching the notations all along this section, we define here all the notations we use:

Let $\mathcal{F}$ be a closure system, and $a, b \in J$ such that $b \notin a^{\Sigma^{\mathcal{F}}}$, i.e. $\Sigma^{\mathcal{F}} \nvdash a \to b$.

- $A = a^{\Sigma^{\mathcal{F}}}$, is the smallest closed set of $\mathcal{F}$ containing a.
- $B = b^{\Sigma^{\mathcal{F}}}$, is the smallest closed set of $\mathcal{F}$ containing b.
- $Q(a) = \{j \in J \mid j^{\Sigma^{\mathcal{F}}} = a^{\Sigma^{\mathcal{F}}}\}$, is the set of elements of J which appear together in any set of $\mathcal{F}$. Without loss of generality, we suppose $Q(a)$ is reduce to the singleton $\{a\}$
- $A_* = A \setminus Q(a)$.
- $\mathcal{A} = \{F \in \mathcal{F} \mid a \in F, b \notin F\}$, is the set of closed sets of $\mathcal{F}$ containing a but not b.
- $\mathcal{A}_* = \{F \in \mathcal{F} \mid A_* \subseteq F, a \notin F, b \notin F\}$, is the set of closed sets of $\mathcal{F}$ containing A_* but not a nor b.

Lemma 1 shows that testing conditions (i) and (ii') is equivalent to testing a bijection between $\mathcal{A}_*$ and $\mathcal{A}$.

Lemma 1. *Consider the mapping* $\varphi : \mathcal{A}_* \to \mathcal{A}$, *with* $\varphi(F) = F \cup \{a\}$. *Then* φ *is a bijection iff the following conditions are satisfied.*

$$(i) \text{ for all } P \to P^{\Sigma^{\mathcal{F}}} \in \Sigma^{\mathcal{F}}_{\downarrow}, \qquad a \in P \quad \Rightarrow b \in P$$
$$(ii') \text{ for all } P \to P^{\Sigma^{\mathcal{F}}} \in \Sigma^{\mathcal{F}}, P^{\Sigma^{\mathcal{F}}} \neq a^{\Sigma^{\mathcal{F}}} \qquad a \in P^{\Sigma^{\mathcal{F}}} \Rightarrow b \in P^{\Sigma^{\mathcal{F}}}$$

Proof. – Suppose $\varphi : \mathcal{A}_* \to \mathcal{A}$, with $\varphi(F) = F \cup \{a\}$ is a bijection, and one of the two conditions is not satisfied.

First suppose condition (i) is not satisfied. There exists $P \to P^{\Sigma^{\mathcal{F}}} \in \Sigma^{\mathcal{F}}_{\downarrow}$ such that $a \in P$ and $b \notin P$. By the bijection φ, we have $P \setminus \{a\} \notin \mathcal{F}$ since $P \notin \mathcal{F}$. Clearly $(P \setminus \{a\})^{\Sigma^{\mathcal{F}}} = P^{\Sigma^{\mathcal{F}}}$ since $(P \setminus \{a\})^{\Sigma^{\mathcal{F}}} \not\subset P$. Let us show that P is not a pseudo-closed set. To do so, we show that $P \setminus \{a\}$ is a quasi-closed set with $(P \setminus \{a\})^{\Sigma^{\mathcal{F}}} = P^{\Sigma^{\mathcal{F}}}$. Let $Y \subset P \setminus \{a\}$ then $Y^{\Sigma^{\mathcal{F}}} \subset P$ or $Y^{\Sigma^{\mathcal{F}}} = (P \setminus \{a\})^{\Sigma^{\mathcal{F}}} = P^{\Sigma^{\mathcal{F}}}$. Suppose $Y^{\Sigma^{\mathcal{F}}} \subset P$. Then $Y^{\Sigma^{\mathcal{F}}} \subset P \setminus \{a\}$ by the bijection φ. We conclude that $P \setminus \{a\}$ is a quasi-closed set and then $P \to P^{\Sigma^{\mathcal{F}}} \notin \Sigma^{\mathcal{F}}_{\downarrow}$.

Second, suppose condition (ii') is not satisfied. We have two cases :
- If $a \in P$, then using condition (i), we have $b \in P$ and so $a, b \in P^{\Sigma^{\mathcal{F}}}$.
- If $a \notin P$, then there exists $P \to P^{\Sigma^{\mathcal{F}}}$ such that $a \notin P$, $a \in P^{\Sigma^{\mathcal{F}}}$ and $b \notin P^{\Sigma^{\mathcal{F}}}$. But $a \notin P$ and $b \notin P^{\Sigma^{\mathcal{F}}}$ implie $a \notin P^{\Sigma^{\mathcal{F}}}$ (since φ is a bijection). This leads to a contradiction.

We conclude that if $\varphi : \mathcal{A}_* \to \mathcal{A} : \varphi(F) = F \cup a$ is a bijection then (i) and (ii) are satisfied.
- Let (i) and (ii') are satisfied.

Suppose $F \in \mathcal{A}$. If $(F \setminus a) \notin \mathcal{F}$ then there exists $P \to P^{\Sigma^{\mathcal{F}}} \in \Sigma^{\mathcal{F}}$ such that $P \subseteq (F \setminus a)$, with $a \in P^{\Sigma^{\mathcal{F}}}$ and $b \notin P^{\Sigma^{\mathcal{F}}}$. This contradicts condition (ii').

Suppose $F \in \mathcal{A}_*$. If $F \cup a \notin \mathcal{F}$ then there exists $P \to P^{\Sigma^{\mathcal{F}}} \in \Sigma^{\mathcal{F}}$ such that $P \subseteq (F \cup a)$. Since $P^{\Sigma^{\mathcal{F}}} \not\subseteq F$, we have $a \in P$. Thus $a \in P$ and $b \notin P$, which contradicts condition (i).   □

Lemma 2 shows that condition (iii') is equivalent to testing if $A \cup (A_* \cup B)^{\Sigma^{\mathcal{F}}} \in \mathcal{F}$.

Lemma 2. *Let $\mathcal{F}$ a closure system such that conditions (i) and (ii') of lemma 1 are satisfied. Then condition (iii') of property 1 is satisfied if and only if $A \cup (A_* \cup B)^{\Sigma^{\mathcal{F}}} \in \mathcal{F}$.*

Proof. Suppose $A \cup (A_* \cup B)^{\Sigma^{\mathcal{F}}} \in \mathcal{F}$ and there exists $P \to (ab)^{\Sigma^{\mathcal{F}}}$ with $a \in P$. By condition (i), we have $\{a, b\} \subseteq P$. Moreover $A \cup (A_* \cup B) \subseteq P$, since P is closed by $\Sigma^{\mathcal{F}}_J$. We distinguish two cases :

- If $a \notin (A_* \cup B)^{\Sigma^{\mathcal{F}}}$, then $(A_* \cup B)^{\Sigma^{\mathcal{F}}} \subset P$ (by definition of pseudo-closed set). We deduce that $A \cup (A_* \cup B)^{\Sigma^{\mathcal{F}}} \subseteq P \subset A \cup (A_* \cup B)^{\Sigma^{\mathcal{F}}}$, which is impossible.
- If $a \in (A_* \cup B)^{\Sigma^{\mathcal{F}}}$, then $A \cup (A_* \cup B) \subseteq P$. Let $Q = P \setminus a$. We have $(A_* \cup B) \subseteq Q$, and $(A_* \cup B)^{\Sigma^{\mathcal{F}}} \subseteq Q^{\Sigma^{\mathcal{F}}}$. Thus, $Q^{\Sigma^{\mathcal{F}}} = (ab)^{\Sigma^{\mathcal{F}}}$.

  Moreover, for all $X \in Q$, if $a \in X^{\Sigma^{\mathcal{F}}}$ then $X^{\Sigma^{\mathcal{F}}} = (ab)^{\Sigma^{\mathcal{F}}}$ (by condition ii). If $a \notin X^{\Sigma^{\mathcal{F}}}$, then $X^{\Sigma^{\mathcal{F}}} \subset P \setminus a$. We conclude Q is a quasi-closed set such that $Q^{\Sigma^{\mathcal{F}}} = P^{\Sigma^{\mathcal{F}}}$, but then, P is not a pseudo-closed set. This is in contradiction with the hypothesis.

Conversely, suppose that condition (iii') is satisfied, and $A \cup (A_* \cup B)^{\Sigma^{\mathcal{F}}} \notin \mathcal{F}$. Then there exists $P' \subset A \cup (A_* \cup B)^{\Sigma^{\mathcal{F}}}$ such that $P'^{\Sigma^{\mathcal{F}}} \subset (A \cup (A_* \cup B)^{\Sigma^{\mathcal{F}}})^{\Sigma^{\mathcal{F}}} = (ab)^{\Sigma^{\mathcal{F}}}$.

Clearly $a \in P'$, otherwise $P'^{\Sigma^{\mathcal{F}}} \subseteq (A_* \cup B)^{\Sigma^{\mathcal{F}}}$. Thus $b \in P'$ (by condition (i)) and $ab \subset P'$. This implies $(ab)^{\Sigma^{\mathcal{F}}} \subseteq P'^{\Sigma^{\mathcal{F}}}$.

From above, we deduce that $P'^{\Sigma^{\mathcal{F}}} = (ab)^{\Sigma^{\mathcal{F}}}$, which is in contradiction with hypothesis. □

Now we are able to give the main theorem of this paper.

Theorem 4. *Let $\mathcal{F}$ and $\mathcal{F}'$ be two closure systems in $\mathcal{C}_\downarrow(\mathcal{F})$, then $\mathcal{F}' \prec \mathcal{F}$ iff the three conditions are satisfied:*

(1) There exist $a, b \in J$ such that $b \notin a^{\Sigma^{\mathcal{F}}}$, $b \in a^{\Sigma^{\mathcal{F}'}}$.
(2) The mapping $\varphi : \mathcal{A}_* \to \mathcal{A}$ with $\varphi(F) = F \cup a$ is a bijection.
(3) $A \cup (A_* \cup B)^{\Sigma^{\mathcal{F}}} \in \mathcal{F}$.

Proof. The proof follow directly from lemmas 1 and 2, which are transcription of conditions in proposition 1, and theorem 3 guarantees the cover relation. □

Theorem 4 gives necessary and sufficient conditions to have $\mathcal{F}' \prec \mathcal{F}$ in $\mathcal{C}_\downarrow(\mathcal{F})$. Next section will illustrate this theorem on an example.

5   Illustration of Theorem 4

Consider the closure system $\mathcal{F} = \{\emptyset, a, \ b, \ c, \ ac, \ bc, \ abcd\}$ on the set $J = \{a, b, c, d\}$. The GD-base of $\mathcal{F}$, is $\Sigma^{\mathcal{F}} = \{ab \to abcd, d \to abcd\}$, with $\Sigma^{\mathcal{F}}_\downarrow = \{ab \to abcd\}$ and $\Sigma^{\mathcal{F}}_J = \{d \to abcd\}$.

To apply theorem 4, we first search for a pair $(x, y) \in J^2$ such that $\Sigma^{\mathcal{F}} \not\vdash x \to y$ (see condition (1)). In this case we have the following implications: $(a, c), (a, b), (a, d), (b, a), (b, c), (b, d), (c, a), (c, b)$ and (c, d). Indeed, the pairs $(d, a), (d, b)$ and (d, c) cannot be chosen since condition (1) is not satisfied.

– Suppose, we choose the implication $a \to c$. First we compute the following sets : $A = \{a\}$, $A_* = \emptyset$, $C = \{c\}$, $\mathcal{A} = \{\{a\}\}$ and $\mathcal{A}_* = \{\emptyset, \{b\}\}$.

  Clearly condition (2) is not satisfied and thus Theorem 4 cannot be applied. Indeed, the addition of the implication $a \to c$, changes $\Sigma^{\mathcal{F}}_\downarrow$, i.e. the pseudo-closed set ab becomes abc.

  Notice that the implication $a \to c$ cannot be reconsidered another time.
– For the implication $a \to b$, we have $A = \{a\}$, $A_* = \emptyset$, $B = \{b\}$, $\mathcal{A} = \{a, ac\}$ and $\mathcal{A}_* = \{\emptyset, \{c\}\}$.

  Condition (2) is satisfied. Now we check condition (3). We have $A \cup (A_* \cup B)^{\Sigma^{\mathcal{F}}} = ab$, which is not closed set of $\mathcal{F}$. Thus condition (3) is not satisfied.

  Here, the addition of the implication $a \to b$, changes the pseudo-closed set ab by a. That is ab remains a quasi-closed set, but not minimal.

– Now, consider the implication $c \to a$. We have $C = \{c\}$, $C_* = \emptyset$, $A = \{a\}$, $\mathcal{C} = \{\{c\}, \{bc\}\}$ and $\mathcal{C}_* = \{\emptyset, \{b\}\}$.

   Conditions (2) and (3) are satisfied, since $C \cup (C_* \cup A)^{\Sigma^{\mathcal{F}}} = ac$ is closed set of $\mathcal{F}$. Thus the closure system $\mathcal{F}' = \{\emptyset, a, b, ac, abcd\}$ is $\Sigma_\downarrow^{\mathcal{F}}$-equivalent and covered by $\mathcal{F}$, with $\Sigma_J^{\mathcal{F}'} = \{d \to abcd, c \to ac\}$.

   Now given $\mathcal{F}'$ we try to apply theorem 4.

– Consider the implication $c \to b$. We have $C = \{ac\}$, $C_* = \{a\}$, $B = \{b\}$, $\mathcal{C} = \{\{ac\}\}$ and $\mathcal{C}_* = \{\{a\}\}$.

   Conditions (2) and (3) are satisfied. Thus the closure system $\mathcal{F}'' = \{\emptyset, a, b, abcd\}$ is $\Sigma_\downarrow^{\mathcal{F}}$-equivalent and covered by $\mathcal{F}'$, with $\Sigma_J^{\mathcal{F}''} = \{d \to abcd, c \to abcd\}$.

   For all remaining pairs, theorem 4 cannot be applied. Thus we conclude that $\mathcal{F}'' = \mathcal{F}_\perp$.

Thus given a closure system $\mathcal{F}$, theorem 4 shows how to compute a $\Sigma_\downarrow$-equivalent closure system $\mathcal{F}'$ such that $\mathcal{F}$ covers $\mathcal{F}'$ in $\mathcal{C}_\downarrow(\mathcal{F})$. To obtain a minimal closure system $\Sigma_\downarrow$-equivalent to $\mathcal{F}$, we repeat the application of theorem 4 following a path in $\mathcal{C}_\downarrow(\mathcal{F})$ until we reach the bottom.

   In the following section we show how to check conditions of theorem 4 using the meet-irreducible elements $\mathcal{M}(\mathcal{F})$, and how compute the meet-irreducible elements of $\mathcal{F}'$ from the meet-irreducible elements of $\mathcal{F}$.

6   Detecting $\Sigma_\downarrow$-Equivalence Using Meet-Irreducible Elements

Notice that for any $X \subseteq J$, $X^{\Sigma^{\mathcal{F}}} = \bigcap \{M \in \mathcal{M}(\mathcal{F}) \mid X \subseteq M\}$. So computing the closure of any set is polynomial in the size of $\mathcal{M}(\mathcal{F})$. Thus checking condition (1) and (3) of theorem 4 can be done in polynomial time.

   Thus, the major difficulty is to test condition (2). Our idea is based on the fact that $\mathcal{A}_* \cup J$ and $\mathcal{A} \cup J$ are closure systems. So checking if φ is a bijection can be done in polynomial time if we can compute polynomial representation of $\mathcal{A}_* \cup J$ and $\mathcal{A} \cup J$ from $\mathcal{M}(\mathcal{F})$.

   Consider the set $\mathcal{M}(\mathcal{F})$ and $a, b \in J$. Our strategy is:

1. Compute a representation of the set $\mathcal{A}_* \cup \mathcal{A}$
2. Compute a representation of the sets $\mathcal{A}_*$ and $\mathcal{A}$ from the representation of $\mathcal{A}_* \cup \mathcal{A}$
3. Check if there is a bijection between $\mathcal{A}_*$ and $\mathcal{A}$ using their representation.

   Consider the following sets:

– $\mathcal{M}_{A_* \backslash B}(\mathcal{F}) = \{M \in \mathcal{M}(\mathcal{F}) \mid A_* \subseteq M, b \notin M\}$, i.e. the set of meet-irreducible elements of $\mathcal{F}$ which belongs to $\mathcal{A}_* \cup \mathcal{A}$.
– $\mathcal{R}(\mathcal{A}_* \cup \mathcal{A}) = \mathcal{M}_{A_* \backslash B}(\mathcal{F}) \cup \{M \cap M' \mid M \in \mathcal{M}_{A_* \backslash B}(\mathcal{F}), M' \in \mathcal{M}(\mathcal{F}), b \in M'\}$, i.e. $\mathcal{M}_{A_* \backslash B}(\mathcal{F})$ union closed sets which may become meet-irreducible when removing closed sets containing b. ($\mathcal{R}$ stands for *representation*)

Proposition 2 shows that $\mathcal{R}(\mathcal{A}_* \cup \mathcal{A})$ is a polynomial representation of $\mathcal{A}_* \cup \mathcal{A}$.

Proposition 2. *The closure of $\mathcal{R}(\mathcal{A}_* \cup \mathcal{A})$ under set intersection is $\mathcal{A}_* \cup \mathcal{A}$. Thus, $\mathcal{R}(\mathcal{A}_* \cup \mathcal{A})$ is a polynomial representation of $\mathcal{A}_* \cup \mathcal{A}$.*

Proof. Let $X \in \mathcal{A}_* \cup \mathcal{A}$, we distinguish two cases:

- $X \in \mathcal{M}_{A_* \setminus B}(\mathcal{F})$ then $X \in \mathcal{R}(\mathcal{A}_* \cup \mathcal{A})$ and it is done.
- $X \notin \mathcal{M}_{A_* \setminus B}(\mathcal{F})$. Clearly, X is not a maximal element of $\mathcal{A}_* \cup \mathcal{A}$ since such elements belong to $\mathcal{M}_{A_* \setminus B}(\mathcal{F})$. Thus X has at least one cover in $\mathcal{A}_* \cup \mathcal{A}$. We distinguish two cases:
  1. X has more than one cover in $\mathcal{A}_* \cup \mathcal{A}$. Suppose Y and Z two covers of X. Inductively, Y and Z are intersections of sets in $\mathcal{R}(\mathcal{A}_* \cup \mathcal{A})$. Then X is the intersection of elements in $\mathcal{R}(\mathcal{A}_* \cup \mathcal{A})$.
  2. X has exactly one cover in $\mathcal{A}_* \cup \mathcal{A}$. Let Y be this cover. Then there exist $M \in \mathcal{M}_{A_* \setminus B}(\mathcal{F})$ and $M' \in \mathcal{F}$ such that $A_* \subseteq M'$ and $b \in M'$, with $X \subseteq M \cap M'$ and $Y \not\subseteq M'$, otherwise $X \in \mathcal{M}_{A_* \setminus B}(\mathcal{F})$. Since $b \notin M \cap M'$, we have $M \cap M' \in \mathcal{A}_* \cup \mathcal{A}$. Furthermore $X \subseteq M \cap M'$ and Y is the unique cover of X then $X \subseteq M \cap M' \subset Y$ and then $X = M \cap M'$.

Conversely, let X be an intersection of some sets in $\mathcal{R}(\mathcal{A}_* \cup \mathcal{A})$. Since the sets of $\mathcal{R}(\mathcal{A}_* \cup \mathcal{A})$ are sets of $\mathcal{F}$ containing A_*, we conclude that $X \in \mathcal{A}_* \cup \mathcal{A}$.   □

From proposition 2, we immediately conclude the following corollary:

Corollary 2. $\mathcal{R}(\mathcal{A}) = \{M \in \mathcal{R}(\mathcal{A}_* \cup \mathcal{A}) \mid a \in M\}$ *is a representation of $\mathcal{A}$.*

Now it remains to compute a representation of $\mathcal{A}_*$. Consider the following sets:

- $\mathcal{M}_{\setminus A}(\mathcal{A}_* \cup \mathcal{A}) = \{M \in \mathcal{R}(\mathcal{A}_* \cup \mathcal{A}) \mid a \notin M\}$
- $\mathcal{R}(\mathcal{A}_*) = \{M \cap M' \mid M \in \mathcal{M}_{\setminus A}(\mathcal{A}_* \cup \mathcal{A}),\ M' \in \mathcal{R}(\mathcal{A}_* \cup \mathcal{A}),\ a \in M'\}$

Using the same proof as in proposition 2, we obtain the following corollary:

Corollary 3. $\mathcal{R}(\mathcal{A}_*) = \{M \cap M' \mid M \in \mathcal{M}_{\setminus A}(\mathcal{A}_* \cup \mathcal{A}),\ M' \in \mathcal{R}(\mathcal{A}_* \cup \mathcal{A}),\ a \in M'\}$ *is a representation of $\mathcal{A}_*$.*

Theorem 5. *The mapping $\varphi : \mathcal{A}_* \to \mathcal{A}$ with $\varphi(F) = F \cup \{a\}$ is a bijection iff the two following mapping are injective :*

- $\psi_* : \mathcal{R}(\mathcal{A}_*) \to \mathcal{A}$ *with* $\psi_*(F) = F \cup \{a\}$
- $\psi : \mathcal{R}(\mathcal{A}) \to \mathcal{A}_*$ *with* $\psi(F) = F \setminus \{a\}$

Proof. Suppose that φ is a bijection. Then ψ_* and ψ are injective, since $\mathcal{R}(\mathcal{A}_*)$ and $\mathcal{R}(\mathcal{A}_* \cup \mathcal{A})$ are elements respectively of $\mathcal{A}_*$ and $\mathcal{A}$.

Conversely, suppose ψ_* and ψ are injective. First we show that φ is injective: Let $X, Y \in \mathcal{A}_*$ with $X \neq Y$. Clearly $\varphi(X) \neq \varphi(Y)$ since $a \notin X$ and $a \notin Y$. It remains to show that $\varphi(X), \varphi(Y) \in \mathcal{A}$. Using corollary 3, the sets X and Y can be obtained as intersection of elements in $\mathcal{R}(\mathcal{A}_*)$, i.e. $X = \bigcap_{i=1,k} X_i$ and $Y = \bigcap_{i=1,k'} Y_i$ with $X_i, Y_i \in \mathcal{R}(\mathcal{A}_*)$. Then $\varphi(X) = \{a\} \cup \bigcap_{i=1,k} X_i = \bigcap_{i=1,k} \psi_*(X_i)$. since $\mathcal{A}_*$ is closed under intersection and $\psi_*(X_i) \in \mathcal{A}$, we conclude that $\varphi(X) \in \mathcal{A}$. In a similar way, we show that $\varphi(Y) \in \mathcal{A}$.

Now, we show that φ is surjective. Let $X \in \mathcal{A}$. We distinguish two cases:

- $X \in \mathcal{R}(\mathcal{A})$. Then $\psi(X) \in \mathcal{A}_*$ and therefore φ is surjective since $\psi(X) = \varphi^{-1}(X)$.
- $X \notin \mathcal{R}(\mathcal{A})$. From corollary 2 we have $X = \bigcap_{i=1,k} X_i$ such that $X_i \in \mathcal{R}(\mathcal{A})$, for $i = 1, \ldots, k$. Moreover $\psi(X) = (\bigcap_{i=1,k} X_i) \setminus \{a\} = \bigcap_{i=1,k}(X_i \setminus \{a\}) = \bigcap_{i=1,k} \psi(X_i)$. Thus $\psi(X) = \varphi^{-1}(X) \in \mathcal{A}_*$ since $\mathcal{A}_*$ is closed under intersection.

This conclude the proof.   □

We deduce from theorem 5, a polynomial time algorithm for checking condition (2), from the meet-irreducible elements of $\mathcal{F}$.

Now it remains to us to show how to compute the meet-irreducible elements of $\mathcal{F}' = \mathcal{F} \setminus \mathcal{A}$. It is straightforward to see that the meet-irreducible elements not in $\mathcal{A}$ remain meet-irreducible in $\mathcal{F}'$ and meet-irreducible elements in $\mathcal{A}$ disappear. But, new meet-irreducible elements might appear when removing $\mathcal{A}$ from $\mathcal{F}$. A closed set becomes a meet-irreducible, if all closed sets that cover it are removed but one. Thus, new meet-irreducible element belong to $\mathcal{A}_*$.

Lemma 3 characterizes closed sets which are new meet-irreducible elements in the new closure system $\mathcal{F}'$.

Lemma 3. *Let $M \in \mathcal{A}_*$ be a new meet-irreducible element in $\mathcal{F}'$. Then $M \cup \{a\}$ is a meet-irreducible element in $\mathcal{A}$.*

Proof. Suppose $M \cup \{a\}$ is not a meet-irreducible element in $\mathcal{F}$. By bijection, we have $M \cup \{a\} \in \mathcal{A}$. We distinguish two cases:

- $M \cup \{a\}$ has two covers in $\mathcal{A}$. Then M has two covers in $\mathcal{A}_*$. So M is not a meet-irreducible element in $\mathcal{F}'$.
- $M \cup \{a\}$ has one cover in $\mathcal{A}$ and another cover F containing b. Thus when removing $\mathcal{A}$, M has one cover in $\mathcal{A}_*$ and F becomes a cover of M.

Theorem 6. *The closure system $\mathcal{F}'$ has less or equal meet-irreducible elements than $\mathcal{F}$.*

Proof. From lemma 3, we deduce that $\mathcal{M}(\mathcal{F}')$ is a subset of $\{M \in \mathcal{M}(\mathcal{F}) \mid a \notin M$ or $b \in M\}$ union $\{M \setminus \{a\} \mid a \in M, b \notin M\}$. Thus $\mathcal{M}(\mathcal{F}')$ has at most the same number of meet-irreducible elements as $\mathcal{M}(\mathcal{F})$.

Corollary 4. *There exists a polynomial time algorithm in $\mathcal{M}(\mathcal{F})$ to compute the meet-irreducible elements of a minimal closure system $\mathcal{F}_\perp$.*

The approach is to find an implication $a \rightarrow b$ wich satisfied the conditions of theorem 4. Conditions of this theorem can be checked in quadratic time in $|\mathcal{M}(\mathcal{F})|$. Then, for an implication wich satisfied these conditions, we compute the set of meet-irreducible elements of the new closure system $\mathcal{F}'$. Again, this can be done in quadratic time.

Once the meet-irreducible elements of $\mathcal{F}'$ are computed, we can apply again the process on $\mathcal{M}(\mathcal{F}')$ while implications may be added.

So, this method may be used in goal to speed-up the computation of a Guigue-Duquenne base. In effect, to compute the Guigue-Duquenne base of a closure system $\mathcal{F}$, given the list of its meet-irreducible elements, classical approachs need to compute all the elements in $\mathcal{F}$. The computation time used to do this is proportional to the number of elements of $\mathcal{F}$.

By a preprocessing step which consist in the transformation of the list of $\mathcal{M}(\mathcal{F})$ in $\mathcal{M}(\mathcal{F}_\perp)$, less closed sets will be computed, and so the practical time may be affected. The following framework may be applied:

1. Compute $\mathcal{M}(\mathcal{F}_\perp)$ from $\mathcal{M}(\mathcal{F})$, store the unit implications added in the process.
2. Compute the Guigues-Duquenne base of $\mathcal{F}_\perp$ from $\mathcal{M}(\mathcal{F}_\perp)$. This step may require the computation of $\mathcal{F}_\perp$, and require a time proportional to $\mathcal{F}_\perp$.
3. Remove from $\Sigma^{\mathcal{F}_\perp}$ unit implications added in step one to obtain the Guigues-Duquenne base of $\mathcal{F}$.

Note that even if the computation of $\mathcal{M}(\mathcal{F}_\perp)$ may be "unefficient", i.e with a polynomial complexity with a degree greater than 2, this complexity may be small in comparison with the time used to compute all elements in $\mathcal{F}$.

7 Conclusion

Consider a closure system $\mathcal{F}$ given by its meet-irreducible elements $\mathcal{M}(\mathcal{F})$. We have proposed a polynomial time algorithm to compute $\mathcal{M}(\mathcal{F}_\perp)$, where $\mathcal{F}_\perp$ is a minimal closure system preserving $\Sigma_\downarrow^{\mathcal{F}}$. Clearly, this step can be used to improve the efficiency of algorithms which compute a minimum base from a context.

Another practical application of our result is for closed sets enumeration algorithms. Indeed, instead of generating closed sets of $\mathcal{F}$ from $\mathcal{M}(\mathcal{F})$, we first generate closed sets of $\mathcal{F}_\perp$ from $\mathcal{M}(\mathcal{F}_\perp)$, and use copies to obtain closed sets of $\mathcal{F}$.

Nation and Pogel[8] have shown that the family of closure systems sharing the same Σ_J is a closure system and has nice properties. Obviously, on can ask, if these properties are satisfied in $\mathcal{C}_\downarrow(\mathcal{F})$? Moreover, it gives us a new approach to study the lattice of all closure systems on a set J, using only their representants, i.e. minimal closure systems.

Acknowledgment. The authors are very grateful to the referees for very helpful comments and their remarks on the gaps in the proof (of the previous version) of proposition 1.

References

1. Eiter, T., Gottlob, G.: Identifying the minimal transversals of a hypergraph and related problems. SIAM Journal on Computing **24(6)** (1995) 1278–1304
2. Eiter, T., Gottlob, G.: Hypergraph transversal computation and related problems in logic and ai. In: European Conference on Logics in Artificial Intelligence (JELIA'02). (2002) 549–564

3. Ibaraki, T., Kogan, A., Makino, K.: Inferring minimal functional dependencies in horn and q-horn theories. Technical report, Rutcor Research Report, RRR 35-2000 (2000)
4. Maier, D.: The theory of relational data bases. Computer Science Press, Rockville (1983)
5. Fredman, M.L., Khachiyan, L.: On the complexity of dualization of monotone disjunctive normal forms. Journal of Algorithms (1996) 618–628
6. Guigues, J., Duquenne, V.: Familles minimales d'implications informatives résultant d'un tableau de données binaires. Math. Sci. hum **95** (1986) 5–18
7. Bordalo, G., Monjardet, B.: The lattice of strict completions of a finite poset. Algebra Universalis **47** (2002) 183–200
8. Nation, J.B., Pogel, A.: The lattice of completions of an ordered set. Order **14** (1997) 1–7
9. Ganter, B.: Two basic algorithms in concept analysis. Technical report, No 831, Technische Hochschule Darmstadt (1984)
10. Beeri, C., Berstein, P.: Computational problems related to the design of normal form relational schemas. ACM Trans. on database systems **1** (1979) 30–59
11. Wild, M.: Implicational bases for finite closure systems. Technical report, No 1210, Technische Hochschule Darmstadt (1989)
12. Shock, R.C.: Computing the minimum cover of functional dependencies. Information Processing Letters **3** (1986) 157–159

Spatial Indexing for Scalability in FCA

Ben Martin[1] and Peter Eklund[2]

[1] Information Technology and Electrical Engineering,
The University of Queensland,
St. Lucia QLD 4072, Australia
monkeyiq@users.sourceforge.net
[2] School of Economics and Information Systems,
The University of Wollongong,
Northfields Avenue, Wollongong, NSW 2522, Australia
peklund@uow.edu.au

Abstract. The paper provides evidence that spatial indexing structures offer faster resolution of Formal Concept Analysis queries than B-Tree/Hash methods. We show that many Formal Concept Analysis operations, computing the contingent and extent sizes as well as listing the matching objects, enjoy improved performance with the use of spatial indexing structures such as the RD-Tree. Speed improvements can vary up to eighty times faster depending on the data and query. The motivation for our study is the application of Formal Concept Analysis to Semantic File Systems. In such applications millions of formal objects must be dealt with. It has been found that spatial indexing also provides an effective indexing technique for more general purpose applications requiring scalability in Formal Concept Analysis systems. The coverage and benchmarking are presented with general applications in mind.

1 Introduction

A common approach to document retrieval using Formal Concept Analysis is to convert associations between many-valued attributes and objects into binary associations between the same objects and new attributes. For example, a many-valued attribute showing a person's income may be converted into three attributes: low, middle and upper which are then associated with the same set of people. The method of associating binary attributes is called either conceptual scaling [10] or logical scaling [20] depending on the perspective chosen.

This is the approach adopted in the ZIT-library application developed by Rock and Wille [22] as well as the Conceptual Email Manager [8, 6]. The approach is mostly applied to static document collections (such as newsclassifieds) as in the program RFCA [7] but also to dynamic collections (such as email) as in MAIL-SLEUTH [2] and files in the Logical File System (LISFS) [19]. In all but the latter two the document collection and full-text keyword index are static. Thus, the FCA interface consists of a mechanism for dynamically deriving binary attributes from a static full-text index. Many-valued contexts are used to materialize formal contexts in which objects are document identifiers.

B. Ganter and L. Kwuida (Eds.): ICFCA 2006, LNAI 3874, pp. 205–220, 2006.

The motivation for the application of spatial structures in this research was for the use of Formal Concept Analysis in a virtual filesystem [11, 19, 16]. In particular the libferris [1] Semantic File System. Spatial indexing has been found to bring similar performance improvements to more general formal concept analysis applications: sometimes referred to as Toscana-systems. We show that the spatial method proposed in this paper has performance which depends on the number of attributes in each query as well as the density and distribution of the formal context.

This paper presents some of the solutions to the problems that arise when integrating Formal Concept Analysis (FCA) [10] into a Semantic File System (SFS) [11, 19, 16] and demonstrates performance improvements that result from the implementation of a spatial indexing structure.

2   Existing Indexing Strategies for Formal Concept Analysis

Two designs dominate current Formal Concept Analysis implementations for the indexing of data: either a single large table in a relational database where objects are rows and their attributes form columns (Toscana) [22] or using inverted files (LISFS) [19]. The libferris design is based on the former with extensions to deal with normalization and the association of emblems [15]. An emblem is a pictorial annotation, usually a small icon, that is associated with an file or directory. An emblem often denotes a category.

Shown in Fig. 1 is an example inverted file index. With an inverted file index values of interest each have a list of the address of the tuples from the origin of the base table. For example, an inverted file index on a *name* column would have a list for the value "peter" with pointers to all the tuples where the name column was "peter". Inverted files work well when there are a limited number of values of interest. Given an inverted file defined such that the values of interest are formal attributes and a concept with intent $\{10000, 01000\}$ one must combine the lists for 10000 and 01000 to list the extent of that concept.

We now focus on systems using relational databases for data storage and indexing.

Assuming, without loss of generality, that the many-valued context is available – denormalized in a single relation which we refer to as the base table. This base table having columns $\{c_1, c_2, ...c_y\}$. As a concrete example, consider a base table with 4 numeric columns $c_1 = size$ and $c_2 = modified$, $c_3 = accessed$ and $c_4 = file\text{-}owner$. Although the modified and accessed columns are numeric they are presented here in a human readable form. As an example consider three ordinal scales on the columns c_1, c_2 and c_3 and a nominal scale on c_4 (see Fig. 3).

More generally for the base relation we consider a formal attribute $\{a_j\}$ to be defined through possible values for one or more columns $\{c_i, ...c_u\}$. It can be convenient to consider the definition of an attribute $\{a_j\}$ as an SQL condition f_j on the values of one or more columns $\{c_i, ...c_u\}$. Thus for all $i \in \{1...j\}$ the formal attribute a_i is defined by the SQL expression f_i on the base table. The

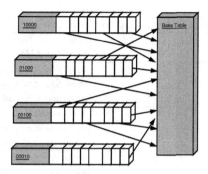

Fig. 1. An inverted file index. For each value of interest there is a list containing all the addresses of tuples which match that value.

object-ID	c_1 size	c_2 modified	c_3 accessed	c_4 file-owner
1	4096	today	today	ben
2	800	yesterday	today	peter
3	400k	1 year ago	last week	ben
...	...	...	...	

Fig. 2. Example base relation containing modification and size data for objects

convenience of using SQL expressions f_j to define the formal attributes a_j is due to the SQL expression returning a binary result. Note that there is a one-to-one correspondence between A and F, every formal attribute is defined by an SQL expression. The number of attributes $|A|$ can vary from the number of columns $|C|$ in the database. The a_x, f_x and c_y are shown in Fig. 3. For example, from in Fig. 3 an attribute a_x might be defined on the columns $\{c_2, c_3\}$ using the SQL expression f_x = `modified < last week AND accessed > yesterday`[1]. Such an attribute would have an attribute extent containing all files which have been accessed today but not modified this week.

Due to the generality of the terms attribute and value some communities use them to refer to specific concepts which are related to the above uses. For example the term attribute in some communities would more naturally refer to the c_i. The above terminology was selected to more closely model Formal Concept Analysis where the formal attributes are binary. Thus the (formal) attributes are modeled as the a_i.

Consider finding the extent of a concept which has attributes $\{a_1, a_3, a_7\}$. The SQL query is formed with an SQL **WHERE** clause as "...where f_1 and f_3 and f_7 ...". For our concrete example, the SQL predicate will be "...where *size* $<= 4096$ and *modified* $<=$ *this week* and *accessed* $<=$ *yesterday* ...".

---

[1] Date values represented as human readable strings in this example.

Attribute	Columns involved	SQL predicate (f_x)
a_1	c_1	*size <= 4096*
a_2	c_1	*size <= 1Mb*
a_3	c_2	*modified <= this week*
a_4	c_2	*modified <= yesterday*
a_5	c_2	*modified <= today*
a_6	c_3	*accessed <= last week*
a_7	c_3	*accessed <= yesterday*
a_8	c_3	*accessed <= today*
a_9	c_4	*file-owner = ben*
a_{10}	c_4	*file-owner = peter*
a_{11}	c_4	*file-owner = foo*
...	...	...

Fig. 3. Ordinal scales on the size, modification and access times of the objects in the base table. Nominal scale on the file-owner.

Current best practice in the Formal Concept Analysis community attempts to assist such queries with B-Tree indexes over subsets of $\{c_1, c_2, ..., c_y\}$. We now discuss how relational databases use B-Tree indexes during query execution.

A common implementation of relational database queries is to check to see if the use of an index is estimated at returning a percent of the base table which is below a given internal threshold [24]. For example, if the use of an index results in 30% of the tuples in the base table being fetched then the database elects not to use that index. If there are no other indexes available for the query then it will sequentially scan the base table to resolve the query. When fetching a large proportion of the base table a sequential scan is usually faster than using the index because the table can be read in order [9]. The estimated ratio of matching tuples is called the *selectivity*. The key to efficient query execution is therefore for the query to be able to use an index which will sufficiently narrow the number of tuples fetched to make index usage attractive.

The selectivity of an index is estimated for the values given in the SQL predicate using statistics of how many tuples will match the given value or value range. For example, if 60% of column c_3 has values below 20 and the SQL predicate is $c_3 < 20$ then an index on column c_3 would be considered unattractive in the resolution of the query because it is not selective enough on average to outperform a sequential scan. The selectivity can be more formally defined as 100×estimated tuple count/size of base table. Thus lower numeric selectivity values are considered "better" in retrieval terms. A relational database's query planner will prohibit the use of all indexes which have an estimated selectivity beyond a predetermined sequential scan cutoff value.

When there are two predicates in the *where* clause commonly the predicate which has an available index with the best selectivity is chosen first. After this initial index selection the other predicate is used as a filter on the tuples as they are read from the base table [24]. This query design strategy works ineffectively

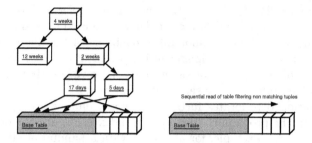

Fig. 4. On the left: B-Tree index on a date column for the base table. Dates in nodes are shown as how long before the current time they represent. The upper nodes are index nodes with the nodes below "12 weeks" omitted. The 17 and 5 days nodes are leaf nodes of the index which point at records in the base table. The B-Tree has a restricted branching factor of two children for illustration purposes. On the right: Resolving the query by a sequential scan filtering out non matching tuples.

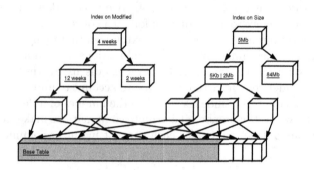

Fig. 5. Two B-Tree indexes on the base table. The left index is on *modified* while the right index is on *size*. Leaf nodes in both indexes point to tuples physically located throughout the base table.

on typical Formal Concept Analysis SQL queries because there is usually more than one predicate joined with a logical *AND*. In the normal case, the selectivity of either predicate will be beyond the query planner's sequential scan cutoff.

When both predicates are considered together a single index over multiple columns may be used in an attempt to achieve better selectivity. For an index created over multiple columns only the leading columns specified in a predicate are considered when computing the number of matching tuples using the index.

Consider an example first. When we seek the size of the concept with intent $\{a_1, a_5, a_{11}\}$ we will have 3 predicates $size <= 4096$, $modified <= today$ and $file-owner = foo$ respectively. These SQL predicates are operating on the columns $\{c_1, c_2, c_4\}$. Assume that an index is created over $\{c_1, c_2, c_3, c_4\}$ to assist this query. Most relational databases do not consider any terms from the predicate which are not contiguous leading terms in the index when calculating the selectivity of an index [24]. Nothing in the query makes reference to c_3 so only the predicates $size <= 4096$, $modified <= today$ will be used to compute selectivity. For this example the index cannot take advantage of the file-owner predicate which may in this case offer a significant improvement to selectivity. Given that the use of the column c_2 will not significantly improve selectivity the use of the whole index deteriorates to the selectivity of c_1 alone.

This situation deteriorates further the more columns are available in the relation due to the probability that leading index terms are not present in the query predicate. For example, for a concept with a handful of attributes in its intent, say $\{\{o_1, o_2\}, \{a_1, a_2, a_3, a_4\}\}$, the chance of having at least one attribute a_x, which happens to have a f_x referring to a column in the index's leading terms, is low. Even with a reference to a leading index term it is unlikely that the reference will be very selective by itself. It is a particular strong point of spatial access methods that they gracefully handle such unreferenced columns on a many column index.

When resolving an SQL query against a base table most relational databases will only consider using a single index [24]. If one considers the possibility of creating a custom index to assist queries for each concept, there are potentially $|C| = 2^{|A|}$ concept intents for a formal context. Given that many f_x will reference the same column, the number of unique combinations of columns from the base table will be less than this number. However, as discussed, the ordering of the columns in the index may have to be taken into account to improve performance. This ordering of columns in indexes will raise the number of indexes needed back towards $|C|$, however, the number of attribute combinations makes it is infeasible to create custom B-Tree indexes for each concept intent or column order.

3   Spatial Indexing for Formal Concept Analysis

We now turn to the application of spatial methods to improve index utilization in query resolution. First we consider using indexes on SQL expressions and then show how spatial methods can be applied to expression indexes to improve performance.

Many relational databases allow the creation of indexes on expressions [3]. For example, given a column *name* an expression index can be created on *lower(name)* to help case insensitive searches. Turning to Formal Concept Analysis one can define an expression index e_x for each respective SQL predicate f_x. Consider again our example from Fig. 3. The expression index e_1 on attribute a_1 is shown in Fig. 6. In an expression index tuples which do not satisfy the index expression are not added to the index.

Turning to the application of expression indexes to Formal Concept Analysis. The indexes $\{e_1, e_2, ..., e_n\}$ having been defined by scales $\{f_1, f_2, ..., f_n\}$ are an implementation artifact which is equivalent to the formal attributes $\{a_1, a_2, ..., a_n\}$ of the formal context. Thus queries on an attribute a_x become queries against the respective index e_x. This allows the materialization of binary attributes from the base table using indexes alone. Creating expression indexes on the f_i expressions does not change the problem of the query planner ignoring such indexes $\cup_1^n e_n$ due to selectivity constraints, highlighted in Section 2. One can however consider indexing structures over the collected $\{e_1, ..., e_n\}$ indexes.

The use of spatial indexing structures over $\{e_1, ..., e_n\}$ can provide substantial increases in FCA query performance. If expression indexes are created for each attribute then above queries such as $Q = \{e_1, e_5, e_{11}\}$ can be classified as a subset query [14] on the expression indexes. In a subset query over $\{e_1, ..., e_n\}$ the objective is to seek all objects with a given subset $S \subseteq \{e_1, ..., e_n\}$ of specified values. For example, the query seeking "... *size* $<= 4096$ and *modified* $<=$ *this week* and *accessed* $<=$ *yesterday* ..." specifies objects matching $S = \{e_1, e_3, e_7\}$.

An indexing structure motivated by the spatial indexing structure, the R-Tree [12, 18], caters for subset queries: the RD-Tree [13, 25]. A particular strong point of these structures is that they index multiple columns in arbitrary order and gracefully handle lookups given a subset of the indexed columns. We first describe the R-Tree followed by the RD-Tree.

The internal nodes in an R-Tree structure contain entries of the form; (bounding n-dimensional box, page pointer), where pages in the subtree reached by page pointers are within the given bounding n-dimensional box (see Fig. 7). This transitive containment relation is the heart of the R-Tree. R-Trees are not limited to 2 or 3 dimensional data but typically use page sizes allowing branching factors much closer to B-Trees than shown in the example.

Searching for a spatial object in the R-Tree starts at the root node and considers all children whose bounding box contains the query object. Searching for the query object in Fig. 7 begins at the root node (R) – the left node (C1) has a bounding box not containing the query object so only the right child (C2) is followed. In turn, the new left node (C2.1) contains the query object and will be followed whereas (C2.2) is not. At the lowest level (the children of C2.1) many nodes may contain the query object and these are followed to retrieve tuples in the base table.

The RD-Tree operates similarly by treating input as an n-dimensional binary spatial area. The R-Tree notion of containment is replaced by set inclusion and

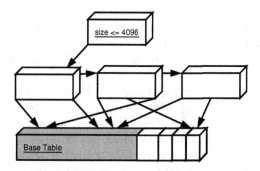

Fig. 6. Expression index on attribute a_1 using f_1, the SQL predicate *size* $<= 4096$

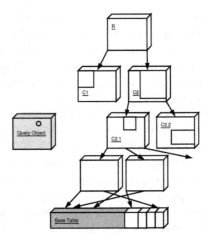

Fig. 7. An example R-Tree with a query object on the left. Each node has a bounding box which fully contains all objects in its child nodes. An implementation stores the bounding box for each child in the parent node. Note the example is limited to 2 dimensional space with a low branching factor for presentation purposes.

the bounding n-dimensional box replaced by a bounding set. The union of a collection of sets forms the bounding set. The bounding set of a child is thus defined as the union of all the elements in the child. The bounding set defined in this way preserves the "containment" notion of the R-Tree during search as a subset relation. When seeking an element which might be in a child it is sufficient to test if the sought element is a subset of the bounding set for the child to know if that subtree should be considered.

An example RD-Tree is shown in Fig. 8. Searching for the query object 01100 starts at the root node discarding (C1) because it does not contain the query object and only following the (C2) child. At (C2) the node (C2.2) is not followed because it does not contain the query object and only (C2.1) is followed. The child (C2.1.2) has a bounding set 00110 which does not contain the query object and is not considered. Only (C2.1.1) matches this query and its contents are tested against the query object to retrieve the results from the base table.

The two main Formal Concept Analysis queries that an RD-Tree can improve are subset and overlap queries [14, 13]. As described above a subset query seeks all objects for which the query object is a subset. For example the query object might be $Q = \{e_1, e_3, e_7\}$ and a matching object $o_i \in O = \{a_1, a_3, a_6, a_7\}'$. For a given set of attributes $A = \{a_1, a_2, ..., a_n\}$ defined by their respective index expressions $E = \{e_1, e_2, ..., e_n\}$ a bitset can be derived $\{b_1, b_2, ..., b_n\}$ such that b_x is set to true when $e_x \in E$ is true. Thus for the example in Fig. 8 we are seeking the query object 01100 which means we want all objects where e_2 and e_3 are true, which is the same as having the formal attributes $\{a_2, a_3\}$.

To resolve a subset query the RD-Tree is walked from the root eliminating any branches with a bounding set which is a subset of the query set. It is apparent

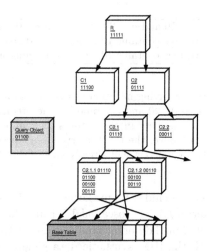

Fig. 8. An example RD-Tree with a query object on the left. Each node has a bounding set associated which fully contains all objects in its child nodes. An implementation would store the bounding set for each child in the parent node. Note that the example is limited to only a small set size with a low branching factor in the tree for presentation.

that the more items from $\{e_1, ..., e_n\}$ specified in the query the less of the index structure will be searched. The trend for RD-Trees is the inverse of that of inverted files. To resolve the above with inverted files the lists for each e_x would have to be fetched and merged. For our same query object 01100 we would have to fetch the lists for 01000 and 00100 to form the set intersection and finally fetch the records from the base table (see Fig. 1).

An overlap query seeks objects which have more than a given number of attributes in common with the query [25]. To efficiently find the contingent size the RD-Tree index must also contain the hamming weight of the binary expression indexes $\cup_1^n e_n$ (ie. formal attributes) which are indexed. The hamming weight for a bitset is the number of bits which are not zero. This is so objects that are in the extent (but not the contingent) can be quickly filtered from the result using the index alone.

The specialized overlap query Q contains the attributes $Q \subseteq \{e_1, e_2, ...e_n\}$ which define the exact attributes sought in the result set. The above subset query would not return object $o_i \in O = \{a_1, a_3, a_6, a_7\}'$ for $Q = \{e_1, e_3, e_7\}$ because attribute a_6 was not specified in the query. It can be seen that o_i would be in the extent of a concept with intent $Q = \{e_1, e_3, e_7\}$ but not in the contingent. An example query translation is shown in Fig. 9.

4   Performance Analysis

The benchmark system is an AMD XP-Mobile running at 2.4GHz with 200Mhz FSB, 1Gb of RAM at 400Mhz dual channel cas222. The software versions which

Normal query	$a < 10$ and $a < 20$ and not $(a < 30)$ and not $(a < 40)$
Simple translation	rd-tree contains 10,20 and not rd-tree contains 30 ...
Custom translation	rd-tree contains 10,20 and hamming-weight(rd-tree) $= 2$

Fig. 9. Translating queries involving negation to take advantage of the RD-Tree. This assumes that the attributes 10, 20 and 30 stand for the predicates $a < 10$ and $a < 20$ and $a < 30$ respectively. The weight function returns the number of RD-Tree predicates a tuple contains. So in the above, the third query doesn't need to negate the 30 and 40 predicates because the weight test will already ensure that 30 and 40 are not set.

may effect performance include Linux kernel 2.6.11rc3, gcc 4.0.0 20050308, PostgreSQL 8.0.1, libferris 1.1.50, ToscanaJ 1.5.1 and Java 1.5.0_01.

Testing was completed on 3 different input data sets: various synthetic formal contexts generated with the IBM synthetic data generator [21], the mushroom and covtype databases from the UCI dataset [4] and a formal context derived from the metadata of 67,000 document files [1].

Also, all columns in the databases had single column B-Tree indexes created on them for every column that might be relevant to query resolution. The mushroom database has 16,832 tuples and the covtype table has 581,012 tuples.

A test consists of lodging a collection of SQL queries against the database as a single batch job. Unless otherwise stated these batch tests were completed after the database was shutdown, the filesystem with the database information was unmounted (and remounted) and finally the database started again. This process flushes internal database buffers and the kernel's disk cache. Where tests were not performed under these cases, the terms "cold cache" refer to a setup where all buffers were flushed and the database restarted as above while "hot cache" mean that the queries were performed with no such flushing or database restart.

For various tests the SQL *explain* was used on each query in the batch to see how many sequential table scans were planned for the batch execution. For small datasets a sequential table scan might prove the fastest method to resolve a query (although performance is bound to be linear and thus will not scale well to larger data sets). Other statistics are shown as well such as the selectivity, mean and standard deviation of a column or table. In order to demonstrate how the spatial indexing performs on various densities of formal context for the synthetic datasets the distribution statistics of the attributes in the formal context is shown.

4.1   Performance on the UCI Mushroom Dataset

The attributes used from the mushroom table are shown in Fig. 10. The two columns *bruises* and *capshape* were used in an attribute list in ToscanaJ. As there are eight binary attributes when each distinct value for these two columns is considered there are a total of 256 SQL queries generated.

Column	Value	Selectivity (count)	Selectivity (% of table)
bruises	NO	10080	59.9
bruises	BRUISES	6752	40.1
capshape	KNOBBED	1680	10.0
capshape	CONVEX	7592	45.1
capshape	FLAT	6584	39.1
capshape	BELL	904	5.4
capshape	SUNKEN	64	0.4
capshape	CONICAL	8	0.05

Fig. 10. Selected attributes for the mushroom table and the number of tuples which have the given attribute-value combination

Test type	Cold cache	Hot cache	Sequential scans
B-Tree only	30	18	4
RD-Tree index simple query translation	10	4	1
RD-Tree optimized query translation	1.6	0.4	0

Fig. 11. Times with hot and cold caches to complete queries for 8 attribute list context. Times are in seconds.

As the relation is relatively small this test was also conducted with explicitly hot caches. This should give an indication of performance differences on small data sets modeling the use case of someone interactively creating and modifying scales. The benchmark was obtained by executing a test multiple times in a row and only taking the last batch time. The results are shown in Fig. 11.

There are 2 versions using an RD-Tree to speed execution: a simple translation and a customized query. The simple translation just substitutes SQL operations to consult the RD-Tree and leaves all other query structure identical. This translation is fairly mechanical and does not fully take advantage of the RD-Tree for query resolution. The custom translation version takes advantage of adding to the RD-Tree the additional information of the hamming weight of the index expressions $\cup_1^n e_n$ as described in Section 3.

4.2 Performance on UCI Covtype Dataset

The UCI covtype database consists of 581,012 tuples with 54 columns of data. For this paper two ordinal columns were used: the slope and elevation. Tests were performed by nesting an ordinal scale on elevation inside an ordinal scale on slope using TOSCANAJ 1.5.1. This nesting produced a total of 378 queries against the database. Given that the primary table is 987Mb tests against a hot cache were not performed. The results are shown in Fig. 12. The RD-Tree index takes 2m:17s to create. As there were no explicit negations there was no gain

Test type	Cold cache (mm:sec)	Sequential scans
B-Tree only	56:16	90
RD-Tree index simple query translation	0:42	0

Fig. 12. Times to complete nested scale queries against the covtype database. The nesting is obtained by generating a nested line diagram in ToscanaJ placing an ordinal scale on elevation inside an ordinal scale on slope.

in producing an RD-Tree optimized SQL query as was done for the mushroom database.

4.3   Performance on Semantic File System Data

An index for part of the libferris filesystem was created for 66,936 files. A formal context based on file name components contains 886 formal attributes for these objects. Formal attributes were packed into a single SQL *bit varying* field making the total size of the formal context only 10Mb. For this formal context there are 488 concepts. Benchmarks for querying the size of the extent of each context is shown for both hot and cold caches in Fig. 13. Without the use of a special purpose index structure the database degenerated to a sequential table scan for almost every query.

Test type	Cold cache	Hot cache	Sequential scans
B-Tree only	80	80	487
RD-Tree index	5	1	0

Fig. 13. Times in seconds with hot and cold caches to complete queries

Query Type	Thousands of trans	Time (128)	Time (64)	Time (32)	Time (16)
B-Tree	1	0.8	0.8	0.7	0.7
RD-Tree	1	0.7	0.6	0.6	0.5
B-Tree	10	3.3	3.3	3.1	3.1
RD-Tree	10	2.4	1.4	0.9	0.7
B-Tree	100	27.6	26.8	26.4	26.2
RD-Tree	100	19.2	10.4	6.7	4.5
B-Tree	1000	6:28	6:14	5:50	5:35
RD-Tree	1000	5:56	2:32	1:30	1:18

Fig. 14. Times for query sets against synthetic databases. SQL Explain shows the B-Tree method always electing to disregard all indexes and perform a sequential scan. The RD-Tree query plan always includes zero sequential scans. The number in brackets below the Time column header is the tlen.

To find the contingent sizes without using an RD-Tree using an SQL query per concept is slower overall than using a single table scan and handling the logic in the client. Using a single table scan from a relational database client takes 70 seconds to find every concept's contingent size. Using an RD-Tree index the same operation takes 2.2 seconds. The use of RD-Trees implies a cost of creating the index, for the above example this is 27 seconds. Index creation can happen faster than a client table scan because it is being done inside the database server process and avoids formatting and copying overhead. So the index can be created and used faster than any other method for finding contingent counts.

4.4 Performance on Synthetic Data

The following use synthetic data generated with the IBM synthetic data generator [21]. Parameters include the number of transactions (ntrans), the transaction length (tlen), length of each pattern (patlen), number of patterns (npat) and number of items (nitems). The number of items was fixed at its minimal value of 1000. The tlen, patlen and npats can be varied to change the density of the resulting formal context while the ntrans is useful for testing the scalability of the query resolution.

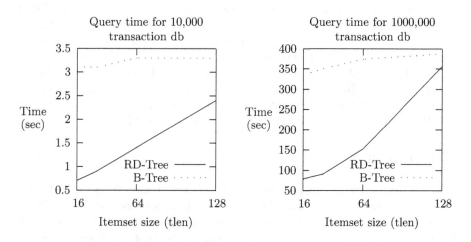

Fig. 15. Execution times for queries using either B-Tree or RD-Tree indexing against databases of varying density

Only the first 32 items were imported into the database. Five values of *tlen* were tested, 256, 128, 64, 32 and 16 at various database sizes ranging from 1,000 to a million transactions.

The query sets were constructed by mining the Closed Frequent Itemsets for the 1,000 transaction database with a minimum support value of 0.01%. The Closed Frequent Itemsets provide the concept intents for an iceberg lattice [23, 17]. This generated 556 concept intents. A query was generated to find the size

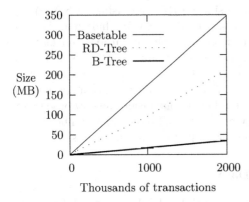

Query	Items/	Max	Mean	Std	Time
Type	pattern	(%)	(%)	Dev	(sec)
B-Tree	256	64	26	19	29.3
RD-Tree					38.9
B-Tree	128	36	13	10	27.6
RD-Tree					19.2
B-Tree	64	19	6	5	26.4
RD-Tree					10.2
B-Tree	32	10	3	3	25.7
RD-Tree					6.7
B-Tree	16	5	1.6	1.4	25.9
RD-Tree					4.8

Fig. 16. (Left) Statistics for the base table and indexes of the synthetic databases. Note that the B-Tree index size is only for a single column whereas the RD-Tree covers all 32 columns. (Right) Effect of formal context density on RD-Tree performance for 100,000 transaction database. The number of items per pattern was reduced in increments from 128 to 16 giving a max, average and standard deviation of set bits in the formal context as shown.

of the extent of each concept. This produced a distribution of 28 single attribute, 224 two attribute, 284 three attribute and 20 four attribute SQL queries. Benchmarks against these datasets are presented in Fig. 14 and graphically in Fig. 15. Size statistics for the DBMS tables and indexes are shown in Fig. 16.

The efficiency of using RD-Trees degenerates as the density of the formal context increases. To measure this effect the number of items per itemset was varied with all other parameters static. Results are shown in Fig. 16. The results with 128 items per itemset are the same as those in Fig. 14.

5   Conclusion

Special indexing structures are essential to FCA systems with large data sets like those encountered in semantic file systems. An index structure derived from spatial indexing for accelerating subset queries has been found to be productive. When the user wishes to list the files matching a concept an RD-Tree permits this within an acceptable time frame for interactive use. The link to spatial indexing structures have not been reported in current best practices elsewhere in the FCA literature [5].

Although the use of spatial indexing was first adopted and implemented to allow Formal Concept Analysis to be applied specifically to the special circumstances encountered by Semantic File Systems, the data structures have also been found to be an advantageous structure for general purpose FCA applications: such as those supported by TOSCANAJ.

The performance of spatial indexing for Formal Concept Analysis in various settings has been examined and shown to provide substantial improvements in many cases. Performance gains from RD-Trees are very effective for sparse formal contexts where queries can be resolved five times faster on large data sets as shown in Section 4.4. The largest benchmark results were found when applied to a large dataset from the UCI collection where the formal context was generated by nesting one conceptual scale inside another. In such an environment queries can be executed over 80 times faster using an RD-Tree than without.

References

1. libferris, http://witme.sourceforge.net/libferris.web/. Visited Nov 2005.
2. Mail-sleuth homepage, http://www.mail-sleuth.com/. Visited Jan 2005.
3. Postgresql, http://www.postgresql.org/. Visited June 2004.
4. Blake, C., Merz, C. UCI Repository of Machine Learning Databases. [http://www.ics.uci.edu/~mlearn/MLRepository.html]. Irvine, CA: University of California, Department of Information and Computer Science, 1998.
5. Claudio Carpineto and Giovanni Romano. *Concept Data Analysis*. Wiley, England, 2004.
6. Richard Cole and Peter Eklund. Analyzing an email collection using formal concept analysis. In *European Conf. on Knowledge and Data Discovery, PKDD'99*, number 1704 in LNAI, pages 309–315. Springer Verlag, 1999.
7. Richard Cole and Peter Eklund. Browsing semi-structured web texts using formal concept analysis. In *Proceedings 9th International Conference on Conceptual Structures*, number 2120 in LNAI, pages 319–332. Springer Verlag, 2001.
8. Richard Cole and Gerd Stumme. Cem: A conceptual email manager. In *7th International Conference on Conceptual Structures, ICCS'2000*. Springer Verlag, 2000.
9. Michael J. Folk and Bill Zoelick. *File Structures*. Addison-Wesley, Reading, Massachusetts 01867, 1992.
10. Bernhard Ganter and Rudolf Wille. *Formal Concept Analysis — Mathematical Foundations*. Springer–Verlag, Berlin Heidelberg, 1999.
11. David K. Gifford, Pierre Jouvelot, Mark A. Sheldon, and James W. Jr O'Toole. Semantic file systems. In *Proceedings of 13th ACM Symposium on Operating Systems Principles*, ACM SIGOPS, pages 16–25, 1991.
12. Antonin Guttman. R-trees: A dynamic index structure for spatial searching. In *Proc. ACM-SIGMOD International Conference on Management of Data*, Boston Mass, 1984.
13. Joseph M. Hellerstein and Avi Pfeifer. The RD-Tree: An Index Structure for Sets, Technical Report 1252. University of Wisconsin at Madison, October 1994.
14. S. Helmer. Index structures for databases containing data items with setvalued attributes, 1997.
15. Ben Martin. File system wide file classification with agents. In *Australian Document Computing Symposium (ADCS03)*. University of Queensland, 2003.
16. Ben Martin. Formal concept analysis and semantic file systems. In Peter W. Eklund, editor, *Concept Lattices, Second International Conference on Formal Concept Analysis, ICFCA 2004, Sydney, Australia, Proceedings*, volume 2961 of *Lecture Notes in Computer Science*, pages 88–95. Springer, 2004.

17. Mohammed J. Zaki, Nagender Parimi, Nilanjana De, Feng Gao, Benjarath Phoophakdee, Joe Urban, Vineet Chaoji, Mohammad Al Hasan and Saeed Salem. Towards generic pattern mining. In Bernhard Ganter and Robert Godin, editors, *Concept Lattices, Third International Conference on Formal Concept Analysis, ICFCA 2005, Proceedings*, Lecture Notes in Computer Science, pages 1–20, Lens, France, 2005. Springer.

18. Hans-Peter Kriegelm Ralf Schneider Norbert Beckmann and Bernhard Seeger. The r*-tree: An efficient and robust access method for points and rectangles. In *Proc. ACM-SIGMOD International Conference on Management of Data*, Atlantic city, N.J., 1990.

19. Yoann Padioleau and Olivier Ridoux. A logic file system. In *USENIX 2003 Annual Technical Conference*, pages 99–112, 2003.

20. Susanne Prediger. Logical scaling in formal concept analysis. In *International Conference on Conceptual Structures*, pages 332–341. Springer, 1997.

21. R. Agrawal, H. Mannila, R Srikant, H. Toivonen and A. Inkeri Verkamo. Fast discovery of association rules. In U. Fayyad et al., editor, *Advances in Knowledge Discovery and Data Mining*, pages 307–328, Menlo Park CA, 1996. AAAI Press.

22. T. Rock and R. Wille. Ein TOSCANA-erkundungssytem zur literatursuche. In G. Stumme and R. Wille, editors, *Begriffliche WissensveraRbeitung: Methoden und Anwendungen*, pages 239–253, Berlin-Heidelberg, 2000. Springer-Verlag.

23. Gerd Stumme, Rafik Taouil, Yves Bastide, Nicolas Pasquier, and Lotfi Lakhal. Computing iceberg concept lattices with titanic. In *J. on Knowledge and Data Engineering (KDE)*, volume 42, pages 189–222, 2002.

24. Dan Tow. *SQL Tuning*. O'Reilly & Associates, Sebastopol, California, 2004.

25. Woo Suk Yang, Yon Dohn Chung, and Myoung Ho Kim. The rd-tree: a structure for processing partial-max/min queries in olap. *Inf. Sci. Appl.*, 146(1-4):137–149, 2002.

Homograph Disambiguation Using Formal Concept Analysis

L. John Old

School of Computing, Napier University, Edinburgh, UK
j.old@napier.ac.uk

Abstract. Homographs are words with identical spellings but different origins and meanings. Natural language processing must deal with the disambiguation of homographs and the attribution of senses to them. Advances have been made using context to discriminate homographs, but the problem is still open. Disambiguating homographs is possible using formal concept analysis. This paper discusses the issues, illustrated by examples, using data from Roget's Thesaurus.

Keywords: Type-10 chains, partitions, components, Roget's Thesaurus, plus operator, word fields, neighbourhood lattices.

1 Introduction

We use formal concept lattices [12] to extract and visualize ambiguous words (homographs) and senses from a lexicon, then use the results to identify whether the ambiguity of the words was resolved by partitions in the lattices. We then compare the results with previous attempts to disambiguate such words using Type-10 components [2], and analyse and discuss identified exceptions.

2 Homographs

Homographs[1] are words with identical spellings but different origins and meanings. These differences are made explicit in lexicons using headword numbers. The senses of a word are then identified under each headword.

1 bat
n. 1. bat [a club]; 2. bat [sports equipment]
vb. bat [to hit using a bat]
2 bat
n. bat [a flying mammal]
...

Also used are etymology (word history) and part-of-speech such as noun (n.) and verb (vb.), as in the example above. It is the lexicographer's judgment as to how many

---

[1] In linguistics the hypernym of homograph is homonym (meaning same name) and includes homophones, words of different meaning but which sound the same when spoken. An example is bore and boar.

R. Missaoui and J. Schmid (Eds.): ICFCA 2006, LNAI 3874, pp. 221–232, 2006.

divisions are made of a headword. The goal is to disambiguate words that look the same but are semantically distinct.

Headwords in lexicons may frequently be found differentiated numerically though they have a common etymological ancestry. In such systems *bat vb.* in the example above would be given a listing as a separate headword. This paper uses a strict definition of a homograph: that two or more words, though spelled the same, have different etymologies. For example, contract (with an emphasis on the first syllable, and meaning an agreement or to make an agreement) and contract (with an emphasis on the second syllable, and meaning to reduce in size), both derive from L. *contractus*, pp. *contrahere*, to draw together. Though these words have quite different meanings, they are not homographs under a strict definition, but cognates (related by descent from the same roots in some ancestral language). Bat1 (club) and bat2 (mammal) in the given example are indeed homographs under the strict definition. Bat1 is derived from L. *battuere* to beat (and, incidentally, cognate with battle and combat), while bat2 comes to English most likely from Old Swedish *nattbakka* (via early Viking settlers [13]).

All words used as examples in this paper were disambiguated etymologically by using their semantic roots. Apart from rare loan words borrowed from non-Indo-European language, the semantic roots used are all Indo-European (IE) roots. Note that all IE roots are hypothetical. They are sometimes referred to as proto-Indo-European roots and prefixed by a *, and are all derived by comparing words and their senses from Indo-European languages such as Latin, Greek and Sanskrit; and modern Icelandic, Hindi, Russian and Iranian (among others).

Note also that more than seventy homographs have more than two expressions. As an example, *pa* has three:

Homograph	Meaning
Pa 1	Protactinium
pa 2	father
pa 3	Maori stronghold

Pa 1 is an abbreviation of a chemical element; pa 2 is an English dialect word synonymous with father; and pa 3 is a foreign loan word. Though homographs with more expressions could be assumed to cause greater difficulties for disambiguation, usually they are rare words with low polysemy and generally easier to differentiate. It is the highly polysemous words (those with many sense variations) that provide the greatest difficulty for disambiguation.

3   Research

Natural language processing (NLP) must deal with the disambiguation of homographs and the attribution of senses to them. Advances have been made by employing context and systems such as N-gram taggers, Bayesian classifiers, vector space models with neural networks, and decision trees ([15],[16]) to discriminate homographs, but the problem is still open. Work has also been done using statistical models of Roget's Thesaurus categories to disambiguate word senses [14]. Treating words as objects and their senses as attributes, identifying words with disjoint sets of senses is possible using formal concept analysis (FCA) [17]. The assumption made and tested in this paper is that this is also applicable to differentiating homographs.

This research manually identified approximately 600 homographs using the strict definition. These were used as a benchmark to compare two partitioning methods, Type-10 chains ([2], [5], [9], [10]) and FCA lattices [12]. The test data was derived from Roget's Thesaurus [1], a semantic dictionary organized by concepts rather than words.

552 homographs were identified in RIT. Of those homographs, 179 either occur in senses by themselves or are the sole representatives of a homograph (the ambiguous partner does not occur in Roget's Thesaurus). For example Nice, a homograph of nice (likeable) appears only in an RIT list called Principle Cities of the World.[2] Not sharing any senses with other words, such homographs are already partitioned and cannot participate in Type-10 chains. Nice could also be easily differentiated using capitalization. The remaining 373 homographs are potentially ambiguous, and eligible for use as test words. These test words were used to compare the results of Type-10 components with neighbourhood lattices regarding effectiveness of discrimination of homographs.

Note that the goal here is to efficiently separate instances of homographs for disambiguation (lead1 from lead2), not to automatically identify and classify all instances of a particular homographic word together (all instances of lead1 together). The latter would indeed be desirable, but at this point entails further problems not yet solved. In other words, it is not yet possible, while partitioning homographs, to automatically group together all instances or senses of a particular word that is a homograph of another word, into a single partition.

4   Definitions

String a sequence of characters representing a word or homograph; an entry in RIT.

Word a disambiguated string, or entry, in RIT. Lead (lead1, guide, not follow) is one word; lead (lead2, the metal) is another, different, word.

Homograph a string for which there are two or more words with the same spelling, but with different etymological origins or roots. Lead1 derives from an Indo-European root, LEIT-2, meaning to guide; lead2 derives from an Indo-European root, EL-1, meaning red (from the colour of the oxide of lead, also known as red lead and sometimes used in primer paint).

Sense a set of words that share a particular meaning or concept; also known as a Synset; (the strings in this set are known as synonyms).

Entry a particular sense of a word; a particular word in a Synset.

Polysemy the number of senses a word has; number of entries representing a particular word in RIT. *Over*, for example, has a polysemy of 22; it can be found listed in 22 senses in RIT.

Paragraph a set of synsets of semantically-related concepts (and part-of-speech) , grouped together in RIT.

Category a set of paragraphs of semantically-related notions, grouped together in RIT.

---

[2] A list in Roget's Thesaurus is a special case where a set of senses each consist of a single word. Each entry in the list is viewed as a separate, but related, concept contributing to the list topic.

Figure 1 shows a sample of the structure of RIT using the word *over*. This represents one sense of over, and this instance of over is one entry in RIT. The entry occurs in Synset 40:10:1, read as RIT Category number 40; Category name, Addition; Paragraph 10; Synset 1. Each of the adjacent synonyms in Synset 40:10:1 is also an entry; a string representing one sense of that word.

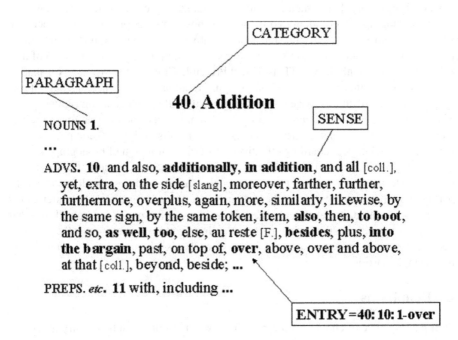

Fig. 1. An example of an entry in Roget's Thesaurus. The word *over* occurs as an entry in Category 40, Paragraph 10, Synset 1. (Bold-type synonyms are considered the most representative words of the sense).

5   Formal Concept Analysis

Several researchers have used so-called neighbourhood lattices to visualize parts of Roget's thesaurus. A semantic neighbourhood is similar to a word field (a set of semantically related words), but also includes the set of the shared senses of the words. A formal context built from a semantic neighbourhood takes the words of a neighbourhood as formal objects and their corresponding senses as formal attributes. The original formalization of neighbourhood lattices was suggested by Wille in an unpublished manuscript. Priss [7] defines neighbourhood lattices as follows:

Instead of using the prime operator ($'$), the plus operator ($+$) retrieves for a set of objects all attributes that belong to at least one of the objects. Formally, for a set G_1 of objects in a context (G, M, I), $\iota^+(G_1) := \{m \in M \mid \exists_{g \in G_1} : gIm\}$. Similarly $\varepsilon^+(M_1) := \{g \in G \mid \exists_{m \in M_1} : gIm\}$ for a set M_1 of attributes. If two plus mappings are applied to a set G_1 it results in a set $\varepsilon^+\iota^+(G_1)$ (with $\varepsilon^+\iota^+(G_1) \supseteq G_1$) which is called the *neighbourhood* of G_1 under I. A neighbourhood of attributes is defined analogously. A neighbourhood context is a context whose sets of objects and attributes are neighbourhoods, such as $(\varepsilon^+\iota^+(G_1), \iota^+\varepsilon^+\iota^+(G_1), I)$. The resulting lattice is called a neighbourhood lattice.

Such lattices are used here to collect and display the senses and synonyms of a topic word. Though there is no limit to the number of times the plus operator can be applied, three times is sufficient to create the neighbourhood of senses with synonyms. The first iteration is to collect the senses of the topic word; the second is to collect the synonyms shared within those senses; and the third to collect the special senses of the synonyms (not shared by the topic word).

6   Type-10 Components

Type-10 chains and Type-10 components derive from the mathematical model of abstract thesauri (of which Roget's Thesaurus is one instantiation), developed by Bryan [2]. The elements in this model are strings and senses (sense definitions, or Synsets), and a relation between them. Bryan defined a series of chains linking entries by word associations, sense associations, or both. If a word appears in two different senses, an association between the senses is implied. If two different words share a sense, an association between the words is implied. The most restrictive, the Type-10 chain, is a double chain. This requires that for any two words sharing a sense, there exists a second sense that that also shares those two words, in order to participate in the chain. The two words plus two senses has been dubbed a *quartet.*

The Type-10 chain restriction is intended to ensure that links are not arbitrary, as happens when two senses are linked by homographs. The assumption is, for example, that there is no second word that accompanies both *lead* (the metal) and *lead* (not follow) for any pair of their senses. Figure 2 illustrates a simple example of RIT entries (the X's) forming quartets.

Type-10 chains are used to form partitions (equivalence relations), called here *components,* on sets of entries. Talburt and Mooney [10] derived all possible Type-10 chain components from the 200,000 entries in RIT in an attempt to automatically separate

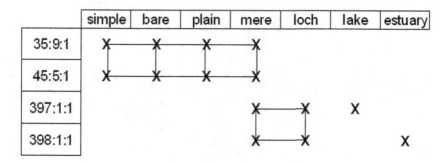

Fig. 2. Quartets formed by some of the synonyms and senses of *mere. 397:1:1 / 398:1:1 / loch / mere* forms one quartet.

all homographs. The largest Component contained more than 22,000 entries, and more than 10,000 components were derived. The effectiveness of this method has never been tested against a complete set of homographs.

Jacuzzi [5] reproduced Talburt and Mooney's work but applied a further constraint: that a quartet can not participate in a component if it shares only one RIT entry with that component. This was because he observed that it was possible for a quartet to satisfy the Type-10 constraint, yet connect to other quartets by only one entry (a string identical to one or more other strings in the component, plus a sense shared by one or more other strings in the component).

Strictly speaking, Jacuzzi's derived components are not partitions because in splitting quartets the offending entry must be included in all derived child components. The components are no longer equivalence relations on the set of all entries in RIT. None-the-less Jaccuzzi's results were chosen for this study as they are more restrictive and the components are smaller and therefore less likely to combine homographs. The maximum sized Jacuzzi component is 1,490 RIT entries.

7   Neighbourhood Lattices

Mere is a homograph. Its main meanings are simple, pure (mere1), and sea, ocean, loch (mere2). Webster's New Collegiate Dictionary [13] defines mere2 as a sheet of standing water: POOL. Figure 3 shows the neighbourhood lattice of the undifferentiated word. The left hand side (the group to the left of the object *mere*) represents mere1. Apart from the concept labelled with *mere*, this forms a partition separating mere1 from senses of mere2. For simplicity, the attribute labels are not shown in this diagram. Two of mere1's attributes, for example, are 35:9:1 Smallness and 45:5:1 Simplicity, Noncomplexity.

Most of the synonyms of mere in Figure 3 are differentiated by senses that are peculiar to them. This causes a more complex structure than is needed to observe partitions between instances of the topic word (the homograph we are interested in).

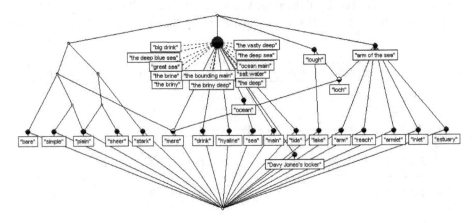

Fig. 3. Formal concept lattice of the semantic neighbourhood of mere (all synonyms and all senses (unlabelled) of the synonyms of mere)

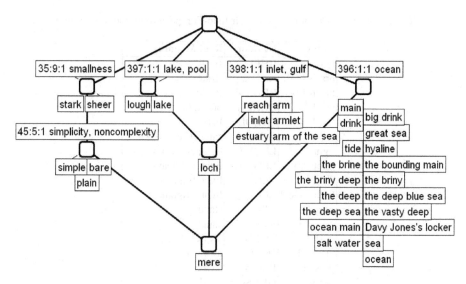

Fig. 4. Formal concept lattice of the semantic neighbourhood of mere (synonyms and senses of mere only)

Figure 4 shows a neighbourhood lattice of mere where the plus operator has been applied only twice. Here, the senses of *mere* alone are used as attributes. Mere1 is identifiable by the two senses 35:9:1 and 45:5:1.

If the top- and bottom-concepts of the lattice are removed from the lattice in Figure 4 we obtain three disjoint graph components; one dealing with mere1 and two dealing with mere2. This shows that there is no overlap of senses or synonyms of mere1 and mere2. The result is called the horizontal decomposition of the lattice and has been used amongst others, by Dekel and Gil [3] to identify component classes in the structure of legacy computer software. The collective lattice of partitions is called the horizontal sum [4]. Thus, Figure 4 is the horizontal sum of the meaningful components.

Figures 3 and 4 illustrate the separation of homographs of a word using the senses as formal attributes (and as differentiae). Of the approximately two-and-a-half-thousand entries in Roget's Thesaurus that represent homographs, all but 22 were found to be differentiable by this method. Those 22 ambiguous entries consisted of 10 homographs. The same entries were differentiable by the Type-10 partitioning (i.e. the 22 were also not differentiable by Type-10 components). The 22 undifferentiated entries (10 homographs) were found among just eight of the Jacuzzi components.

8   The Exceptions

The 10 problem homographs fell into three cases. Each case involved a word shared by each of the homographs (a synonym in common). This shared word provided a bridge between the homographs and prevented the formation of a partition. The 10 homographs

Table 1. Ambiguous homographs, their Indo-European roots and root meanings

Entry	IE Root	Root Meaning
brash 1	BHEL-2	swell, blow
brash 1	KAU-2	strike, hew
brash 1	RE-1	Backward
brash 2	BHREG-	break, breach
fell 1	P(H)OL-	Fall
fell 2	GHEL-2	shine, bright
light 1	LEGWH-	light, not heavy
lightsome 1	LEGWH-	light, not heavy
light 2	LEUK-	light, brightness
lightsome 2	LEUK-	light, brightness
post 1	STA-	stand
poster 1	STA-	stand
post 2	(A)PO-	away, off
poster 2	(A)PO-	away, off
press 1	PER-5	strike
press 2	GHESOR-	hand
rash 1	KAU-2	strike, hew
rash 1	RE-1	backward
rash 2	RED-	scrape, scratch, gnaw
set 1	SED-1	sit
set 2	SEKW-1	follow
set 3	N/A	Egyptian god
slug 1	SLAK-	strike
slug 2	SLEU-	sluggish, slow

were *brash, fell, light, lightsome, post, poster, press, rash, set* and *slug*. These are listed in Table 1, along with their Indo-European Roots and each roots' meanings[3].

The first case involved homographs that each shared a supposedly unambiguous (non-homographic) word as a synonym. As it happens these shared words had many, very diverse senses (25 in the first instance). One of those senses of the shared word overlapped with the meaning of one homograph, while a second sense of the word overlapped with the meaning of the second homograph. Such cases will be referred to as the ambiguous synonyms category.

As an example, press1 generally relates to printing or pressure, and is derived from an IE root, PER-5 (meaning, to strike); while press2 relates to drafting into military service or being at hand, and derives from GHESOR- (meaning, hand). Both press1 and press2 share *call* as a synonym. The sense of the word call shared with press1 has to do with calling *on* someone, as in: to pressure someone for money or a sale. The second sense of call, shared with press2, has to do with calling *up* someone (call up has one meaning of ordering someone to report for military duty). It is clear that these

---

[3] Brash1 comes from bold + rash1; rash1 is a variant of rush1 (hurry); rush1 and rash1 come from RE-1 (backward) + KAU-2 (strike, hew); and bold comes from BHEL-2 (swell, blow). Brash1, therefore, derives from the three IE roots BHEL-2 + RE-1 + KAU-2. Hence the three entries in the table for brash1, and the two entries for rash1.

Table 2. Category one of undifferentiated homographs involves ambiguous synonyms

Hom1	Meaning1	Hom2	Meaning2	Comp#
Press1	**call** on	Press2	**call** up	[VJ2 2478]
Set1	**head** for	Set2	**head** of (collection)	[VJ2 184]
Set1	**suit**, attune (to match)	Set2	**suit** (matching attr.s)	[VJ2 9323]

senses of two, otherwise distinct and unambiguous, homographs live close to each other in the semantic universe.

The other homograph in this category, *set*, has two instances of this problem. Considering that the word *set* has 53 senses, 51 of which were disambiguated, this is not a dismal result. The ambiguous entries in Roget's Thesaurus for the first category are listed in Table 2. The last column shows the component numbers from Jaccuzzi's Type-10 chain components.

Figure 5 illustrates the restricted neighbourhood lattice of the homographs of *press*. To reduce complexity the formal context was restricted to those synonyms that occur in more than one sense of press. The majority of senses belong to press1. Press2 has the three senses located to the extreme right. *Call* is a shared synonym of both press1 and press2, and prevents the formation of a partition between the two homographs.

The second category of undifferentiated homographs is identifiable by shared synonyms that are themselves ambiguous homographs. These are, again, highly polysemous words. They have a wide range of senses that allow them to overlap semantically, as synonyms, with many other strings-including the strings that constitute the instances of these topic homographs. Table 3 shows the second category.

Down2 is an expanse of rolling, grassy, treeless upland used for grazing (a moor) in the context of the undifferentiated RIT fell entry. It is often used in the plural, in this sense, as part of a place-name, as in: *Watership Downs*. A fell (fell2) is a type of flat

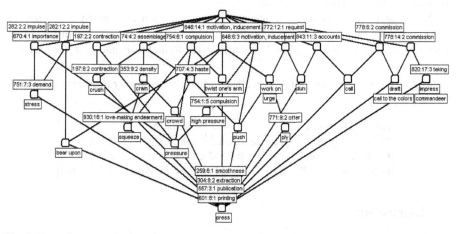

Fig. 5. Formal concept lattice of the restricted semantic neighbourhood of press (all synonyms sharing more than one sense, plus senses of press only). Press2 has the three senses located to the extreme right. Call is a shared synonym of both press1 and press2.

Table 3. Category two of undifferentiated homographs involves ambiguous homographs

Hom1	Meaning1	Hom2	Meaning2	Comp#
Fell1	fall, Down2	Fell2	moor, a Down2	[VJ2 4573]
Slug1	slowpoke, Poke1	Slug2	slog, Poke1	[VJ2 1501]

land, synonymous with moor. Down2 is also the adverb, *down* (as in: up, down, forward and backwards). It is an extension of this second persona of down2 which is found as a synonym of fell1 (to down something; to chop down or drop, as in: to fell a tree). Down1, incidentally, is the down derived from goose feathers.

Poke1 is an abbreviation of an Americanism, slowpoke [US], from cowpoke [US], a cowpuncher [US] or cowboy. Punching cows involved prodding (or poking) them with poles to make them enter railroad cars. Slug1 is the slimy invertebrate and slowpoke means one who moves slowly. Slug2 means hit; strike. (Poke2, not listed, is a pocket, as in: a pig in a poke).

Category three involves three cases where synonyms that are differentiable homographs co-occur with related homographs that are themselves differentiable homographs, but that match with homographs accompanying their homographic nemeses. The consequence is that there is a catch-22; each homograph relies on the other to differentiate it. Because they have identical spellings, and their cohort homographs have identical spellings, they match in a rare undifferentiated quartet. Referring to Table 4, light1 (not heavy) and light2 (bright), both have *lightsome* as a synonym. Lightsome1 means weightless (also cheerfulness and caprice) while lightsome2 means full of light. Lightsome (1 & 2) is an infrequent and archaic word in English.

Likewise, *brash* and *rash* can both mean impulsive or indicate a collection of red spots that are symptoms of a disease. And *post* or *poster* can both mean put up an announcement and also be the names for the role of a mailman. Brash2, poster1, and poster2 are all rare words in English.

Note that in category three, lightsome, poster and rash are equally ambiguous to light, brash, and post. So this represents six, not just three, instances of undifferentiated homographs. Table 4 was restricted to emphasizing light, brash, and post, for brevity.

Table 4. Category three of undifferentiated homographs involves other undifferentiated homographs

Hom1	Meaning1	Hom2	Meaning2	Comp#
Light1	weightless, Lightsome1	Light2	bright, Lightsome2	[VJ2 7393]
Brash1	impulsive, Rash1	Brash2	disease, Rash2	[VJ2 2183]
Post1	(put up a) Poster1	Post2	mailer, (a) Poster2	[VJ2 8564]

9   Discussion

It is well known by linguists that words with otherwise distinct etymologies can influence each other to trade meaning and blend, and eventually come to have the same connotations. This may well have happened with press1, press2, light1 and light2, and possibly some of the other homographs unsuccessfully disambiguated here by either the FCA lattices or the Type-10 components methods. Also, many of these ambiguous synonyms and homographs are highly polysemous, giving ample opportunity for instances of semantic overlap.

Finally, in some of these cases one of the senses or one of the words is rare. It should not be surprising that these are difficult to disambiguate. Languages tend to discard or modify ambiguous words, but rare instances have less opportunity for scrutiny. In fact many of the successfully disambiguated homographs suggested a pattern. That if a homograph came from a specialty area, such as a branch of science, it was more likely to have a matching common word as a homograph. An example is abbreviations for chemical elements. Be: Beryllium; He: Helium; As: Arsenic; In: Indium; At: Astatine.

It may appear to be a contradiction that both highly polysemous and rare-sense words should both contribute to ambiguity, but there appears to be a balance required for disambiguation in language-sufficient context to differentiate but not so much as to cause confusion.

10   Conclusion

We have compared more than two-and-a-half-thousand semantically ambiguous entries in Roget's Thesaurus using two methods, FCA neighbourhood lattices and Type-10 chain components. The ambiguity amongst entries was caused by homographs-words of identical spelling but with different origins and meaning. We conclude that, given a lexicon and set of homographs in common, FCA neighbourhood lattices can discriminate homographs as well as Type-10 chain components. Furthermore, while Type-10 components may contain up to 1,500 thesaurus entries, semantic neighbourhoods are constrained to the senses and synonyms of the topic word. Consequently, the partitions formed around homographs using FCA make the data more tractable and human-accessible.

Ten of the 373 homographs used in this study had senses that were undifferentiable by either of the two methods. These cases involved senses where homographs of completely different origins (by definition) overlapped semantically via words in common. These words-in-common were other homographs in all but three instances. The failed instances involve very rare or very common (highly polysemous) words, and may represent the boundaries for discrimination of homographs. They may also indicate the range and combination of frequency and rarity necessary to disambiguate polysemous words in human conceptual processing.

Future work should examine the effectiveness of combining previously documented methods of homograph disambiguation with FCA neighbourhood lattices to disambiguate homographs with a view to improving effectiveness. Furthermore, while neighbourhood lattices are effective at partitioning senses of homographs, at the same time they may partition senses within the set of senses. For completeness, a method should be developed which classifies together all of the senses of any homograph to which the partitioning method is applied.

Acknowledgements

The lattices used in this paper were developed using Conexp [17] and Anaconda [11]. The electronic version of Roget's International Thesaurus was used with kind permission from Dr W. A. Sedelow Jr. and Dr S. Yeates Sedelow.

References

1. Berrey, L. (Ed.). (1962). *Roget's international thesaurus (3rd ed.)*. New York: Crowell.
2. Bryan, R. M. (1973). Abstract thesauri and graph theory applications to thesaurus research. In S. Y. Sedelow (Ed.), *Automated language analysis, report on research 1972-73* (pp. 45-89). Lawrence, KS: University of Kansas.
3. Dekel, U., and Yossi, G. (2003). Revealing class structure with concept lattices. *The Tenth Working Conference on Reverse Engineering,* IEEE Computer Society Press, pp. 353-365.
4. Ganter, B., and Wille, R. (1999). *Formal concept analysis: Mathematical foundations.* Berlin-Heidelberg-New York: Springer. ISBN 3-3540-62771-5.
5. Jacuzzi, V. (1991, May). *Modeling semantic association using the hierarchical structure of Roget's international thesaurus.* Paper presented at the Dictionary Society of North America Conference, Columbus, Missouri.
6. Miller, G., Beckwith, R., Fellbaum, C., Gross, D., Miller, K., and Tengi, R. (1993). *Five papers on WordNet.* Technical Report. Princeton, N.J: Princeton University.
7. Priss, U. (1996). *Relational concept analysis: Semantic structures in dictionaries and lexical databases.* (Doctoral Dissertation, Technical University of Darmstadt, 1998). Aachen, Germany: Shaker Verlag.
8. Priss, U., and Old, L. J. (2004). Modelling lexical databases with formal concept analysis. *Journal of Universal Computer Science, 10*(8), 967-984.
9. Sedelow, S.Y. (1991). Exploring the terra incognita of whole-language thesauri. In R. Gamble and W. Ball (Eds.), *Proceedings of the Third Midwest AI and Cognitive Science Conference* (pp. 108-111). Carbondale, IL: Southern Illinois University.
10. Talburt, J. R., and Mooney, D. M. (1990). An evaluation of Type-10 homograph discrimination at the semi-colon level in Roget's international thesaurus. *Proceedings of the 1990 ACM SIGSMALL/PC Symposium*, 156-159.
11. Vogt, F. (1996). *Formale Begriffsanalyse mit C++: Datenstrukturen und Algorithmen.* Springer-Verlag, Berlin-Heidelberg, 1996, ISBN: 3-540-61071-5.
12. Wille, R. (1982). Restructuring lattice theory: an approach based on hierarchies of concepts. In I. Rival, (Ed.), *Ordered sets* (pp. 445-470). Dordrecht: Reidel.
13. Woolf, B. H. (Ed.). (1976). *Webster's new collegiate dictionary*, G. & C. Merriam Company, Springfield, Massachusetts.
14. Yarowsky, D. (1992). Word-sense disambiguation using statistical models of Roget's categories trained on large corpora. *Proceedings of 14th International Conference on Computational Linguistics, COLING-92.* Nantes, pp. 454-460.
15. Yarowsky, D. (1996). Homograph disambiguation in text-to-speech synthesis. In J. van Santen, R. Sproat, J. Olive, and J. Hirschberg, (Eds.), *Progress in Speech Synthesis*: Springer, New York, 1996.
16. Yarowsky, D. (2000). Word sense disambiguation. In R. Dale, H. Moisl and H. Somers (Eds.) *The Handbook of Natural Language Processing.* New York: Marcel Dekker, pp. 629-654.
17. Yevtushenko, S. (2005). Conexp (Concept Explorer) 4-Beta. *SourceForge.net.* Available via http://sourceforge.net/projects/conexp

Using Concept Lattices to Uncover Causal Dependencies in Software

John L. Pfaltz

Dept. of Computer Science, Univ. of Virginia,
Charlottesville, VA 22904-4740
jlp@virginia.edu

Abstract. Suppose that whenever event x occurs, a second event y must subsequently occur. We say that x "causes" y, or y is causally dependent on x. Deterministic causality abounds in software where execution of one routine can necessarily force execution of a subsequent sub-routine. Discovery of such causal dependencies can be an important step to understanding the structure of undocumented, legacy code.

In this paper we describe a methodology based on formal concept analysis that uncovers possible causal dependencies in execution trace streams. We first walk through the process using a small synthetic, but easily comprehensible, example. Then we illustrate its potential using 57 threads involving 18,969 executed operations that were monitored in an open source middleware system.

1 Introduction

Since its first application as "concept analysis" [22], Galois closure [13] has proven to be a valuable tool for the analysis of various phenomena. Many examples can be found in *Formal Concept Analysis* [5] and the reader is assumed to be familiar with this fundamental work.

To our knowledge the first effort to apply closure concepts to software engineering was by Gregor Snelting who used formal concept analysis to analyze legacy code [12, 18]. Siff and Reps [17] published shortly after. Snelting's goal was to reconstruct the overall system structure by determining which variables (columns) were accessed by which modules (rows). It was hoped that the concept structure would become visually apparent as it does in all of Ganter and Wille's examples [5]. Unfortunately, the resulting concept lattice shown on page 356 of [12] is little more than a black blob. Visual interpretation of closure concepts does not seem to scale well. In [1], Ball specifically proposes using concept analysis to establish the relationship between individual test runs and procedure executions in a red-black tree system as shown in Figure 1. He, then goes on to visually identify which procedures dominate others — that is, force their execution. Given the small size of R and $\mathcal{L}$, dominance is visually derivable. But, in a larger system this might be unwieldy. Unfortunately, Ball does not seem to have done any further work on this concept based approach to dynamic software analysis. In this paper we will push this kind of analysis a bit further.

R. Missaoui and J. Schmid (Eds.): ICFCA 2006, LNAI 3874, pp. 233–247, 2006.

Test	Procedures						a1rMSD
------	------	-----------	------	------	------	--------	
	(a)dd	(1)Rotate	(r)em	(M)in	(S)ucc	(D)elFix	
t1	X		X			X	
t2	X	X	X			X	
t3	X	X	X	X	X		
t4	X	X	X	X	X	X	
t5	X	X	X	X	X	X	

Fig. 1. Execution of sub-procedures in a red-black tree program under various test configurations

```
1    a ··· u ··· c ··· q ··· s
2    u ··· e ··· b ··· r ··· t ··· p ··· b ··· r ··· t ··· p
3    b ··· d ··· a ··· s ··· r ··· q
4    c ··· a ··· t ··· u ··· s ··· q ··· a ··· t ··· s
5    a ··· c ··· s ··· q ··· t
6    d ··· e ··· c ··· t ··· s ··· u
7    p ··· b ··· u ··· e ··· t ··· r ··· e
8    a ··· u ··· q ··· e ··· p
```

Fig. 2. 8 event sequences extracted from simulated trace data

Let $a, b, c, d, e, p, q, s, t, u$ denote specific software events, and let the 8 sequences of Figure 2 depict relevant portions of trace data from 8 executions of a single software system. Our goal will be to analyze necessary dependencies between these events, if any. This will constitute a running example through Section 3. It provides a clearer introduction. In Section 4, we will turn to the analysis of real trace data consisting of 57 separate threads comprised of 18,969 invocations of 77 different operators.

2   Discrete Deterministic Data Mining

The first step in our analysis of software execution employs the discrete deterministic data mining (DDDM) system we have developed at the Univ. of Virginia. As described in [15, 16] this system extracts all the logical dependencies between attributes, or properties, of observed objects as recorded in a binary relation $R(O, A)$. We let $\mathcal{L}_R$ denote the concept lattice generated by $R(O, A)$. Each closed concept, together with its generator(s) determine a logical dependency. More specifically, if a subset at is a generator of the closed set $acqst$ of attributes then, as shown in [15], at logically implies $acqst$. The dependency, or implication, can be expressed in first-order notation as

$$(\forall o \in O)[(a(o) \wedge t(o)) \vee (q(o) \wedge t(o)) \rightarrow a(o) \wedge c(o) \wedge q(o) \wedge s(o) \wedge t(o)] \quad (1)$$

which by letting concatenation denote conjunction and suppressing the universal quantifier, we abbreviate as simply

$$at \vee qt \rightarrow acqst \quad (2)$$

These expressions implicitly indicate that the closed set $acqst$ has two generators. The data mining performed by the DDDM system is deterministic because
these implications *must* occur. We sometimes call this closed set data mining to
distinguish it from more customary *apriori*, or frequent set, data mining which
yields statistical associations between the attributes, properties, or items.

The value of identifying closed sets in order to minimize redundant associations in traditional *apriori* type data mining has been rather thoroughly explored
in [10, 26, 27]. To get a feel for the power of focusing on closed sets, we observe
that our DDDM system yielded 2,641 closed concepts, and thus logical implications, when applied to a relation, $R(O, A)$, consisting of 8,124 objects, or rows,
and 39 attributes, or columns, that enumerated the horticultural properties of
mushrooms. Admittedly, most of these 2,641 implications were either trivial or
useless. Nevertheless, 2,641 closed set concepts is an order of magnitude less
than the 25,210 frequent set associations generated by an open source *apriori*
algorithm on the same $8,124 \times 39$ data set, using reasonable support and confidence parameters. Emphasizing Galois closed sets in data mining can have a
huge performance payoff. And, a rather simple filter was able to reduce this
mass of implications to only 37 relatively simple rules for determining whether
a mushroom is *edible* or *poisonous*. We might consider these to be the most
important attributes of any mushroom. Obtaining deterministic identification
rules is very desirable here!

We say that the DDDM approach is *discrete* because the universal quantifi-
cation can only be over the finite domain O comprising the relation $R(O, A)$.

The DDDM system constructs the concept lattice $\mathcal{L}_R$ incrementally in a manner that was first described by Godin and Missaoui in [7, 8, 9] and refined a bit
more in [20, 21]. Incremental construction of the concept lattice facilitates incor-
poration of new data into an existing set of formal concepts without rereading
the earlier data. The actual implementation of our system is more fully described
in [16].

Given the sequence data of Figure 2 we first create the boolean relation
$R(O, A)$ shown as Figure 3(a). Here (n, x) is true if event trace sequence n
contains an occurrence of event x. Observe that x may, and often does, appear repeatedly in a single trace. Setting (n, x) to true only indicates that x
has appeared at least once in trace n. Our DDDM system then incrementally
generates the lattice of 26 concepts shown as Figure 3(b). To determine, and
possibly modify, generators incrementally we use the technique proven in [11]
which determines generating sets by examining covering relationships between
closed concepts in $\mathcal{L}_R$. Let Z be a closed concept that covers the concepts Y_i in
$\mathcal{L}_R$. If $X \subseteq Z$ is a minimal subset such that $X \cap (Z - Y_i) \neq \emptyset$ for all i, then X
is a generator of Z.

Figure 4 more accurately illustrates the output generated by the DDDM process. The program does not actually draw the concept lattice $\mathcal{L}_R$; it is too hard
to do well. Instead we list the attributes comprising the closed concepts, the gen-
erator(s) of these concepts, and the support of each concept (objects involved in
the Galois closure), together with a concept identifier and list of covering lattice

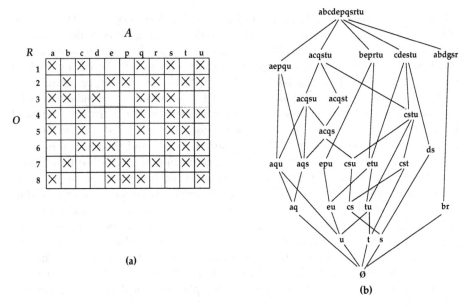

Fig. 3. The resulting concept lattice

edges (which we have not shown). To conserve space, we have only listed those 20 concepts of Figure 3(b) which are supported by at least two observations. Observe that it is concept #10, $acqst$ with generators $\{at, qt\}$ that is the source of the logical implications presented earlier as (1) and (2). The reader should also verify that these logical implications are, in fact, true for the discrete domain O. For example, any object, or row, in which both a and t appear will also contain c, q and s, as indicated by concept #10.

3   Causal Dependency

Logical implication is not equivalent to causal dependency. The accepted concept of "causality" involves time, whereas logic does not. If at is the precedent, as in concept #10, or expression (2), then we expect that the conjunction of these two events must precede the occurrence of the events c, q and s of the consequent if we are to say that at "causes" qst as a consequence. Causal dependence is assumed to be strictly anti-symmetric with respect to time. Since concept #10 is supported by traces 4 and 5, we examine each more closely. Event t is the last event in trace 5. In no way could it be considered to have any causal effect on the preceding events c, q or s. So similarly, the conjunction of events ct cannot possibly be a causal agent.

Now consider concept #1 which can be logically expressed as
$$acu \vee asu \vee cqu \vee qsu \ \longrightarrow \ acqsu.$$
Its support is traces 1 and 4. Examination shows that, in both traces, the conjunction of events acu always precedes both q and s. So, $acu \Rightarrow qs$ is a reasonable

id#	closed concept	support	generators
1	acqsu	1,4	acu, asu, cqu, qsu
2	beprtu	2,7	bu, ru, bt, rt, be, er, bp, pr, pt
3	u	1,2,4,6,7,8	u
5	aqs	1,3,4,5	as, qs
7	br	2,3,7	b, r
9	tu	2,4,6,7	tu
10	acqst	4,5	at, qt
11	acqs	1,4,5	ac, cq
12	t	2,4,5,6,7	t
14	etu	2,6,7	et
15	ds	3,6	d
16	s	1,3,4,5,6	s
17	cstu	4,6	ctu, stu
18	csu	1,4,6	cu, su
19	cs	1,4,5,6	c
20	cst	4,5,6	ct, st
22	epu	2,7,8	p
23	eu	2,6,7,8	e
24	aq	1,3,4,5,8	a, q
25	aqu	1,4,8	au, qu

Fig. 4. Selected concepts generated by the relation $R(O, A)$ of Figure 3(a)

causal hypothesis, while none of the other logical disjuncts can be. We use $\rightarrow$ to denote logical implication and $\Rightarrow$ to denote causal dependence.

We deliberately use the word "hypothesis" in the preceding sentence. We cannot establish that the conjunction of events a, c and u actually cause events q or s to occur. We can only establish that they satisfy the necessary conditions for "causality". We will discuss this further in Section 6.

Because we record the support for each closed concept along with its generators, it is not hard to re-examine the appropriate trace data sequences to verify, or exclude, specific generators as possible causal precedents. Applying this procedure to the 20 concepts of Figure 4 we get the following list of 6 possible causal dependencies shown in Figure 5.

The 6 dependencies of Figure 5 represent a rather significant reduction in the sheer number of concepts that are typically created.

$$
\begin{array}{lll}
\#1 & acu \Rightarrow & acqsu \\
\#7 & b \Rightarrow & r \\
\#11 & ac \Rightarrow & qs \\
\#15 & d \Rightarrow & s \\
\#19 & c \Rightarrow & s \\
\#24 & a \Rightarrow & q \\
\end{array}
$$

Fig. 5. 6 possible causal dependencies

It is easy to show in a first-order logic, if $a \rightarrow x$ and $b \rightarrow y$ then $ab \rightarrow xy$. Such rules of inference are common place. But, they need not be valid in causal dependence. For example, we have #19 $c \Rightarrow s$ and the trivial dependency #3 $u \Rightarrow u$. Yet, $cu \not\Rightarrow csu$ because in trace 6, $s < u$. Consequently, given the dependencies #24 $a \Rightarrow q$ and #19 $c \Rightarrow s$ we cannot logically infer that $ac \Rightarrow qs$, even though in this case #11 is, in fact, true.

Similarly, in first-order logic it is customary to declare x and y to be equivalent, $x \equiv y$ if $x \rightarrow y$ and $y \rightarrow x$. However, a concept of causal equivalence in which x causes y and y causes x does not appear to make semantic sense. Nevertheless, such apparent patterns are common in our trace data where we have repeated sections of code, or loops, as in

$$\cdots a \cdots b \cdots a \cdots b \cdots a \cdots b \cdots .$$

Such repeating patterns can be exposed by techniques developed in [24, 23], but even here, some form a priori knowledge of what kind of pattern is being sought must be applied.

4   A Real Example

To test these ideas, the author used trace data down loaded from *JBoss*, an open source, professional middleware company which is accessible through www.jboss.com. All of the method entrance events of the transaction management module in JBoss 1.4.2 were instrumented by my colleagues, Jinlin Yang and David Evans. They then ran the entire JBoss regression test suite to collect the traces [25]. A small sample of this trace data from a single thread is shown below in Figure 6.

```
3     TxManager.getTransaction()Ljavax/transaction/Transaction;
2     TxManager.getThreadInfo()Lorg/jboss/tm/TxManager$ThreadInfo;
4     TxUtils.isActive(Ljavax/transaction/Transaction;)Z
1     TxManager.getStatus()I
2     TxManager.getThreadInfo()Lorg/jboss/tm/TxManager$ThreadInfo;
1     TxManager.getStatus()I
2     TxManager.getThreadInfo()Lorg/jboss/tm/TxManager$ThreadInfo;
3     TxManager.getTransaction()Ljavax/transaction/Transaction;
2     TxManager.getThreadInfo()Lorg/jboss/tm/TxManager$ThreadInfo;
4     TxUtils.isActive(Ljavax/transaction/Transaction;)Z
5     TxManager.suspend()Ljavax/transaction/Transaction;
2     TxManager.getThreadInfo()Lorg/jboss/tm/TxManager$ThreadInfo;
1     TxManager.getStatus()I
2     TxManager.getThreadInfo()Lorg/jboss/tm/TxManager$ThreadInfo;
1     TxManager.getStatus()I
2     TxManager.getThreadInfo()Lorg/jboss/tm/TxManager$ThreadInfo;
3     TxManager.getTransaction()Ljavax/transaction/Transaction;
```

Fig. 6. A representative fragment of an operator sequence

Preprocessing consisted of taking 57 such threads; scanning each operation; and, if new, assigning it an identifying integer. The integers to the left in Figure 6 are examples. This preprocessing had several benefits. First, it insures that the closed sets are extracted without using any embedded semantic information. Second, it permits us to display a set of operations as a set of integers, which we will see has definite *display* benefits. Third, because identifying integers are assigned in sequence as operators are scanned we have an interesting artifact in which related operators often appear as a number sequence. Our programs make no use of this artifact, but human inspection can reveal interesting structures that are not uncovered by the Galois closure itself.

The trace fragment of Figure 6 would be perceived by our DDDM software as

$$... 3\ 2\ 4\ 1\ 2\ 1\ 2\ 3\ 2\ 4\ 5\ 2\ 1\ 2\ 1\ 2\ 3\ ...$$

From now on we consider only discrete integer representations. We observe that many operations are repeated in definite patterns, but our analysis is set based. All we can assert is that operations { 1,2,3,4,5 } all occur somewhere in this trace.

We analyzed 57 distinct traces consisting of 77 distinct operations. The shortest trace consisted of no more than 6 operations; the longest trace involved 1,393 operations.

The *set* of operations comprising each trace were input incrementally to our DDDM system. Its output is illustrated in Figures 7 and 8.

We were more than a little surprised. Only twenty seven non-trivial closed sets of operations emerged.

As we indicated in Section 2, the lattice shown in Figure 7 is hand drawn. For each concept in the lattice our software really outputs a sequence of concepts consisting of the items in the closed set of the concept, the set of generating sets, the set of supporting rows and the set of concepts covered by the concept. The latter facilitates drawing the lattice and maneuvering through the lattice. This table of closed concepts, in Figure 8, and their generators requires a bit of interpretation; particularly since we use the hyphen (-) in two different ways.

The first column denotes the concept number. They correspond to the concept numbers in Figure 7. The reader can verify that the closed sets of concepts in the lattice below any specific concept, say concept #2, are contained in the closed set of #2. The closed set of concept #2 consists of operations 1 through 5 and operations 12 through 64. Here, rather than enumerating every operation id, separated by commas, we use the hyphen as an "extended and".

Concept #2 has many generators. Operations 33 and 34 are each singleton generators. If either operation occurs in a trace then every operation of the closed set must also occur. The combination { 4, 12-32 } is also listed as a generator. This means that { 4, 12 }, { 4, 13 } ... { 4, 32 } are each generating sets. Here we are using the hyphen as an "extended or", that is, operation 4 in combination with operation 12 or operation 13 or ... is a generator.

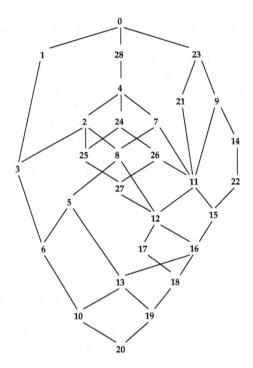

Fig. 7. Lattice, $\mathcal{L}_{ops}$, of closed operator sets

Concepts #17 and #18 are of interest because all of their generators are singleton; but there are many of them. For example, from concept #17, if any of the operations 46 through 49 appear in the trace then all of the operations in the closed set will occur in the trace.

The final column displays the number of traces in which this concept can be found, that is its size of "support". We felt that enumerating this support, as in Figure 4 would be overkill.

It is interesting to note that the infimum set { 3, 2 } of operations (concept #20) is generated by the empty set. These two operations occur in every trace.

5   Establishing Dominance

Execution of a procedure, or operator, often depends on a conjunction of conditions. Figure 8 lists many such conjunctive generators. However, analysis of such generators is beyond the capabilities of our current software. Consequently, we will restrict ourselves to analyzing only those concepts with singleton generators. Our DDDM software makes identification of these concepts quite easy.

If a singleton set, such as { 1 }, generates the Galois closed set { 1, 2, 3, 5 } as in concept #6 of Figure 8 then we can logically assert that

$$1 \rightarrow 1235 \tag{3}$$

Concept number	Closed set	Generators	Size of support
1	{1-11}	{6-11}	1
2	{1-5, 12-64}	{33-34},{4,12-32},{4,35-64},{1,25-32},{1,45}	6
3	{1-5}	{4}	7
4	{1-5,12-65}	{4,65},{33-34,65},{1,25-32,65},{1,35-38,65}, {1,45,65}	4
5	{1-3,5,13-14}	{1,13-14}	27
6	{1-3,5}	{1}	28
7	{1-3,5,12-24,39-44,46-65}	{1,65}	10
8	{1-3,5,12-24,39-44,46-64}	{1,12},{1,15-24},{1,39-44},{1,45-64}	12
9	{2-3,5,12-24,39-44,46-72}	{46-49,66-72},{51-52,66-72},{60,66-72}, {62,66-72}	7
10	{2-3,5}	{5}	56
11	{2-3,5,12-24,39-44,46-65}	{46-49,65},{51-52,65},{60,65},{62,65}	23
12	{2-3,5,12-24,39-44,46-64}	{5,46-49},{12,46-49},{5,51-52},{12,51-52}, {5,60},{12,60},{5,62},{12,62}	25
13	{2-3,5,13-14}	{5,13-14}	55
14	{2-3,5,12-24,39-44,50, 53-59,61,63-72}	{66}	10
15	{2-3,5,12-24,39-44,50, 53-59,61,63-65}	{65}	27
16	{2-3,5,12-24,39-44,50, 53-59,61,63-64}	{12},{5,16-24},{5,39-44},{5,50},{5,53-59} {5,61},{5,63-64}	29
17	{2-3,13-24,39-44,46-64}	{46-49},{51-52},{60},{62}	26
18	{2-3,13-24,39-44,50, 53-59,61,63-64}	{15-24},{39-44},{50},{53-59},{61},{63-64}	30
19	{2-3,13-14}	{13-14}	56
20	{2-3}	{}	57
21	{2-3,5,12-24,39-44,46-65, 73-75}	{73-75}	2
22	{2-3,5,12-24,39-44,46-72}	{67-72}	11
23	{2-3,5,12-24,39-44,46-75}	{66-72,73-75}	1
24	{2-3,5,12-32,35-65}	{25-32,65},{35-38,65}	5
25	{2-3,5,12-32,35-64}	{25-32},{35-38}	7
26	{2-3,5,12-32,39-65}	{45,65}	6
27	{2-3,5,12-32,39-64}	{45}	8
28	{1-5,12-65,76-77}	{76-77}	1

Fig. 8. Output from DDDM system

or equivalently, "if 1 appears in a trace then 2, 3, and 5 must also appear". But, this does not necessarily imply that the execution of operation 1 "causes" the execution of the other operators. There need be no causal dependency. Such logical implication is only one necessary condition.

A second necessary condition for causal dependency is that the generator must precede the consequent(s) in *all* traces. In the 28 traces supporting concept #6 this is not always true, so (3) cannot be rewritten as a causal dependency.

For each singleton generator, {gen_op}, of a concept, we re-examine each of the trace sequences supporting the concept. If an operator op in the closed set of the

concept precedes the first occurrence of *gen_op* in *any* trace, then *gen_op* $\not\Rightarrow$ *op*.
If *op* always follows at least one occurrence of *gen_op* then we say that *gen_op*
dominates *op* and *gen_op* $\Rightarrow$ *op* becomes plausible.

This reasoning is particularly applicable when there are several singleton generators. They cannot all be causally equivalent. In the following analysis procedure we first resolve domination among multiple singleton generators, if any,
and then analyze domination among the other operators of the closed concept
set.

```
operator_domination (LATTICE L, TABLE Dominates)
   // Analyze the concepts in 'L' to create the
   // table of operator domination
   {
   ELEMENT       dg, fg,       // generators -- dominating, first
                 o;
   SET           SG, OP;
   CONCEPT       c;
   OP_SEQUENCE seq;

   for_all c in L.concepts do
       {
       SG <- singletons (c.generators);
       if empty(SG)
           continue;
                           // concept 'c' has singleton generators
       dg <- null;         // as yet no dominating generator
       for_all seq in c.support do
           {     // examine every sequence supporting this concept
               // get the first singleton generator in this sequence
           fg <- first_element (SG, seq);
           if dg = null
               dg <- fg;    // now, check that 'fg' is always first
           else
               dg <- null;
               break;
           }
       if dg = null          // no generator in SG is dominating
           continue;
       else
           add [dg -> SG] to 'Dominates';
               // now, check if 'dg' dominates other operators in 'c'
       OP <- c.closed_set not c.SG;
       for_all o in OP do
           {     // verify that this 'o' follows 'dg' in all
               // supporting sequences
           for_all seq in c.support do
```

```
                    {
                    if o precedes dg in seq
                        {
                        OP <- OP not {o};
                        break;
                        }
                    }
            if not empty(OP)
                add [dg -> OP] to 'Dominates';
            }
    }
```

Application of this `operator_domination` procedure to the output of our DDDM software, as illustrated in Figure 8, is shown in Figure 9. We observe that concepts #3, #6, #10, #27 make no contribution to this list of possible causal dependencies, even though all have singleton generators. They fail the operator dominance test.

```
#1      6 ⟹ 7 ⟹ 8 ⟹ 9 ⟹ 10 ⟹ 11
#2      33 ⟹ 34 ⟹ { 4, 35 ... 38 }
#14     66 ⟹ { 24, 50, 53 ... 59, 61, 63, 64, 67 ... 72 }
#15     65 ⟹ { 24, 50, 53 ... 59, 61, 63, 64 }
#16     12 ⟹ { 13 ... 15 }
#17     46 ⟹ 47 ⟹ 48 ⟹ 49 ⟹ { 51, 52, 64 }
#18     16 ⟹ 17 ⟹ { 24, 50, 53 ... 59, 61, 63, 64 }
        18 ⟹ ,.. ⟹ 23 ⟹ { 24, 39 ... 44, 50, 53 ... 59, 61, 63, 64 }
        24 ⟹ { 50, 53 ... 55, 58, 59, 61, 63, 64 }
        39 ⟹ ,.. ⟹ 44 ⟹ { 50, 53 ... 59, 61, 63, 64 }
        50 ⟹ {53 ... 55, 58, 59, 61, 63, 64 }
        53 ⟹ 54 ⟹ 55 ⟹ {61, 63, 64 }
        56 ⟹ 57 ⟹ {61, 63, 64 }
        61 ⟹ { 63, 64 }
#19     13 ⟹ 14
#21     73 ⟹ 74 ⟹ 75 ⟹ { 24, 46 ... 73 }
#25     25 ⟹ ,.. ⟹ 32 ⟹ { 35 ... 64 }
#28     76 ⟹ 77
```

Fig. 9. Possible causal dependencies

There are numerous instances of multiple generator sequences such as

$$46 \Rightarrow 47 \Rightarrow 48 \Rightarrow 49$$

in concept #17. We surmise that these traces come from software employing a stack architecture and that these represent iterated invocations down through the stack. Concept #18 gives rise to many apparent dependencies. Readily we could simplify the enumeration considerably. Applying causal transitivity we could simply write, for instance,

$$16 \Rightarrow 17 \Rightarrow \{24, 56\}$$
$$18 \Rightarrow 19 \Rightarrow \{24, 39\}$$

Events 24, 39 and 56 in turn generate more of the closed set comprising concept #18. Such condensation makes certain differences more evident, but our preliminary, proof of concept software is not, as yet, capable of doing it.

6   Considerations

The technique described in the preceding sections has four distinct steps. It: (1) identifies software events of interest; (2) extracts them from trace data to form a relation $R(T, E)^1$; (3) creates a concept lattice $\mathcal{L}_R$ embodying a number of logical implications of the form $< generator > \to < closed\ concept >$; and (4) retains only those implications for which the $< generator >$ precedes the remainder of the consequent $< concept >$ in all supporting trace sequences in T. This approach works. But, there are still a number of issues to be considered.

First, the prior identification of software events of interest can be awkward. If the events denote entrance, and exit, from modules, procedures or other bodies of code as in Ball [1] and this paper, then this step is fairly straight forward. But, there are other kinds of "events" that are of interest in software analysis. Prime examples are "conditions" such as "$x + y > 100 * z$". Typically such conditions form the basis of triggers, or guards. Uncovering the various relationships between conditions and the events they may trigger is a key to finding the "likely invariants" that describe a body of software [1, 3, 14].

Michael Ernst, in particular, has been a leader in identifying likely invariants from dynamic trace data [3, 14]. Causal dependencies are a form of software invariant. So, this paper can be considered to be an extension of his work. But, neither Ernst nor we know how to discover what conditional relationships might participate in a likely software invariant without first identifying them *a priori*. It is a significant outstanding problem that we are currently investigating.

Second, given a set of causal dependencies such as Figure 5 we would like to be able to reason about them. Some rules, such as the transitive law, *if* $x \Rightarrow y$ *and* $y \Rightarrow z$ *then* $x \Rightarrow z$ remain true in a causal logic. But, as we saw in Section 3, others do not.

There is a considerable body of literature concerning "temporal logic" which has been studied since the early 70's as an analytic tool associated with finite state controllers, reactive devices and parallel systems [2, 4, 6, 19]. Most varieties of linear time logic (LTL) introduce 4 additional temporal operators, X, U, F, G where, given boolean expressions α, β, we have

$X\alpha$ denotes "next α"
$F\alpha$ denotes "eventually α"
$G\alpha$ denotes "generally (or always) α".
$\alpha U \beta$ denotes "α until β"

---

[1] It seems appropriate in this application to relabel the relation $R(O, A)$ as $R(T, E)$, where T denotes the set of traces and E denotes the set of events.

Causal dependencies can be expressed in terms of the X and F operators. Unfortunately valid derivations within a temporal logic are rare. Temporal logics encounter the same kinds of issues we have illustrated with our causal dependencies.

The third issue we must consider is the last step, operator domination, in which we winnow out those logical implications which cannot represent causal dependencies. As was pointed out in Section 3, we can use the support for each concept to limit the number of trace sequence that must be examined to verify the temporal precedence properties. But, this would seem to negate much of the advantage obtained by incrementally creating the concept lattice in the manner of Godin and Missaoui. We will have to keep the entire set of trace data on hand and possibly re-examine hundreds of trace sequences as each new concept is entered into the lattice.

Fortunately, this is only an apparent problem caused by our rough and ready, proof of concept software. One can incrementally create a "precedes" relation, $<$, as shown in Figure 10. Here, $x < y$ if x precedes y in any trace t. Readily, $<$ is only a pre-order, since it is transitive but not antisymmetric. If $x < y$ then y cannot causally determine x. By creating a precedence relation in parallel with the concept lattice, one can incrementally uncover likely causal dependencies on the fly with no need to re-examine earlier trace data, or even to retain it. Of course, if we do not keep the original trace data we will then lose the opportunity to look more carefully at any particular trace to see why this is, or is not, a case of causal dependency.

$<$	a	b	c	d	e	p	q	r	s	t	u
a			×		×	×	×		×	×	×
b	×			×	×	×	×	×	×	×	×
c	×					×		×	×	×	
d	×		×		×		×	×	×	×	×
e		×	×			×		×	×	×	×
p	×			×				×		×	×
q			×	×				×	×		
r			×	×					×		
s							×	×			
t						×	×	×	×		×
u	×	×		×	×	×	×	×	×	×	

Fig. 10. The precedes relation from Figure 2

But possibly a precedence relation such as Figure 10 can do more. Why not use this relation to directly indicate causal dependencies? For example, we see in Figure 10 that $b < a$ while $a \not< b$. Thus we know that b precedes a in at least one trace, and that a never precedes b. This strict anti-symmetry is one property that we have postulated for causal dependence. It is a necessary condition. But, it is not sufficient. Our understanding of causal dependence $b \Rightarrow a$ is that whenever b occurs then a must also always follow. This is not true in the sequences 2 and 7 of Figure 2. This second necessary condition whose logical expression is

$$(\forall t \in T)[b(t) \rightarrow [b(t) < a(t)]] \tag{4}$$

seems to be a fundamental property of causal dependence that cannot be derived from simple precedence relations.

This author believes that the key to discovering causal dependencies from observed software behavior must involve the use of Galois closure, which is the basis of formal concept analysis. Only by adopting a formal concept methodology can we derive an expression such as (4). It seems to have been a key piece that has been missing in the search for "likely software invariants".

Finally, we observe that our procedure still only reveals "likely" causal dependencies. Because we find that $a \Rightarrow q$ in the set T of trace data, we cannot literally say that the event a "causes" the event q as a consequence. One can only base such a claim on examination of the code itself. But, without having likely dependencies to specifically look for, such examination is extremely difficult; and in the case of legacy systems without source code it is essentially impossible.

The principles of formal concept analysis have an important application in software analysis and software engineering.

References

1. Thomas Ball. The Concept of Dynamic Analysis. *Proc. Seventh European Software Engineering Conf.*, pages 216–234, Sept. 1999.
2. E. Allen Emerson.   Temporal and Modal Logic . In *Handbood of Theoretical Computer Science*, pages 997–1071. Elsevier Science, 1990.
3. Michael D. Ernst, Jake Cockrell, William G. Griswold, and David Notkin. Dynamically Discovering Likely Program Invariants to Support Program Evolution. *IEEE Trans. Software Eng.*, 27(2):1–25, Feb. 2001.
4. Michael Fisher. A Model Checker for Linear Time Temporal Logic. *Formal Aspects of Computing*, 4(3):299–319, 1992.
5. Bernhard Ganter and Rudolf Wille. *Formal Concept Analysis - Mathematical Foundations*. Springer Verlag, Heidelberg, 1999.
6. Rob Gerth, Doron Peled, Moshe Y. Vardi, and Pierre Wolper. Simple on-the-fly Automatic Verification of Linear Temporal Logic. In Piotr Dembinski and Marek Sredniawa, editors, *Protocol Specification, Testing and Verification XV, Proc. of 15th IFIP Workshop*, pages 3–18, Warsaw, Juue 1996.
7. R. Godin and Hafedh Mili. Building and Maintaining Analysis-Level Class Hierarchies Using Galois Lattices. In *ACM Conf. on Object-Oriented Programming Systems, Languages and Applications (OOPSLA'93)*, pages 394–410, Washington, DC, 1993.
8. Robert Godin and Rokia Missaoui. An Incremental Concept Formation Approach for Learning from Databases. In *Theoretical Comp. Sci.*, volume 133, pages 387–419, 1994.
9. Robert Godin, Rokia Missaoui, and Hassan Alaoui. Incremental Concept Formation Algorithms Based on Galois (Concept) Lattices. *Computational Intelligence*, 11(2):246–267, 1995.
10. Karam Gouda and Mohammed J. Zaki. Efficiently Mining Maximal Frequent Item Sets. In *1st IEEE Intern'l Conf. on Data Mining*, San Jose, CA, Nov. 2001.
11. Robert E. Jamison and John L. Pfaltz. Closure Spaces that are not Uniquely Generated. *Discrete Appl Math.*, 147:69–79, Feb. 2005. also in Ordinal and Symbolic Data Analysis, OSDA 2000, Brussels, Belgium July 2000.
12. Christian Lindig and Gregor Snelting. Assessing Modular Structure of Legacy Code Based on Mathematical Concept Analysis. In *Proc of 1997 International Conf. on Software Engineering*, pages 349–359, Boston, MA, May 1997.

13. Oystein Ore. Galois Connexions. *Trans. of AMS*, 55:493–513, 1944.
14. Jeff H. Perkins and Michael D. Ernst. Efficient Incremental Algorithms for Dynamic Detection of Likely Invariants. *Proc. SIGSOFT'04/FSE-2*, pages 23–32, Nov. 2004.
15. John L. Pfaltz and Christopher M. Taylor. Closed Set Mining of Biological Data. In *BIOKDD 2002, 2nd Workshop on Data Mining in Bioinformatics*, pages 43–48, Edmonton, Alberta, July 2002.
16. John L. Pfaltz and Christopher M. Taylor. Concept Lattices as a Scientific Knowledge Discovery Technique. In *Workshop on Discrete Mathematics and Data Mining, 2nd SIAM International Conference on Data Mining*, pages 65–74, Arlington, VA, Apr. 2002.
17. Michael Siff and Thomas Reps. Identifying Modules via Concept Analysis. In *Intn'l Conf. on Software Maintenance*, pages 170–179, Bari, Italy, Oct. 1997.
18. Gregor Snelting and Frank Tip. Reengineering Class Hierarchies Using Concept Analysis. In *Proc. ACM SIGSOFT 6th International Symposium on Foundations of Software Engineering, FSE-6*, pages 99–110, Lake Buena Vista, FL, 1998.
19. Paulo Tabuada and George J. Pappas. Linear Time Logic Control of Discrete-time Linear Systems. *IEEE Trans. on Automatic Control*, page (in revision), 2005.
20. Petko Valtchev, Rokia Missaoui, and Robert Godin. A Framework for Incremental Generation of Frequent Closed Itemsets. In Peter Hammer, editor, *Workshop on Discrete Mathematics & Data Mining, 2nd SIAM Conf. on Data Mining*, pages 75–86, Arlington, VA, April 2002.
21. Petko Valtchev, Rokia Missaoui, Rouane Hacene, and Robert Godin. Incremental Maintenance of Association Rule Bases. In *Proc. Workshop on Discrete Mathematic and Data Mining*, San Francisco, CA, 2003.
22. Rudolf Wille. Restructuring Lattice Theory: An approach based on hierarchies of concepts. In Ivan Rival, editor, *Ordered Sets*, pages 445–470. Reidel, 1982.
23. Jinlin Yang and David Evans. Automatically Inferring Temporal Properties for Program Evolution. In *15th IEEE Symposium on Software Reliability Engineering (ISSRE 2004)*, Saint-Malo, France, Nov. 2004.
24. Jinlin Yang and David Evans. Dynamically Inferring Temporal Properties. In *Proc. ACM SIGPLAN-SIGSOFT Workshop on Program Analysis for Software Tools and Engineering (PASTE 2004*, Washington, DC, June 2004.
25. Jinlin Yang, David Evans, Deepali Bhardwaj, Thirumalesh Bhat, and Manuvir Das. Terracotta: Mining Temporal API Rules from Imperfect Traces. In *28th Internl. Conf. on Software Engineering (ICSE 2006)*, page (submitted), Shanghai, China, May 2006.
26. Mohammed J. Zaki. Generating Non-Redundant Association Rules. In *6th ACM SIGKDD Intern'l Conf. on Knowledge Discovery and Data Mining*, pages 34–43, Boston, MA, Aug. 2000.
27. Mohammed J. Zaki and Ching-Jui Hsiao. CHARM: An Efficient Algorithm for Closed Association Rule Mining. In Robert Grossman, editor, *2nd SIAM International Conf. on Data Mining*, pages 457–473, Arlington, VA, April 2002.

An FCA Interpretation of Relation Algebra

Uta Priss

School of Computing, Napier University, Edinburgh, UK
u.priss@napier.ac.uk
www.upriss.org.uk

Abstract. This paper discusses an interpretation of relation algebra and fork algebra with respect to FCA contexts. In this case, "relation algebra" refers to the DeMorgan-Peirce-Schroeder-Tarski algebra and not to the "relational algebra" as described by Codd. The goal of this interpretation is to provide an algebraic formalisation of object-relational databases that is based on binary relations and thus closer to FCA and formal contexts than the traditional formalisation based on Codd. The formalisation provides insights into certain symmetries (among quantifiers) and the use of ternary relations and part-whole relations for building relational databases.

1 Introduction

Algebras of relations, such as Codd's (1970) relational algebra (RLA) or Peirce-Tarski's relation algebra (RA)[1], have been studied by logicians since the mid 19th century. But apart from the use of RLA in relational databases, relational methods have not been in the mainstream for more than a hundred years, even though they have promising applications. Only during the past 15 years, there has been an increased interest in "Relational Methods in Computer Science" as evidenced by the creation of a new journal in this area[2].

Relational methods can be considered a "paradigm" that is different from some set-based logical formulas because a relational representation abstracts from elements and certain quantifiers. Programming languages that are based on relational methods tend to be more of a non-functional, list processing character. Users sometimes find such languages or formalisms difficult to read - as has been documented with respect to the relational database language SQL (eg. Hansen & Hansen (1988)). This may explain why relational methods are only slowly gaining more popularity. Nevertheless, relational methods have interesting applications and because of the recent interest in relational methods in computer science and because RA and FCA share common structures, we believe that a detailed discussion of FCA and RA is of interest to the FCA community and provides links to this newly emerging research area.

A combination of RA and FCA can be used to analyse formal aspects that underpin relational and object-relational databases. Current RLA-based implementations of databases are highly optimised with respect to functionality and efficiency. But RA can

---

[1] In this paper "RLA" is used as an abbreviation for Codd's relational algebra and "RA" for a Tarski-style relation algebra.

[2] http://www.jormics.org

R. Missaoui and J. Schmid (Eds.): ICFCA 2006, LNAI 3874, pp. 248–263, 2006.

provide new insights into the structural properties of relational databases, such as into certain symmetries (among quantifiers) and the use of ternary relations and part-whole relations for relational databases. Currently, there does not exist a widely accepted extension of RLA to object-relational databases (cf. Atkinson et al. (1989)). A broader approach using a variety of algebras (including RA and RLA) may lead to such a formalisation of object-relational databases and formal ontologies. With respect to FCA, this paper shows that RA is sufficiently expressive to represent basic FCA notions and a fork extension of RA is sufficient to represent many-valued contexts and power contexts.

This paper presents a continuation and elaboration of some ideas which were presented in a preliminary form by Priss (2005). But in contrast to Priss (2005), this paper adds a more detailed mathematical presentation, a use of fork algebras, a distinction between a "named" and an "unnamed" perspective and a more detailed elaboration of relational schemata.

2 Algebras of Relations: Codd Versus Tarski

The most influential algebra that is currently used in computer science is probably RLA because it serves as the foundation of relational databases. It is not a trivial task to mathematically formalise RLA in detail with respect to relational databases - as indicated by the fact that at least four different types of suggestions for such formalisations of RLA exist (Abiteboul et al., 1995). Because of this and because RLA uses n-ary relations, it is also not trivial to combine relational databases directly with Formal Concept Analysis (FCA). Wille's (2002) notion of power context families incorporates n-ary relations into FCA, but it does not cover all the detail of relational databases and RLA. Hereth (2002) has made some progress with respect to RLA and FCA.

Although Codd (1970) is usually quoted as the inventor of RLA (and he certainly advocated the practical use of it), a more detailed and comprehensive description of algebras of relations was provided by Tarski in the 1940s (cf. Van den Bussche (2001) for an overview). Tarski described two types of algebras: RA and Cylindric Set Algebra, which according to Imielinski & Lipski (1984) is closely connected to RLA. The idea of RA can be traced back from Tarski (1941) to Peirce and de Morgan and Schröder (cf. Pratt (1992) for an overview). In contrast to RLA which has expressive power equivalent to first order logic, the expressive power of RA is only equivalent to first order logic with at most 3 distinct variables (cf. Van den Bussche (2001)). Thus RA is much less powerful than RLA. But there is an extension of RA called "fork algebra", which is equivalent to first order logic. Because of a close relationship between RA and binary relations, it is of interest to consider RA and fork algebra together with FCA.

3 Relation Algebra: Definition and Overview

The following definition follows Tarski and is adapted from Brink et al. (1992):

Definition 1. A *relation algebra* is an algebra $(R, +, \cdot, ', 0, 1, ;, \smile, e)$ satisfying the following axioms for each $r, s, t, \in R$:

R1 $(R, +, \cdot, ', 0, 1)$ is a Boolean algebra   R5 $(r + s); t = r; t + s; t$

R2 $r; (s; t) = (r; s); t$ R6 $(r + s)^\smile = r^\smile + s^\smile$

R3 $r; e = r = e; r$ R7 $(r; s)^\smile = s^\smile; r^\smile$

R4 $r^{\smile\smile} = r$ R8 $r^\smile; (r; s)' \leq s'$.

If R in Definition 1 is a set of binary relations, then the following can be defined:

Definition 2. A *proper relation algebra* (RA) is an algebra $(R, \cup, {}^-, one, \circ, {}^d, dia)$ where for a set A and an equivalence relation $one \subseteq A \times A$, R is a set of binary relations equal to the powerset of one and $dia := \{(x, x) \in one\}$ and for $I, J \in R$: $I \cup J := \{(x, y) \mid (x, y) \in I \text{ or } (x, y) \in J\}$; $\overline{I} := \{(x, y) \mid (x, y) \in one, (x, y) \notin I\}$; $I \circ J := \{(x, y) \mid \exists_{z \in A} : (x, z) \in I \text{ and } (z, y) \in J\}$; $I^d := \{(x, y) \mid (y, x) \in I\}$.

Different authors use different symbols for RA operations. It is common to list the Boolean operators before the non-Boolean ones in the signature of the algebra. It is also common to use Tarski's notation $(+, \cdot, ; , 0, 1)$ for relation algebras and $\cup, \cap, {}^-$ and other symbols for proper relation algebras. The left column of table 1 shows the notations used in this paper. For the purposes of this paper, a mapping from an equational class (as in definition 1) to an algebra which has elements with set-theoretically defined structure (as in definition 2) is called an "interpretation". Apart from interpreting relation algebras with respect to binary relations, they can also be interpreted with respect to FCA contexts, as shown in later sections of this paper. Pratt (1993) observes that only the non-Boolean operations $(\circ, {}^d)$ make use of the inner structure of the elements of the relations (such as inverting the pairs using ^{d}). For the Boolean operations $(\cup, \cap, {}^-)$, relations are just sets. It has been shown by Lyndon (1950) that there are interpretations of relation algebras which are not isomorphic to proper relation algebras. In this paper, only proper relation algebras are considered and their notational symbols are used. "RA" stands for proper relation algebra in the remainder of this paper. Based on definition 2, a representable relation algebra (RRA) is usually defined as a subalgebra of a proper relation algebra. The class RRA forms a variety (Tarski, 1955).

Table 1. Overview of basic RA operations and some extensions

RA	Tarski's	name	basis	definitions
$\cup$	$+$	union	yes	
$-$	$'$	negation (complement)	yes	
$\cap$	$\cdot$	intersection		$I \cap J := \overline{\overline{I} \cup \overline{J}}$
$\circ$	$;$	composition	yes	
d	$\smile$	inverse (dual)	yes	
$\bullet$	$\dagger$	de Morgan compl. of $\circ$		$I \bullet J := \overline{\overline{I} \circ \overline{J}}$
nul	0			$nul := \overline{one}$
one	1	universal relation	yes	
dia	e	diagonal	yes	
$\subseteq$				$I \subseteq J :\Leftrightarrow I \cap J = I$
$=$		equality		$I = J :\Leftrightarrow I \subseteq J$ and $J \subseteq I$
$\subset$		containment		$I \subset J :\Leftrightarrow I \subseteq J$ and not $I = J$
trs		transitive closure	(yes)	$I^{trs} := I \cup I \circ I \cup I \circ I \circ I \cup \ldots$
$\star$		refl. trans. closure		$I^* := dia \cup I^{trs}$

The top half of table 1 shows how other common operations and elements of RA can be derived from the basis operations and elements. Numerous mathematical (or logical) properties can be proven for RA (cf. Maddux (1996), Pratt (1992), Pratt (1993), Van den Bussche (2001), Kim (1982)). There are numerous applications for RA, which obviously include and extend applications of Boolean algebras. Apart from applications in logic, RA has been used for the semantics of programming languages (Maddux, 1996). Pratt (1993) explains that RA is very similar to Chu spaces.

The bottom half of table 1 shows some extensions of RA: equality and transitive closure. For this paper, equality and containment is assumed to be defined for RA. Transitive closure is not a first order logic property and cannot be derived from the other RA operations. It can be useful in some applications to have transitive closure available. For example, if I represents the incidence matrix of a graph, then I^{trs} shows all transitive paths between the graph nodes. We believe that a major reason for the recent popularity of XML for ontologies and other tree-like structures is because the calculation of paths in a tree is natural for XML but difficult in SQL. In fact, only the more recent SQL standard (SQL3) contains a suggestion for a recursion operator that can be used for calculating paths in a tree. Unfortunately, the implementation of this operator is inconsistent among different database vendors (Wagner, 2003). The reasons for this may be that transitive closure is missing from RLA and that it can be computationally expensive to calculate transitivity. Nevertheless, we believe, that if transitive closure had been added to SQL at an earlier stage, the history of XML as a format for representing ontologies might have been different. This brief excursus on XML and databases should indicate the significance of the presence or absence of a transitive closure operation. In this paper, we assume that transitive closure is available as needed.

4   RA Interpretations as FCA Contexts

4.1   Active Domains

In analogy to relational database theory, an "active domain" (ACT) is introduced for the purposes of this paper. In relational database theory, an active domain is the finite set of actually occurring values and value combinations, which is a subset of the infinite "universe" (U) of possible values. For example, the complement of a relation in relational databases is usually calculated with respect to the active domain to avoid the use of infinite sets. Relational databases contain finite sets of data at any point in time, but a fork operation as introduced in section 5.2 requires an infinite set of elements at all times. To cope with the infinity of the fork operation, the following two sets are defined in this paper: a set of identifiers containing a finite set of even-numbered identifiers (EVN) and an infinite set of odd-numbered identifiers (ODD). Even-numbered identifiers are used for actual, persistent or "important" data and odd-numbered ones are used for potential, transient or "un-important" data. This distinction follows the practise of object-relational databases which automatically generate "object identifiers" for instances of tables. It also follows the distinction between "persistent" data (for data that may need to be reused in other applications and should be stored) and "transient" data (which can be forgotten, such as the values of a counter) by Atkinson et al. (1989). It should be noted that even though the set of even identifiers is finite and fixed at any point in time, it can change over time if new data is added to an application (eg. if database tables are updated). In addition to the sets of identifiers, there is also a finite set N of named elements of a relational database (i.e., names of tables, columns, values, etc).

Definition 3. A *universe* of possible elements is a set $U := N \cup EVN \cup ODD$ where N is a finite set of names, EVN is a finite set of even-numbered identifiers and ODD is an infinite set of odd-numbered identifiers and N, EVN, and ODD are pairwise disjoint. An *active domain* is a finite subset of U defined as $ACT := N \cup EVN$. For practical purposes, ACT is assumed to have a fixed linear order.

4.2 The Unnamed Perspective

There are potentially numerous ways for using RA with respect to FCA. Because formal contexts are usually represented as cross tables, for the rest of this paper binary relations are viewed as Boolean matrices (or binary matrices or cross-tables) in the sense of Kim (1982). In analogy to a distinction made in relational database theory (Abiteboul, 1995), we distinguish between an "unnamed perspective" and a "named perspective". In the unnamed perspective, all data of an application, i.e., all formal contexts of an application, are represented as (possibly large) matrices of the same dimension[3]:

Definition 4. In the unnamed perspective, with $|\mathcal{A}| := |ACT| \times |ACT|$, an *active domain* $\mathcal{A}$ is the set of all binary $|\mathcal{A}|$-dimensional matrices I so that semantically for all elements in ACT, the nth element in ACT corresponds to the nth row and column in I. It is then said that I *is based on* $\mathcal{A}$ denoted by $I_{\mathcal{A}}$.

The subscript $\mathcal{A}$ in $I_{\mathcal{A}}$ can be omitted if it clear from context. Obviously, it would be impractical for most applications to actually construct such large matrices. The unnamed perspective is mainly used to define some operations in a somewhat more context-independent manner, which can be useful for certain context compositions in the named perspective. Otherwise, the unnamed perspective is mostly of theoretical value. Each object or attribute of any formal context relating to a single application is uniquely identified by its position in $\mathcal{A}$. Row and column permutations change the values of rows and columns but not their semantic correspondence to elements in $\mathcal{A}$ (thus may not be meaningful operations). Even though, the names of the elements in ACT are not strictly required, it is usually more convenient to use them instead of using row and column numbers.

The next definition assumes the usual operations for Boolean matrices (cf. Kim (1982)), i.e., with $(i, j)_I$ denoting the element in row i, column j in matrix I and $\vee, \wedge$ and $\neg$ denoting Boolean OR, AND and NOT: $(i, j)_{I \cup J} := (i, j)_I \vee (i, j)_J$; $(i, j)_{\overline{I}} := \neg(i, j)_I$; $(i, j)_{I \circ J} := 1$ iff $\exists_k : (i, k)_I \wedge (k, j)_J$; $(i, j)_{I^d} := (j, i)_I$. A matrix is symmetric if $I = I^d$, reflexive if $dia \subseteq I$, transitive if $I^2 \subseteq I$. Because matrix operations and operations on binary relations are so similar we use a set-theoretic notation for both. The distinction between sets and matrices is made using typeface (see footnote 3).

Definition 5. A *matrix-RA based on* $\mathcal{A}$ is an algebra $(R, \cup, ^-, one, \circ, ^d, dia())$ where $one \in R$ is a reflexive, symmetric and transitive matrix; R is a set of Boolean matrices

---

[3] In the rest of this paper, typewriter font (A, B, etc) is used for subsets and elements of ACT and uppercase italics (I, J etc) for matrices and binary relations. If elements of ACT are used in names of matrices, then they are written in italics but underlined (but not if used as subscript).

based on $\mathcal{A}$ with $I \in R \Leftrightarrow I \subseteq one$; and $\cup, ^-, \circ, ^d$ are the usual Boolean matrix operations; and for any set $S \subseteq \mathtt{ACT}$ and $\mathtt{a(n)}$ denoting the nth element in $\mathtt{ACT}$, $dia(S)$ is defined by $(i,j)_{dia(S)} = 1$ iff $i = j$ and $\mathtt{a}(i) \in S$ (but only if $dia(S) \subseteq one$).

Table 2 summarises the definition and introduces some further operations (with $\mathtt{G, M, S} \subseteq \mathtt{ACT}$). The operation $\cap$ is still the de Morgan complement of $\cup$. nul can be derived from dia or via $nul = \overline{one}$. Because binary relations can be equivalently represented as sets of pairs or as binary matrices it follows that:

Table 2. The unnamed perspective: A matrix-RA based on $\mathcal{A}$

notation	definition	basis
$\cup$	component-wise $\vee$	yes
$^-$	component-wise $^-$	yes
$\circ$	binary matrix multiplication	yes
^{d}	matrix transposition (mirrored along diagonal)	yes
nul	$:= \overline{dia \cup \overline{dia}}$	
dia	$:= dia(\mathtt{ACT})$	
$dia(S)$	has 1's according to S	yes
$dia(I_\uparrow)$	$:= dia \cap (one \circ I) = I^d \circ I \cap dia$	
$dia(I_\rightarrow)$	$:= dia \cap (I \circ one) = I \circ I^d \cap dia$	
$sqr(\mathtt{G, M})$	$:= dia(\mathtt{G}) \circ one \circ dia(\mathtt{M})$	

Lemma 1. Definitions 2 and 5 are equivalent: every RA is a matrix-RA and vice versa.

It is not necessary to use sets $(S, \mathtt{ACT})$ in definition 5. Instead of defining $dia(S)$, one could define dia and then derive $dia(I_\rightarrow), dia(I_\uparrow)$ and state that every matrix $J \subseteq dia$ corresponds to a set. Thus the algebra in definition 5 is not truly two sorted. But the use of $dia(S)$ is convenient with respect to the named perspective. The other definitions from table 2 can be explained as follows: the matrices $dia(I_\uparrow)$ and $dia(I_\rightarrow)$ represent column-wise and row-wise projections of a matrix I onto the diagonal. For $dia(I_\uparrow)$ this means that for each column in I that contains at least one 1, $dia(I_\uparrow)$ contains a 1 in that position on the diagonal. A matrix $sqr(\mathtt{G, M})$ contains a 1 for each cell whose row name is in $\mathtt{G}$ and whose column name is in $\mathtt{M}$. $sqr(\mathtt{G, M})$ is an encoding of an empty cross table of a formal context based on $\mathcal{A}$. A formal context can now be represented as $(sqr(\mathtt{G, M}), I)$ where $\mathtt{G, M} \subseteq \mathtt{ACT}$ and I is a matrix based on $\mathcal{A}$ with $I \subseteq sqr(\mathtt{G, M})$.

Definition 6. A *context-RA based on* $\mathcal{A}$ for a set of formal contexts is the smallest matrix-RA based on $\mathcal{A}$ that contains these contexts.

This means that for a context $(sqr(\mathtt{G, M}), I)$ the context-RA contains all contexts that have any subsets of $\mathtt{G} \cup \mathtt{M}$ as sets of objects and attributes. It should be noted that a smaller RRA could be constructed that contains a set of contexts, if R in definition 5 was not required to contain all matrices $I \subseteq one$. But since the prime operator $(')$ in FCA is normally applicable to all subsets of objects or attributes, definition 5 (which allows the formation of matrices corresponding to subsets) seems reasonable. Subsets of $\mathtt{G}$ and $\mathtt{M}$ can be represented as diagonal matrices or as matrices which contain identical rows (eg. $sqr(\mathtt{ACT, S})$) or identical columns (eg. $sqr(\mathtt{S, ACT})$). These three ways are equivalent because the matrices can be converted: $dia(S) = sqr(ACT, S) \cap dia = sqr(S, ACT) \cap dia$ and $sqr(ACT, S) = one \circ dia(S)$.

Lemma 2. In the unnamed perspective the basic FCA operations can be represented as summarised in table 3.

Table 3. Basic FCA operations in the unnamed perspective

standard FCA	RA: unnamed perspective
gIm	$sqr(\{g\}, \{m\}) \subseteq I$
$g' := \{m \in M \mid gIm\}$	$dia(g') = dia(g^+) := \overline{dia \cap sqr(ACT, \{g\})} \circ I$
$H' := \{m \in M \mid \forall_{g \in G} : g \in H \Longrightarrow gIm\}$	$dia(H') = dia \cap sqr(ACT, H) \circ \overline{I}$
$H^+ := \{m \in M \mid \exists_{g \in G} : g \in H \text{ and } gIm\}$	$dia(H^+) = dia \cap sqr(ACT, H) \circ I$

The equivalence of the expressions in the left and right columns in table 3 follows directly from the definitions. But a further explanation of the table is required: a context $(sqr(G, M), I)$ is assumed with $H \subseteq G$; $N \subseteq M$; $g \in G$; $m \in M$. The plus $(+)$ operator, which is somewhat dual to the prime $(')$ operator originates from use in lexical databases (cf. Priss(1998) and Priss & Old (2004)). The operations for sets of attributes are analogous to the ones for sets of objects in table 3.

4.3 The Named Perspective

In contrast to the unnamed perspective where all matrices of an application are of dimension $|\mathcal{A}|$, in the named perspective matrices can have different dimensions and may not even be square.

Definition 7. In the named perspective, the *active domain* ACT is linearly ordered. A formal context (G, M, I) *based on* ACT consists of two sets $G, M \subseteq$ ACT, which are linearly ordered using the by ACT-induced ordering, and of a binary matrix I of dimension $|G| \times |M|$ where the ith row corresponds to the ith element in G and the jth column corresponds to the jth element in M. This can be denoted as $I_{G,M}$.

Semantically, this implies a unique name assumption because if the same name is used in different formal contexts or in a single context both as an object and as an attribute, then these elements are semantically indistinguishable because they refer to the same element in ACT. The unique name assumption ensures that the operations $\cup$ and $\circ$ can be meaningfully generalised to contexts of different dimensions as follows:

Definition 8. For formal contexts $\mathcal{K}_1 := (G_1, M_1, I)$ and $\mathcal{K}_2 := (G_2, M_2, J)$ the following *context operations* are defined:

$\mathcal{K}_1 \sqcup \mathcal{K}_2 := (G_1 \cup G_2, M_1 \cup M_2, I \sqcup J)$ with $gI \sqcup Jm :\Longleftrightarrow gIm$ or gJm
$\mathcal{K}_1 \sqcap \mathcal{K}_2 := (G_1 \cup G_2, M_1 \cup M_2, I \sqcap J)$ with $gI \sqcap Jm :\Longleftrightarrow gIm$ and gJm
$\mathcal{K}_1 \diamond \mathcal{K}_2 := (G_1, M_2, I \diamond J)$ with $gI \diamond Jm :\Longleftrightarrow \exists_{n \in (M_1 \cap G_2)} : gIn$ and nJm
$\overline{\mathcal{K}_1} := (G_1, M_1, \overline{I})$; $\mathcal{K}_1^d := (M_1, G_1, I^d)$.

Table 4 shows some further operations that can be defined for formal contexts in the named perspective. Most of the operations are essentially the same as in the unnamed

Table 4. Further context operations

$\mathcal{K}_1 \cup \mathcal{K}_2$	$:= \mathcal{K}_1 \sqcup \mathcal{K}_2$ if $G_1 = G_2, M_1 = M_2$
$\mathcal{K}_1 \circ \mathcal{K}_2$	$:= \mathcal{K}_1 \diamond \mathcal{K}_2$ if $M_1 = G_2$
$\dfrac{\mathcal{K}_1}{\mathcal{K}_2}$	$:= \mathcal{K}_1 \sqcup \mathcal{K}_2$ if $G_1 \cap G_2 = \emptyset$ and $M_1 = M_2$
$\mathcal{K}_1 \vert \mathcal{K}_2$	$:= \mathcal{K}_1 \sqcup \mathcal{K}_2$ if $G_1 = G_2$ and $M_1 \cap M_2 = \emptyset$

$dia_G(S_\rightarrow)$	see definition 9
$dia_M(S_\uparrow), dia_G(I_\rightarrow), dia_M(I_\uparrow)$	analogous to $dia_G(S_\rightarrow)$
$set_G(I), set_M(I)$	see definition 9, $set_G(I) = set_G(dia_G(I_\rightarrow))$
$nul_{G,M}$	$:= \overline{I_{G,M} \cup \overline{I_{G,M}}}$
$red_{G,M}(J)$	$:= dia_G \diamond J_{G_1,M_1} \diamond dia_M$
$col_G(S)$	$:= dia_G(S_\rightarrow) \circ one_{G,\{x\}}$
$row_M(S)$	$:= one_{\{x\},M} \circ dia_M(S_\uparrow)$

perspective. Because dia is square, one set as subscript is sufficient ($dia_G := dia_{G,G}$). In addition to operations which convert a set into a matrix ($dia_M(S_\rightarrow)$), there are also operations which convert a matrix into a set: $set_G(I)$. Union potentially enlarges the dimension of the original matrices. A reduction operation $red_{G,M}(J)$ eliminates all rows and columns from a matrix J which do not correspond to elements in G and M, respectively. The following holds for context composition: $\mathcal{K}_1 \diamond \mathcal{K}_2 = (G_1, M_1 \cup G_2, I \sqcup nul_{G_1,G_2}) \circ (M_1 \cup G_2, M_2, J \sqcup nul_{M_1,M_2})$

Definition 9. A *context algebraic structure (CAS) based on* ACT is a three sorted algebra $(R_1, R_2, R_3, \sqcup, ^-, \diamond, ^d, dia(), set(), (,,))$ where R_2 is a set of subsets of ACT, R_3 is a set of Boolean matrices, R_1 is a set of formal contexts based on ACT and constructed using the partial function $(,,) : R_2^2 \times R_3 \rightarrow R_1$; $\sqcup, ^-, \diamond, ^d$ are according to definition 8; $set_G(I) := \{g \in G \mid \exists_{m \in M} : gIm\}$; $set_M(I) := \{m \in M \mid \exists_{g \in G} : gIm\}$; and $dia_G(S_\rightarrow)$ is defined by $(i,j)_{dia_G(S_\rightarrow)} = 1$ iff $i = j$ and for the ith element in G: $g(i) \in S$.

The algebra in definition 9 is not a RA because formal contexts have different nul elements and composition from the left and the right may require a different dia element. But presumably a homomorphism can be constructed that maps each context (G, M, I) onto a pair $(sqr(G, M), I)$, that maps all diagonal matrices onto dia and all null matrices onto nul, and that maps the other operations accordingly resulting in a context-RA. (The details of this are left to future research.)

Lemma 3. In the named perspective the basic FCA operations can be represented as summarised in table 5.

Table 5. Basic FCA operations in the named perspective

standard FCA		CAS
gIm		$g^d \circ I \circ \underline{m}^d = (1)$
$g' := \{m \in M \mid gIm\}$		$\underline{g'} = \underline{g}^+ := \underline{g}^d \circ I$
$H' := \{m \in M \mid \forall_{g \in G} : g \in H \Longrightarrow gIm\}$		$H' := H^d \circ \overline{I}$
$H^+ := \{m \in M \mid \exists_{g \in G} : g \in H \text{ and } gIm\}$		$H^+ := H^d \circ I$

The following conventions are used in table 5: for a context (G, M, I); $H \subseteq G$; $N \subseteq M$; $g \in G$; $m \in M$; $H := col_G(H)$; $g := col_G(\{g\})$. As declared in footnote 3, the matrix names derived from elements of $\widehat{ACT}$ are underlined (such as $\underline{g}$). In the first row, (1) is a 1×1 matrix with element 1. In the named perspective, sets are best represented as row or column matrices. Because G and M need not be disjoint, it can be ambiguous whether $\underline{g}$ is a row or column. In that case, the notations $\underline{g}_c := col_G(\{g\})$ and $\underline{g}_r := row_M(\{g\})$ can be used. In table 5, H' is a row matrix but $\underline{H}$ and H'' are column matrices. H'' is calculated dually to H' by composition with I from the left: $H'' = \overline{\overline{I} \circ H^d \circ \overline{I}^d} = \overline{\overline{I} \circ \overline{I}^d \circ H}$. The notations from the unnamed and named perspective are compatible with each other and can be used together.

4.4 Eight Quantifiers

The use of negation and composition in the calculation of H' and H^+ raises the question as to whether other combinations of negation and composition are of interest. Table 6 summarises all 8 possible combinations of negation and composition for a context (G, M, I) and a set $N \subseteq G$. The third column in that table provides a rough linguistic description, which should be taken with caution because words such as "only" are fairly ambiguous in natural languages. In many applications, these 8 quantifiers result in 8 different sets, which together describe the relationship of N and G in some detail. In the next section, an example of a lattice construction is provided that summarises all 8 quantifiers in one diagram. It should be noted that with respect to relational databases, it can be quite challenging to formulate these 8 quantifiers in SQL because the "ALL" quantifier (corresponding to the so-called relational division in RLA) is not a primitive operation in SQL. In fact to represent this "ALL" quantifier in SQL, two sub-select statements are required (cf. Priss (2005) for an example).

4.5 Compositional Schemata

To represent more complex data than just a single relation using RA some kind of canonic means for translating complex data into binary relations is needed.

Definition 10. A *compositional schema* consists of a set of 4 or 9 formal contexts, which are arranged in a tabular manner, (cf. figure 1) so that some of the contexts can be derived from adjacent contexts using composition.

Table 6. Eight Quantifiers

N^+	$I \circ N$	at least one, some
$G \setminus N^+$	$\overline{\overline{I} \circ N}$	none
N'	$\overline{\overline{I} \circ N}$	relates to all
$G \setminus N'$	$\overline{I} \circ N$	does not relate to all
$(M \setminus N)^+$	$I \circ \overline{N}$	relates to those that are not only
$G \setminus ((M \setminus N)^+)$	$\overline{I \circ \overline{N}}$	relates to those that are only
$(M \setminus N)'$	$\overline{\overline{I} \circ \overline{N}}$	relates to all outwith
$G \setminus ((M \setminus N)')$	$\overline{I} \circ \overline{N}$	does not relate to all outwith

	A	B
A	1	2
C	3	4

	A	B	C
B	1	2	3
A	4	5	6
D	7	8	9

	A	B	C
B			L
A		J	$J \circ L$
D	I	$I \circ J$	$I \circ J \circ L$

Fig. 1. A compositional schema

The idea of compositional schemata is not new. There have been many papers on FCA which use such schemata explicitly or implicitly (eg. Ganter & Wille (1999), Priss (1998), Faid et al. (1997)). By identifying certain types of compositional schemata, their properties can be described in a general manner.

The numbering of the four or nine cells as presented in the left hand side of figure 1 is used in the remainder of this paper. In the case of nine cells, the compositional schema is built from the formal contexts $K_J := (A, B, J)$; $K_I := (D, A, I)$ and $K_L := (B, C, L)$. Because K_I and K_J share the set A, a context $K_{I \circ J} := (D, B, I \circ J)$ can be formed. Similarly, a context $K_{J \circ L}$ can be formed. Instead of the existence quantifier used in the construction of the matrices in cells 6, 8 and 9, any of the other seven quantifiers from table 6 can be used. A further context $K_{I \circ J \circ L}$ can be formed in cell 9 to complete the schema. It should be noted that while $K_{J \circ L}$ and $K_{I \circ J}$ are formed by composing the context to the left with the one above, $K_{I \circ J \circ L}$ is formed by composing the context to the left with the context two steps above (or the context two steps to the left with the one above). An exception is if J is a reflexive, transitive relation, in which case $J \circ J = J$ and $I \circ J \circ L = I \circ J \circ J \circ L$. Depending on the application, cells 1, 2, 4 can be filled with *nul* or *dia* or something else. In many cases, it may not be necessary to calculate a lattice for the context consisting of all 9 cells, but instead only for cells 5, 6, 8, 9 or just for individual cells. For identifying where objects and attributes are located in the schema, row and column matrices representing the sets A, B, and so on can be used. For example, an element g is an object in K_J if $g_c \subseteq A_c$.

Figure 2 provides two examples of compositional schemata. The first example can be constructed for any concept lattice. This example shows a formal context $(\{a, b, c, d\}, \{1, 2, 3, 4\}, I)$ where I is the matrix in cell 4 of the schema. After computing the set of concepts of this lattice, $(\{A, B, C, D, E\}$ without the top and bottom concept), cell 1 is filled with the conceptual hierarchy I_{sub}; cell 2 is filled with the intension relation between attributes and concepts (here called I_{attr}); cell 3 is filled with the extension relation between objects and concepts (called I_{inst}). Because I_{sub} is reflexive and transitive: $I = I_{inst} \circ I_{sub} \circ I_{attr} = I_{inst} \circ I_{sub} \circ I_{sub} \circ I_{attr}$. In this case also $I_{inst} = I_{inst} \circ I_{sub}$ because $\exists_{c_1} : g I_{inst} c_1, c_1 I_{sub} c_2 \iff g I_{inst} c_2$ and the same for I_{attr}.

The bottom half of figure 2 shows a lattice that visualises all 8 quantifiers from table 6. The compositional schema is constructed by inserting *dia* into cell 1, I into cell 3, the $\in$ relation into cell 2, and $I \circ N$ into cell 4. The lattice diagram shows that $I \circ N$ and $\overline{I} \circ N$ are the intensions of the join and meet of the elements in N. $I \circ \overline{N}$ and $\overline{I} \circ \overline{N}$ are the intensions of the join and meet of $M \backslash N$ which is $\overline{N}$. The other 4 quantifiers need not correspond to single concepts in the lattice but are the set-complements of the first 4 quantifiers. This lattice has the original lattice (G, M, I) as a sublattice. If this lattice was produced as an answer to a query about elements in N, it would answer many questions simultaneously: whether elements are at least in N, not in N, just in N, and so on.

	A	B	C	D	E	1	2	3	4
A	x				x				
B		x				x			
C			x					x	
D	x			x		x		x	
E	x	x			x	x	x		
a	x	x			x	x	x		
b	x			x		x		x	
c			x					x	
d		x						x	

	1	2	3	4	N	M\N
1	x				x	
2		x			x	
3			x			x
4				x		x
a	x	x			x	
b	x		x		x	x
c				x		x
d		x			x	

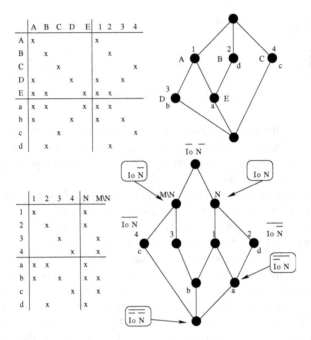

Fig. 2. Two examples of compositional schemata

5   Relational Schemata and Fork Algebra

5.1   Relational Schemata

This section covers schemata that represent the table structure of a relational database but without the actual values that are stored in the database and without showing which tables correspond to what is called "entities" and what is called "relations" in relational databases. More complex schemata with values and relations are covered in the next section. Relational schemata are relevant not just for relational databases but also for object-relational databases. There are some differences between the implementations (and thus the underlying formalisations) of object-relational databases among different vendors. For this paper, object-relational databases are considered to have a subtype relation among tables, i.e., one table can be defined to be a subtype of another table. This subtype relation is declared to be reflexive, acyclic and transitive. A subtype table inherits all columns (attributes) from its supertypes and the instances (rows) of a subtype are also assigned to its supertypes after deleting non-applicable columns. With this definition, a relational database is an object-relational database where the subtype relation is the identity (each table is only subtype of itself).

Definition 11. A *relational schema (of an object-relational database) based on* ACT is a CAS using a compositional schema according to figure 3 where Tbls is a set of table names, Inst a set of instances and Attr a set of column names with Inst $\subseteq$ ACT; Tbls, Attr $\subseteq$ N, and the sets are pairwise disjoint and linearly ordered according to ACT. The subschema consisting of the cells 5, 6, 8, 9 is denoted by $\mathcal{DB}$.

	Tbls	Tbls	Attr
Tbls			I_{attr}
Tbls		I_{sub}	$I_{sub} \circ I_{attr}$
Inst	I_{inst}	$I_{inst} \circ I_{sub}$	$I_{isba} := I_{inst} \circ I_{sub} \circ I_{attr}$

Fig. 3. The basic relational schema for an object-relational database

It is normally assumed that I_{attr} has no empty columns and that I_{inst} has no empty rows because having instances or attributes which are not in relationship to anything else is strange. The relational schema $\mathcal{DB}$ is basically the same as the first example of figure 2 because I_{sub} is reflexive and transitive and thus, for example, $I_{inst} \circ I_{sub} = I_{inst} \circ I_{sub} \circ I_{sub}$. In the line diagram of the lattice of $\mathcal{DB}$ the name of a table is attached to a node both as an object and as an attribute or in other words:

Theorem 1. For each table $\mathtt{t} \in \mathtt{Tbls}$ there exists a formal concept $c(\mathtt{t})$ in the concept lattice of $\mathcal{DB}$ which has $\mathtt{t}$ in its contingent extent and in its contingent intent. If $\mathtt{t_1} \neq \mathtt{t_2}$ then $c(\mathtt{t_1}) \neq c(\mathtt{t_2})$.

Proof: As before $\mathtt{t_r}$ denote a table among the attributes and $\mathtt{t_c}$ the same table among the objects. $\mathtt{t'_r} = \{\mathtt{s_c} \in \mathtt{Tbls} \mid \mathtt{s_c} I_{sub} \mathtt{t_r}\} \cup \{\mathtt{i} \in \mathtt{Inst} \mid \mathtt{i}(I_{inst} \circ I_{sub}) \mathtt{t_r}\}$ and $\mathtt{t''_c} = \{\mathtt{s_r} \in \mathtt{Tbls} \mid \forall_{\mathtt{y} \in \mathtt{t'_c}} : \mathtt{s_r}(I_{sub} \mid I_{sub} \circ I_{attr})\mathtt{y}\} \cup \{\mathtt{i} \in \mathtt{Inst} \mid \forall_{\mathtt{y} \in \mathtt{t'_c}} : \mathtt{i}(I_{inst} \circ I_{sub} \mid I_{isba})\mathtt{y}\}$. Because of $\mathtt{s} I_{sub} \mathtt{t_r} \iff \forall_{\mathtt{y} \in \mathtt{t'_c}} : \mathtt{s}(I_{sub} \mid I_{sub} \circ I_{attr})\mathtt{y}$ and $\mathtt{i}(I_{inst} \circ I_{sub})\mathtt{t_r} \iff \forall_{\mathtt{y} \in \mathtt{t'_c}} : \mathtt{s}(I_{inst} \circ I_{sub} \mid I_{isba})\mathtt{y}$ it follows that $\underline{\mathtt{t}}'_r = \underline{\mathtt{t}}''_c$. Same for $\underline{\mathtt{t}}'_c = \underline{\mathtt{t}}''_c$. This implies that $(\mathtt{t''_c}, \mathtt{t'_c})$ and $(\mathtt{t'_r}, \mathtt{t''_r})$ describe the same concept which has $\mathtt{t}$ in its contingent. Because the subtype relation I_{sub} is assumed to be acyclic (i.e., a table cannot be subtype of a second table which is itself a subtype of the first table), there is a different formal concept in $\mathcal{DB}$ for each different table.

The concept $c(\mathtt{t})$ has all attributes of $\mathtt{t}$ in its intension and all instances of $\mathtt{t}$ in its extension. Thus the concept lattice of $\mathcal{DB}$ summarises important information about the tables of an object-relational database. Another feature of $\mathcal{DB}$ is that different types of inheritance can be defined and analysed (cf. Priss (2005)). The formal context C_t for a table $\mathtt{t}$ can be derived as $C_t = (\mathtt{set_{Inst}}(\underline{\mathtt{t_c}}'), \mathtt{set_{Attr}}(\underline{\mathtt{t_r}}'), red_{G_t, M_t}(I_{inst} \circ dia_{\mathtt{Tbls}}(\underline{\mathtt{t}}) \circ I_{attr}))$.

5.2   Fork Algebraic Definitions

Tarski showed that RA is equivalent to first order logic with three variables (cf. Van den Bussche (2001)). An indication for why three variables are sufficient is given by Van den Bussche's example: $\{(x, y) \mid \exists_z (\exists_y (\exists_z (R(x, z) \wedge R(z, y)) \wedge R(y, z)) \wedge R(z, y))\}$. Tarski further showed that what is missing from RA is a form of "pairing", i.e., a means for combining two elements into a pair which then itself behaves like a primitive element. This pairing is required to build ternary relations. Different methods for adding a "pairing axiom" to RA have been suggested (eg. Jain, Mendhekar & Van Gucht (1995)). The approach which seems to be most widely used and which indeed has expressive power equivalent to first order logic is called "fork algebra" (Frias et al., 2004). It was developed in the area of programming language semantics for the purpose of dealing with non-deterministic algorithms. To our knowledge, applications in the area of databases as we are suggesting in this paper have not been discussed before.

The following two definitions are adapted from (Frias et al., 2004). For the purposes of this paper the usual operation (∇), is replaced by its relational dual denoted by $\triangle$.

Definition 12. A *fork algebra* is an algebra $(R, +, \cdot,', 0, 1, ;, \smile, e, \triangle)$ so that $(R, +, \cdot,', 0, 1, ;, \smile, e)$ is a relation algebra and for all $r, s, t, u \in R$:

F1 $r \triangle s = ((e \triangle 1); r) \cdot ((1 \triangle e); s)$ F3 $(e \triangle 1)^d \triangle (1 \triangle e)^d \leq e$
F2 $(r^d \triangle s^d)^d; (t \triangle u) = (r; t) \cdot (s; u)$

Definition 13. A *pre-proper fork algebra* (FRA) is a two sorted algebra $(R, U, \cup, \cap, \bar{\ }, nul, one, \circ,^d, dia, \triangle, \mathtt{frk}())$ where $(R, \cup, \cap, \bar{\ }, nul, one, \circ,^d, dia)$ is a RA on a set U; a binary function $\mathtt{frk} : U \times U \to U$ is injective on the restriction of its domain to *one*; the operation $\triangle$ is defined as $I \triangle J := \{(\mathtt{frk}(x, y), z) \mid (x, z) \in I; (y, z) \in J\}$ and R is closed under $\triangle$.

Proper fork algebras are defined somewhat more abstractly than pre-proper ones, but they are not required for this paper. Unless $one = dia$, the $\mathtt{frk}$ operation in definition 13 requires an infinite set of elements because of the injectivity. It should be noted that $\mathtt{frk}(\mathtt{frk}(x, y), z) \neq \mathtt{frk}(x, \mathtt{frk}(y, z))$. With respect to active domains, the $\mathtt{frk}$ operation in this paper has the purpose of assigning unique identifiers.

Definition 14. A *context-FRA based on* $\mathcal{A}$ is a context-RA based on $\mathcal{A}$ with a FRA on ACT which fulfills the following: with $\mathtt{frk} : U_1 \times U_2 \to U_3$: if $U_1 = U_2 = \mathtt{ACT} \Rightarrow U_3 = \mathtt{ODD} \cup \mathtt{EVN}$ and if $U_1 = \mathtt{ODD}$ or $U_2 = \mathtt{ODD} \Rightarrow U_3 = \mathtt{ODD}$. For $x, y \in \mathtt{EVN}$: $\mathtt{frk}(x, y) \neq x$ and $\mathtt{frk}(x, y) \neq y$. The following restrictions to ACT are defined: $I \triangle|_{\mathtt{EVN}} J := \{(\mathtt{frk}(x, y), z) \mid (x, z) \in I; (y, z) \in J; \mathtt{frk}(x, y) \in \mathtt{EVN}\}$ and $lft := dia_{\mathcal{A}} \triangle|_{\mathtt{EVN}} one_{\mathcal{A}}$ and $rgt := one_{\mathcal{A}} \triangle|_{\mathtt{EVN}} dia_{\mathcal{A}}$ and $prt := rgt \sqcup lft \sqcup (rgt \diamond lft) \sqcup (lft \diamond rgt) \sqcup (rgt \diamond rgt) \sqcup (lft \diamond lft) \ldots$ and $end := dia(prt_{\uparrow}) \cap dia(prt_{\rightarrow})$.

It can be shown that F1 and F3 (but not F2) from definition 12 still hold for $\triangle|_{\mathtt{EVN}}$. lft and rgt are projections because $lft = \{(\mathtt{frk}(x, y), x) \mid \mathtt{frk}(x, y) \in \mathtt{EVN}\}$. According to F1, all of the information about $\triangle|_{\mathtt{EVN}}$ is contained in lft and rgt. The matrices lft and rgt are fixed at any point in time according to EVN. Calculations with $\triangle|_{\mathtt{EVN}}$ are thus reduced to look-ups in lft and rgt together with ordinary RA operations. It should be noted that to calculate the parts, prt requires some sort of transitive closure (thus is not strictly an RA operation). Frias et al. (2004) do not discuss the need for transitive closure, but it is not known to us whether they do not require it or whether they have overlooked the problem.

5.3 Relational Schemata Using Fork Algebra

Using the fork algebraic extensions from the previous section, it is now possible to define a complete relational schema for an object-relational database that contains both simple and composite tables with all their values. Simple tables are traditionally called "entity tables". They collect instances, such as "employee" or "project". Instances (or rows) in such tables are usually identified by a single key, which is a column of the table and contains unique values, such as "employee number" or "project number". Composite tables are traditionally called "relations", which are built using the keys from simple tables as "foreign keys". For example, a relation "work" can be built from tables "employee" and "project" using the keys "employee number" and "project number". Such a table represents a database relation between employees and projects.

Definition 15. A *complete relational schema based on* ACT is a relational schema based on ACT with a context-FRA based on $\mathcal{A}$ and with sets: Keys $\subseteq$ Attr; Nkey := Attr $\setminus$ Keys; Simp $\subseteq$ Inst with Simp := $\{s \in \text{Inst} | \neg \exists_y : s\, prt\, y\} \cup \{s \in \text{ACT} | \exists_x : x\, end\, s\}$ and Comp := Inst $\setminus$ Simp; so that simple instances have at most one key: for $s \in$ Simp, $k_1, k_2 \in$ Keys: $s I_{isba} k_1, s I_{isba} k_2 \Rightarrow k_1 = k_2$ and the keys of composite instances correspond exactly to their fork algebraic end parts: for $c \in$ Comp: $\exists_s : c\, end\, s \Longleftrightarrow \exists_{k \in \text{Keys}} : c\, end \diamond I_{isba}\, k$; for $c \in$ Comp, $k \in$ Keys: $c\, end \diamond I_{isba}\, k \Longleftrightarrow c I_{isba} k$ and $|\mathsf{set}((\underline{c} \diamond end)_\uparrow))| = |\mathsf{set}_{\text{Keys}}((\underline{c} \diamond I_{isba})_\uparrow)|$.

Definition 15 does not allow for the same attribute to be used more than once as a foreign key. This is only a problem if these attributes are in a many-to-many relation because otherwise the relation does not require a separate table. But even then it is possible to generate a generic key attribute using identifiers and treating the other attributes as non-key attributes.

Remark 1. Definition 15 translates the database notion that instances are uniquely identified by keys into a fork algebraic part-whole relationship! This is significant because in RLA, keys form just another set and instance pairs are not structurally different from value pairs. But in the fork algebraic formalisation, the special nature of keys is structurally represented.

The conditions in definition 15 can also be expressed relationally: simple instances do not have parts: $dia(\text{Simp}) \diamond prt = nul$. End parts are simple: $dia(\text{Simp}) \supseteq dia(end_\uparrow)$ or $end = prt \diamond dia(\text{Simp})$. Simple instances have exactly one key: $red_{\text{Simp,Keys}}(I_{isba})$ is a permutation matrix[4]. Keys of composite instances correspond to the fork algebraic end parts of the instances: for each $c \in$ Comp: $dia_{\mathsf{set}}((\underline{c} \diamond end)_\uparrow) \diamond I_{isba} \diamond dia_{\text{Keys}}((\underline{c} I_{isba})_\uparrow)$ is a permutation matrix.

So far, instances can be constructed and attributes can be assigned to instances – but only in a binary manner showing which instance has which attribute but not which value belongs to an instance with respect to an attribute. In the following definition, Vals stands for instance-value pairs. Attribute values are often drawn from potentially infinite domains (such as the set of real numbers). This is why attribute values do not usually correspond to keys. According to Definition 15 simple instances do not need to have a key. Simple instances without keys are attribute values. But not all attribute values need to be listed as instances in a complete relational schema.

Definition 16. A *value assignment context* for a complete relational schema based on ACT is a formal context $(\text{Vals}, \text{Nkey}, I_{vals})$ where Vals $\subseteq$ ACT with $dia(\text{Vals}) \subseteq dia(lft_\rightarrow)$ so that $((dia(\text{Vals}) \diamond lft)_\uparrow) \subseteq dia(\text{Inst})$ and I_{vals} is a binary matrix so that for each attribute a: the matrix $dia((I_{vals} \diamond \underline{a})_\rightarrow) \diamond lft = dia(\mathsf{set}(\underline{a}')) \diamond lft$ has at most 1 cross per column.

The conditions in definition 16 ensure that elements of Vals are pairs where the left element is an instance. The instance-value relation for each attribute can be retrieved via

---

[4] A binary relation I represents an exact correspondence, if it contains exactly one 1 in each row and column (i.e., $I \circ I^d = dia$). Such a matrix is called a permutation matrix.

$lft^d \diamond dia(\mathtt{set}(\underline{a}')) \diamond rgt$. The last condition in definition 16 ensures that each instance has exactly one value for each attribute. In relational database terms this means that the tables are in first normalform. Definition 16 does not only require this for each single table, but instead across the whole database. If an instance has a value for an attribute, then it has the same value in all tables in which this instance and this attribute occur. This means that a multiple inheritance anomaly (Priss, 2005) is avoided. From an implementation viewpoint, this can always be achieved by renaming attributes if these attributes have table-specific values.

The construction in definition 16 is in principle similar to the treatment of "many-valued contexts" in traditional FCA (Ganter & Wille, 1999) and to Wille's (2002) power context families. The difference is, however, that in definition 16 a single formal context is used for all attributes of all database tables of an application. The information about which instances and values belong to which database tables is coded into the fork algebraic part-whole structure of the elements in $\mathtt{Vals}$. The fork algebraic construction also uses a more restrictive set of operations. Instead of the use of a Cartesian Product, pairs of instances or of instances and values are added to lft and rgt on an as-needed basis. Another restriction (but not limitation) is that n-ary relations are built stepwise from binary relations.

6   Conclusion

This paper describes a formalisation of object-relational databases using RA and fork algebra. An advantage for this approach is that in contrast to traditional RLA, the mathematisation is mainly based on binary relations and thus closer to FCA, which provides easy access to visualisations in form of FCA line diagrams. Compared to RLA, the basis of the algebraic operations that are required is quite similar. Both RA and RLA use union and complement. The RLA operations of projection and selection are achieved in RA by using composition, dual and a selection via composition with $dia()$. The RLA operation of cross (or Cartesian) product corresponds to fork algebraic constructions together with RA operations that allow to convert between sets and matrices. The similarities and differences between the two approaches provide insights into the structure of relational databases. It is hoped that in the future an implementation can be developed that explores practical applications of the RA/fork algebraic structures.

References

1. Abiteboul, Serge; Hull, Richard; Vianu, Victor (1995). *Foundations of databases*. Addison Wesley.
2. Atkinson, M.; Bancilhon, F.; DeWitt, D.; Dittrich, K.; Maier, D.; Zdonik, S. (1989). *The Object-Oriented Database System Manifesto*. In: Proceedings of the First International Conference on Deductive and Object-Oriented Databases, Kyoto, Japan. p, 223-240.
3. Brink, C.; Britz, K.; Schmidt, R. (1992). *Peirce Algebras*. Max-Planck-Institut für Informatik. MPI-I-92-229.
4. Codd, E. (1970). *A relational model for large shared data banks*. Communications of the ACM, 13:6.

5. Faid, M.; Missaoui, R.; Godin, R. (1997). *Mining Complex Structures Using Context Concatenation in Formal Concept Analysis.* In: Mineau, Guy; Fall, Andrew (eds.), Proceedings of the Second International KRUSE Symposium. p. 45-59.
6. Frias, Marcelo; Veloso, Paulo; Baum, Gabriel (2004). *Fork Algebras: Past, Present and Future.* Journal on Relational Methods in Computer Science, 1, p. 181-216.
7. Ganter, B.; Wille, R. (1999). *Formal Concept Analysis.* Mathematical Foundations. Berlin-Heidelberg-New York: Springer, Berlin-Heidelberg.
8. Hansen, Gary; Hansen, James (1988). *Human Performance in Relational Algebra, Tuple Calculus, and Domain Calculus.*, International Journal of Man-Machine Studies, 29, 503-516.
9. Hereth, J. (2002). *Relational Scaling and Databases.* In: Priss; Corbett; Angelova (Eds.) Conceptual Structures: Integration and Interfaces. LNCS 2393, Springer Verlag. p. 62-76.
10. Imielinski, T.; Lipski, W. (1984). *The relational model of data and cylindric algebras.* Journal of Computer and Systems Sciences, 28, p. 80-102.
11. Jain, M.; Mendhekar, A.; Van Gucht, D. (1995). *A Uniform Data Model for Relational Data and Meta-Data Query Processing.* International Conference on Management of Data.
12. Kim, K. H. (1982). *Boolean Matrix Theory and Applications.* Marcel Dekker Inc.
13. Lyndon, R. (1950). *The representation of relational algebras.* Ann. of Math., 2, 51, p. 707-729.
14. Maddux R. (1996). *Relation-algebraic semantics.* Theoretical Computer Science, 160, p. 1-85.
15. Pratt, V.R. (1992). *Origins of the Calculus of Binary Relations.* Proc. IEEE Symp. on Logic in Computer Science, p. 248-254.
16. Pratt, V.R. (1993). *The Second Calculus of Binary Relations.* Proc. 18th International Symposium on Mathematical Foundations of Computer Science, Gdansk, Poland, Springer-Verlag, p. 142-155.
17. Priss, U. (1998). *Relational Concept Analysis: Semantic Structures in Dictionaries and Lexical Databases.* (PhD Thesis) Verlag Shaker, Aachen 1998.
18. Priss, U.; Old, L. J. (2004). *Modelling Lexical Databases with Formal Concept Analysis.* Journal of Universal Computer Science, Vol 10, 8, 2004, p. 967-984.
19. Priss, U. (2005). *Establishing connections between Formal Concept Analysis and Relational Databases.* In: Dau; Mugnier; Stumme (eds.), Common Semantics for Sharing Knowledge: Contributions to ICCS 2005, p. 132-145.
20. Tarski, A. (1941). *On the calculus of relations.* Journal of Symbolic Logic, 6, p. 73-89.
21. Tarski, A. (1955). *Contributions to the theory of models.* Indag. Math, 17, p. 56-64.
22. Van den Bussche, Jan (2001). *Applications of Alfred Tarski's Ideas in Database Theory.* Proceedings of the 15th International Workshop on Computer Science Logic. LNCS 2142, p. 20-37.
23. Wagner, Stefan (2003). *Transitive closure in relational database systems.* On-line available at http://www.stefan-wagner.info/cs/trans_clos.php.
24. Wille, Rudolf (2002). *Existential Concept Graphs of Power Context Families.* In: Priss; Corbett; Angelova (Eds.) Conceptual Structures: Integration and Interfaces. LNCS 2393, Springer Verlag, p. 382-395.

Spring-Based Lattice Drawing Highlighting Conceptual Similarity

Tim Hannan and Alex Pogel

Physical Science Laboratory, New Mexico State University,
Las Cruces, NM 88003, USA
{Tim.Hannan, Alex.Pogel}@psl.nmsu.edu

Abstract. This paper presents a spring-based lattice drawing method that uses natural spring lengths determined by assigning a dissimilarity value, the size of symmetric difference, to every pair of concept extents. This extends previous work on incorporating support structure in a concept lattice diagram, in which the support weight function was applied to modify any layout. That work was a partial advance toward the goal of viewing high confidence association rules via the lattice diagram in a way that naturally extends the traditional viewing of implications in a diagram, but also caused the appearance of nearly horizontal edges. The spring-based method solves this problem by placing concepts in the ambient space such that the distance between concepts is proportional to the size of the symmetric difference of the extents of the respective concepts. Besides meeting the proportionality criteria, the algorithm yields highly symmetric diagrams in cases where it is expected.

1 Introduction

In most applications of Formal Concept Analysis, the analysis activity is driven by human interaction with a diagram of the formal concept lattice, a labeled lattice derived from binary-valued tabular data. Consequently, the lattice layout problem is of singular importance. As stated in [StW], "A serious problem is how to represent graphically concept lattices such that the semantical relationships within the data become mostly transparent." The goal of this lattice drawing activity is to extend what is the most basic use of a lattice diagram, the recognition of implications, to the recognition of near implications, i.e. high confidence association rules, and to do so naturally, so that the fundamental rule "implication reads upwards" is extended to "*near* implication *nearly* reads upwards". We are not concerned with reading exact confidence values in a diagram, since the inclusion of such values for every pair of concepts would overload the lattice diagram with information, and because these values are always available via simple user requests. This goal is intended to enhance knowledge discovery applications of Formal Concept Analysis in which association rules are important, including epidemiology [OPH] and analysis of multi-agent computer simulations with prominent stochastic components: in both these domains implications are rare, but near implications are numerous.

R. Missaoui and J. Schmid (Eds.): ICFCA 2006, LNAI 3874, pp. 264–279, 2006.

This effort extends the results in [PHM] that focused on expressing the support structure of a concept lattice via the support weight function. There, concepts were presented at heights that were relative to their support, and this created diagrams in which pairs of concepts that were order related and had nearly the same support values - the hallmark of a high confidence association rule - were sometimes presented near one another, clearly extending the usual paradigm for reading a line diagram, while other times were presented far from each other, such that nearly horizontal lines appeared in the diagram. To address the latter problem, this paper is motivated by a more specific goal: we want to create a lattice drawing method that makes conceptual similarity transparent. Intuitively, we think two concepts are *conceptually similar* if their extents are nearly equal, that is, if there are only a few objects distinguishing the two. We are purposefully leaving "only a few objects", and thus the ultimate meaning of conceptual similarity, to the interpretation of the analyst, but the important point is that we can capture the intuitive notion quantitatively by computing the symmetric difference of the extents of the respective concepts. The core idea of this paper is that the goal of expressing conceptual similarity is accomplished by drawing concept lattices such that a *proportionality criteria* is met: for each pair of concepts, the distance between concepts is proportional to the size of the symmetric difference of the extents of the respective concepts.

Given the proportionality criteria, in a particular drawing of a lattice there may be pairs of concepts that will be too far apart given how small the symmetric difference of their extents is, and there may be pairs of concepts that will be too close given how large the symmetric difference of their extents is. In either case, a spring whose natural length is the size of the symmetric difference would act to correct the displacement from the natural length. Thus, a spring-based algorithm is appropriate to achieve the proportionality criteria, and this paper presents a spring-based lattice drawing method that *iteratively improves* a diagram through reference to a natural spring length between concepts that is determined by the size of the symmetric difference of the respective extents.

Now we describe the structure of the paper. First we present details of the algorithm. Next we consider practical applications of the results of the algorithm, via lattice diagrams created from the Zoological data and Mushroom data available at the UCI KDD repository [UCI]. Then we discuss the algorithm's performance on the well-known Knives concept lattice, also demonstrating that the algorithm can be considered a layout method in its own right, not merely a modification method. Finally we display quantitative evidence we have gathered to show that the proportionality criteria is more satisfied by this method than other vector-based methods, even when the vector-based methods are combined with the support weight function.

2   The Spring-Based Algorithm

We use the standard nomenclature of Formal Concept Analysis, as found in [FCA]. Various concept lattice layout techniques are available at this time, including the geometrical method, the additive line diagram method, and the nested line diagram method, as discussed in [FCA], along with the upper covers method [PHM], the order-based improvement algorithm in LatDrawWin [LDW], and the tension-repulsion algorithm in GaLicia [GaL]. In this paper, we will use two vector-based drawing methods, which may be modified via a height scaling that uses the support weight function. The first method, V1 (vector-based method 1), is a three-dimensional vector-based method inspired by the traditional additive line diagram method, while V2 is the Upper Covers vector-based method. Whenever method V1 or V2 is modified by the support weight function, we denote the resulting drawing method by VS1 or VS2 (Vector+Support1 and Vector+Support2).

Our general view of lattice drawing is that some mapping $graph : \mathbf{L} \to \mathbf{R}^3$ provides a strictly order-preserving map from $\mathbf{L}$ into $\mathbf{R}^3$, where order in the codomain is determined by the z-coordinate. We use $\mathbf{R}^3$ in place of $\mathbf{R}^2$ because this allows a user to rotate the lattice about a central axis (as in [LDW]) to determine a "best" projection into $\mathbf{R}^2$ (onto the computer screen). Experience with [LDW] shows this is an effective, and computationally cheap, solution. Thus our focus is on producing a desirable mapping of a lattice into 3D space (meeting the proportionality criteria), with the manipulation controls (rotating, translating, and zooming) being used to explore the lattice.

Now we explain how a model of spring lengths is used to create an improvement algorithm. Our use of springs is consistent with their traditional application in graph drawing algorithms [Dea], but is now motivated by concept-based applications. We assume we are given a concept lattice with each concept placed at some (x, y, z) coordinate. For each concept in the lattice, we compute a force vector that is based on forces computed when all of the concepts are treated as point masses of equal mass with a spring connecting each pair of concepts. As we model Hooke's Law, every single concept applies a force on every other concept that is proportional to the displacement from the natural length of the spring and the force acts in the direction that restores the spring to its natural length. Each spring is given a natural length equal to some (static) measure of the dissimilarity of its two concepts. We used the dissimilarity measure $D(C_i, C_j) = |extent(C_i) \triangle extent(C_j)|$. There are clearly other possibilities for dissimilarity measures, and some are discussed in the last section, but our choice of this particular dissimilarity measure is made to meet our goal of expressing conceptual similarity in a diagram. From $D(C_i, C_j)$ and $d(C_i, C_j)$, the distance between the current positions (x_i, y_i, z_i) and (x_j, y_j, z_j) of concepts C_i and C_j respectively, we compute a force on concept C_i that is due to concept C_j, and then we sum all those forces:

$$F(C_i) = \sum_{\{j : j \in \{1,2,\dots,n\},\ j \neq i, d(c_i, c_j) \neq 0\}} \overline{v} * [d(c_i, c_j) - D(c_i, c_j)]$$

In the summand expression, the vector $\overline{v}$ on the left is the unit vector

$$\frac{(x_j - x_i, y_j - y_i, z_j - z_i)}{d(c_i, c_j)}$$

pointing from concept C_i to concept C_j. The quantity in the brackets is the spring's displacement from its natural length, and the sign of this value determines whether the force is attractive or repulsive. Thus, for every concept C_i in the lattice, we compute the force it feels from every other concept, and we add up all of these force contributions to get the total force felt by C_i.

Notice that the force $F(C_i)$ can be very large if the concept lattice cardinality is large. This has potential to induce instability in the improvement process.[1] Therefore, once the total force vectors are computed for every concept, we simply use some fraction f of the force acting on each concept to determine a new position for the concept. We employ the fraction $f = \frac{1}{4*|\mathfrak{B}(\mathbb{K})|}$, and believe other expressions of the form $f = \frac{1}{k*|\mathfrak{B}(\mathbb{K})|}$ will work as well with $k > 1$. Larger k values slow the improvement. In principle, the algorithm is that simple. In practice, implementing the algorithm has required minor variations of the core idea, but none drastic enough to explain further.

3   Practical Results of the Improvement Algorithm

Here we examine the utility of the improvement algorithm in viewing high confidence association rules and in expressing conceptual similarity. We begin with a brief review of the support and confidence functions often used to evaluate association rules.

Given a set M of attributes, an association rule is a pair (X, Y), often written $X \to Y$, with X, Y subsets of M, interpreted to say "in (some) cases where X holds, Y also holds" (near implication), or "in the event of X, event Y also occurs" (conditional event). Two functions used to formulate evaluation criteria for association rules of a formal context $\mathbb{K} = (G, M, I)$, are $conf_{\mathbb{K}}(-)$ (confidence) and $supp_{\mathbb{K}}(-)$ (support), given by

$$conf_{\mathbb{K}}(\,(X,Y)\,) = \frac{|X' \cap Y'|}{|X'|} \quad and \quad supp_{\mathbb{K}}(\,(X,Y)\,) = \frac{|X' \cap Y'|}{|G|}$$

We define a function $Csupp_{\mathbb{K}} : \mathfrak{B}(\mathbb{K}) \to [0,1]$, called the *concept support* function, by assigning to each concept $D = (P, B)$ of $\mathbb{K} = (G, M, I)$ the value $Csupp_{\mathbb{K}}(D) = \frac{|P|}{|G|}$. Then the support of an association rule (X, Y) is the concept support of the concept generated by $X \cup Y$, that is

$$supp_{\mathbb{K}}(\,(X,Y)\,) = Csupp_{\mathbb{K}}((X', X'') \wedge (Y', Y'')) = Csupp_{\mathbb{K}}(\,(X' \cap Y', (X' \cap Y')')\,).$$

Also note that the support of a valid implication $X \to Y$ (i.e. an association rule (X, Y) with 100% confidence) is the concept support of the concept (X', X'') generated by the premise X.

For a first example, consider the zoological dataset available at [UCI]. The dataset contains 101 animals and 17 attributes. Of the 17 attributes, 15 are

---

[1] We have seen a similar instability when larger lattices are visualized (roughly 75 concepts and larger) in LatDrawWin [LDW].

binary and 2 are numerical. From this dataset we create two subdatasets. The first dataset has all 101 animals and 4 of the binary attributes for which we have especially strong pre-existing interpretations, chosen as an illustrative example, while the second includes all 15 of the binary attributes. In the first dataset, the attributes "hair" and "milk" are chosen because to most people they would seem very similar conceptually. Most animals with hair nurse their young and most animals which give milk also have hair. The attribute "eggs" was chosen because it is nearly complementary to the attributes "hair" and "milk". Finally, a fourth attribute, "predator", was chosen because most people would consider that to be independent of the other three attributes. This subdataset with these four attributes gives a clear illustration of the *purpose* of the improvement algorithm.

The smaller dataset has 11 concepts. The first figure shows the initial drawing of the lattice of this small dataset drawn, at left, using the VS1 method. Notice the nearly horizontal lines from "milk" and "hair" to their meet. As discussed in [PHM], these nearly horizontal lines show near implications between the concepts "milk" and "hair".

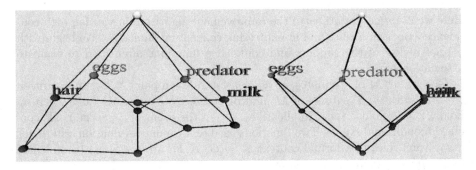

Fig. 1. At left, an unimproved lattice drawn using the VS1 method. At right, the improved version, with the "hair" and "milk" attribute concepts drawn so close together (at far right edge) that they are nearly indistinguishable.

The lattice diagram at right in Figure 1 shows the improved version of this lattice, and we see that the concepts "milk" and "hair" are drawn very near each other (so much so that the attribute names are laid over one another), while the concept "eggs" is repulsed from the pair, now appearing at the opposite (left end) of the right diagram. Also, the concept "predator" has no preference between "eggs" and the cluster with "milk" and "hair", so it ends up somewhere in the middle. Evaluating the rules "milk" → "hair" and "hair" → "milk", we find they have respective confidence values of 95.1% and 90.6%. Perhaps the most important result produced by the algorithm is its visual clustering of concepts, thus providing an indication that, according to the data, multiple concepts are minor variations on a single concept.

Now we turn to the concept lattice of the larger dataset, consisting of 101 animals and all 15 of the binary attributes in the zoological dataset. This lattice has 238 concepts, and the improvement algorithm converged on a stable diagram within 10 seconds of its initial layout via the VS2 method. We have seen similar speed on many other lattices in the same size range (roughly 150 to 300) , and our experience is that the improvement algorithm can easily handle large lattices. Note in the Figure that the sublattice generated by "hair", "milk", "eggs", and "predator" is drawn almost exactly as it was when isolated in its own diagram (right side of Figure 1).

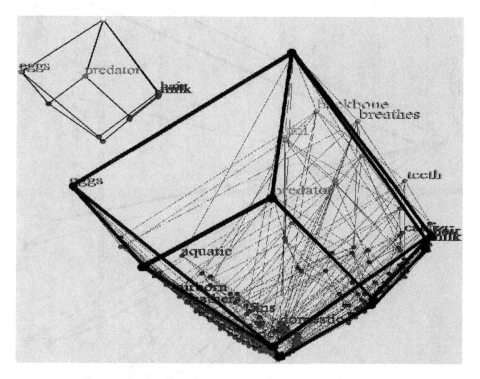

Fig. 2. Full lattice of the context containing all 15 binary attributes. Notice the highlighted small sublattice has the same shape as the improved diagram at right in Figure 1, shown again at upper left. A much higher resolution image is available [PSL].

Next we turn our attention to the well-known Mushroom Database from the UCI KDD Repository [UCI]. Consider the iceberg lattice [St1] created using a 0.55 support cutoff. In the next Figure, we see that the spring-based improvement algorithm clusters together the concepts that are nearly identical in terms of extent, while pushing apart those with less overlap.

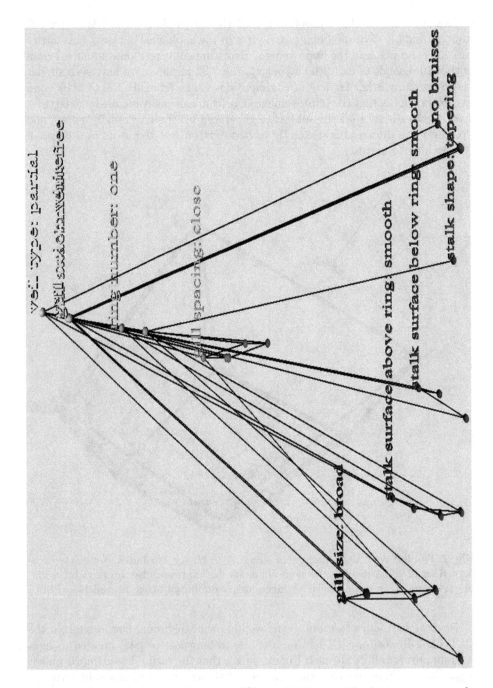

Fig. 3. Mushroom iceberg lattice with 55% cutoff, drawn with the improvement algorithm, after an initial (poor) layout via VS2. A higher resolution image is available [PSL].

We have two key points to make regarding the iceberg lattice of the Mushroom database. First, note that the improvement algorithm result in Figure 3 has apparently depicted 23 concepts in place of the 30 that are actually present, by clustering those concepts that have nearly equal extents (in fact, the additional 7 concepts can be easily seen by shrinking the size of the nodes in the diagram and zooming in to the relevant sections). In this case, there is a 25% reduction of the number of concepts, so the algorithm has automatically produced a leaner diagram that is comparably expressive to the human-supervised vector-based diagram for the same iceberg lattice that is shown in [St1].

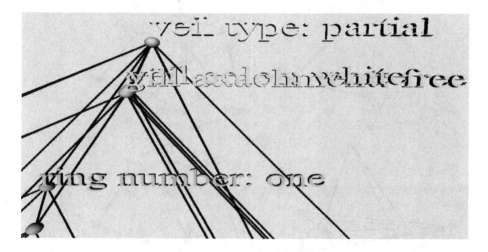

Fig. 4. Zoom in to top of mushroom iceberg lattice with 55 percent cutoff

We see in Figure 4 that two attribute concepts at the top of the lattice are laid over one another – "Veil color: white" and "Gill attachment: free" – indicating that there is (nearly) a single concept, namely the meet of these two, that is representative of both. We would expect a mycologist (mushroom expert) to know this concept immediately, perhaps already aware of a single category characterized by these two features, in which case the lattice diagram reflects the expert knowledge.

4   Further Qualities of the Resulting Diagrams

Here we display the concept lattice of the challenging Knives context (a clarified version of pocket knives and the tools they contain, as in [Lea]), created by the improvement algorithm. Starting with any of the four distinct layout methods in the left column of Figure 5 (VS1 method [resulting in a slightly twisted diagram]; VS2 method [with slight horizontal distension]; a random placement of concepts in the space between the top and the bottom elements of the lattice, which is *not* a Hasse diagram; and a layout in which all non-extremal concepts are placed

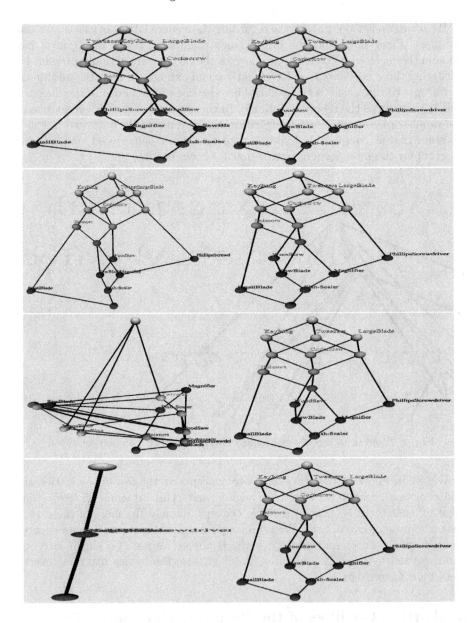

Fig. 5. Four initial layouts for the Knives lattice are shown in the left column, and the four improved lattices appear in the right column. Each of these four diagrams is available [PSL] in much higher resolution.

at the midpoint between the top and bottom element, which is also *not* a Hasse diagram), the algorithm yields the single layout shown in Figure 6.

We mention again that the latter two layout methods of the four in Figure 5 should almost never produce Hasse diagrams. Nonetheless, the improvement

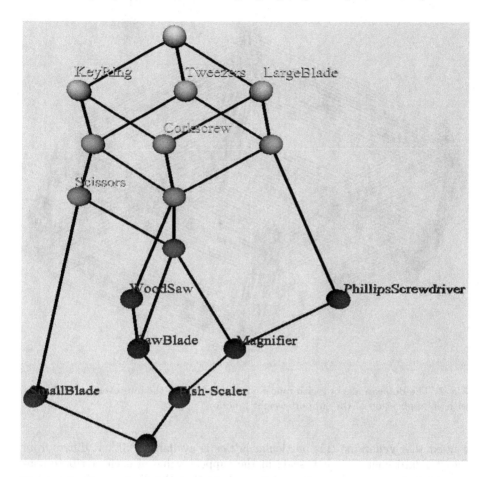

Fig. 6. The Knives lattice drawn using the improvement algorithm. Applying the improvement algorithm iteratively to each of the four diagrams mentioned above leads to this same layout.

algorithm manages to return indistinguishable (to the human eye) diagrams from each of the four layout methods. It appears there is similar stability for nearly all small lattice diagrams, since for any lattice with roughly 100 elements or less (that we have considered), the algorithm produces indistinguishable diagrams from any initial layout[2]. The evidence we have gathered regarding small lattices indicates that the algorithm can be viewed as an initial layout method itself. Various images supporting this argument are available [PSL].

Next we consider a nine-dimensional Boolean algebra, drawn in Figure 7 by the improvement algorithm with the support weight function applied. The actual

---

[2] Remark: In lattices with symmetries in the horizontal component, the uniqueness is up to symmetry, e.g. the Boolean algebra with three atoms can be drawn with two orientations determined by the three atoms, but the diagrams will look the same.

Fig. 7. The Boolean algebra with nine coatoms, drawn via the improvement algorithm with an application of the support weight function

context that generated this particular lattice is available [PSL]. It differs from the standard contranominal scale in the support value of each attribute concept (it is neither clarified nor reduced). Note that the diagram indicates that this Boolean algebra *should be understood* as a product of a 3-coatom Boolean algebra and a 6-coatom Boolean algebra, and further that the inner 6-coatom Boolean algebra is best understood as a product of two 3-coatom Boolean algebras. There is minor twisting evident in the long lines connecting the 6-coatom Boolean algebras. Nonetheless, regardless of the initial configuration, the same local clusters are produced, since the forces generated from the spring lengths enforce this outcome. Various before-and-after images showing the results generated from various initial layouts, including point-line configurations that are *not* Hasse diagrams and variations on the input data, are available [PSL].

5   Quantitative Evaluation of Proportionality Criteria

In this section, we present evidence that the algorithm presented here does create lattices that satisfy the proportionality criteria better than in the initial layout that was input to the algorithm. To show that this is the case, for each pair of concepts we plot the Euclidean distance between the pair versus the size of the

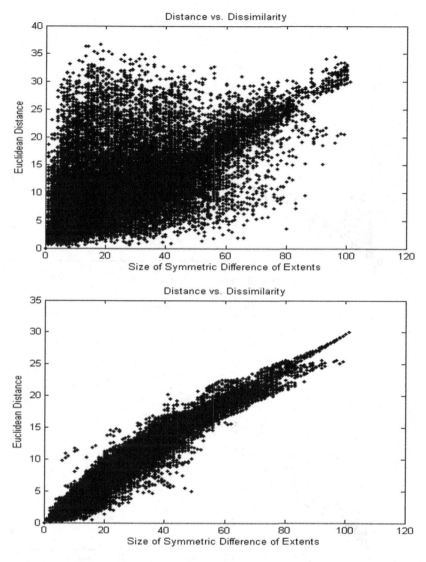

Fig. 8. A plot of Euclidean distance vs. size of symmetric difference of extents for each pair of concepts in the Zoological dataset prior to and after improvement. Notice the nearly linear relationship in the bottom image.

symmetric difference of the extents of the concepts. Ideally, these values would be equal, but because we often change the spatial scale used in presenting a concept lattice, whereas the symmetric differences do not change, the best we could hope for is a linear relationship.

Consider the Zoological dataset from the UCI KDD repository whose lattice was viewed in Section 3. The following figures show the plot of Euclidean distance

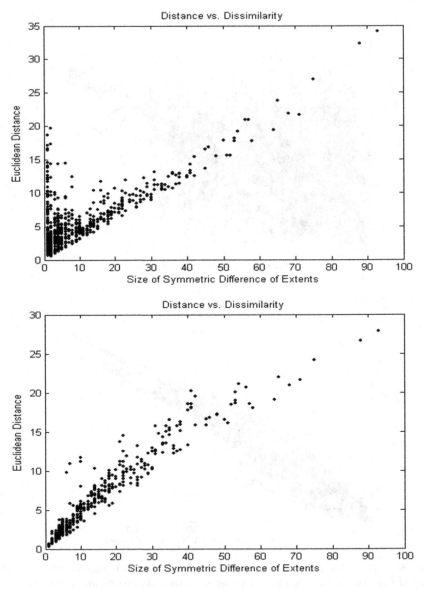

Fig. 9. A plot of Euclidean distance vs. size of symmetric difference of extents for each covering relation in the Zoological dataset prior to and after improvement. Again, the nearly linear relationship appears. Notice the change at the left edge.

vs. symmetric difference of extents for each pair of concepts in the lattice diagram before the improvement and after the improvement.

We also created the same plots for the set of pairs of concepts that are in the covering relation of the concept lattice, so in simpler terms, each point in the plot corresponds to an edge in the diagram.

We have created these plots for many other lattices and have seen no variation from the principle exhibited here: the proportionality criteria is better fulfilled in an improved diagram. Further quantitative evidence is available [PSL].

6   Future Work

The second largest lattice diagram (Figure 2 with 238 concepts) we considered in this paper shows that the improvement algorithm has a drawback from the usability point of view, i.e. it puts very many of the lower support concepts close to one another, as shown in Figure 10, which displays a zoom into the bottom section of the lattice (all the concepts with support 25% or less). This clustering at the bottom is not surprising, as it is clearly due to the fact that the symmetric difference of small support concepts will be small, thus making their spring lengths very small – and this fact makes it very difficult to see structure near the bottom of the lattice, particularly in the big picture, i.e. when the whole lattice is viewed at once. We suspect that removing any edges that lead to concepts not pictured in the current frame (we call them unspecified edges) will drastically improve the readability of the portion of the lattice showing in this frame. This issue does not arise very often in small lattice diagrams, e.g. in this paper none of the small examples had many concepts clustered around the bottom of the lattice.

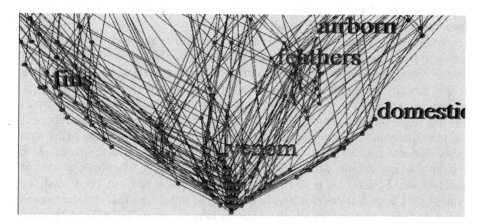

Fig. 10. A zoom into the bottom quarter (concepts with 25% support or less) of the Zoological lattice. A higher resolution image is available at [PSL].

In Figure 10 we have changed the scale in which the lower section of the Zoological lattice is viewed. Imagining a cleaned up version of Figure 10 with no unspecified edges (especially in the higher resolution version available from the authors), it appears that the improvement algorithm is having the same effect at the bottom of the lattice that it had at the higher support values already considered in this paper: some concepts appear very near one another (clusters are clear even in this messy picture), and further zooming can reveal details if desired, while others that are nearly complementary ("fins and "domestic") are drawn as far apart as they can be given the constraints arising from the higher support concepts. Thus, one of our key steps for future versions of the environment in which the spring-based drawing method is used will be the removal of all unspecified edges from the frame in which a portion of the lattice is viewed.

In our original work toward a useful algorithm, we generated a variety of natural spring length values from the symmetric difference dissimilarity measure. First, we tried a dissimilarity measure which is the original dissimilarity measure multiplied by the ratio of the average degree of the two concepts and the average degree of all concepts in the lattice. This was an attempt to spread out the areas with the most edges to make the concepts easier to see. Another attempt involved using the dissimilarity measure

$$|extent(C_1) \triangle extent(C_2)| * [(\frac{|extent(C_1)| + |extent(C_2)|}{|G|} - 1)^2 * k + 1] \, ,$$

expecting it would leave the middle of the lattice alone and stretch out the bottom and the top horizontally by a factor of no more than $k + 1$. (It will only slightly change the top in most cases since there are often very few concepts with support > 0.5.) So far, in each of these attempts to spread out the bottom of the lattice diagram, the improvement algorithm became unstable, and it would not settle toward an ideal picture, so neither of these alternative dissimilarity measure have given desirable results.

We plan to continue testing alternative dissimilarity measures, particularly a measure that involves the size of the intersection of the two concepts' extents (the size of the extent of their meet) as part of the determination of natural spring length size, in some combination with symmetric difference. We also expect to consider modifying the existing diagrams via contraction (expansion) to (from) the line segment, from the top to the bottom of the lattice, depending on support value (expansion when support is low, contraction when support is high), and other variants on this idea. The strategy here is to maintain the nearness of concepts whose extents are nearly equal while disallowing the dense clutter at the bottom. Finally, we note that while optimization has not been used, we believe that large lattices can lead to multiple equilibria, so this certainly opens the door for us to consider which of the various equilibria (as defined by some stopping condition) are best, i.e. which equilibria optimize some criteria.

Another question that arose during our work on this algorithm was whether there existed pathological initial configurations of a lattice in which the algorithm cannot improve the configuration, specifically so that the proportionality criteria is not met. Such pathological examples were never witnessed in normal usage of the improvement algorithm on real datasets, but we were able to use our understanding of the algorithm to construct a lattice and artificial initial configuration such that the proportionality criteria fails. This example involves a $\{0, 1\}$-gluing of four lattices (three copies of $1 \oplus \overline{2}^4 \oplus 1$ and one copy of $\overline{2}^3$, where $\overline{2}$ is the two-element chain), but is too complex to present here, especially since the corresponding context is an important part of the argument, and we feel there are more surrounding issues to present. We will report on this topic in a future paper.

Acknowledgement. The authors thank Physical Science Laboratory programmers Wesley Varela, Lance Miller, Arturo Mayorga, and Jon W. Newton for their creation of supporting software.

References

[Dea]   Di Battista, G., Eades, P., Tamassia, R., and Tollis, I.G., Graph Drawing: Algorithms for the Visualization of Graphs, Prentice Hall, NJ, 1999.

[Col]   Cole, R.J.: Automatic Layout of Concept Lattices using Layer Diagrams and Additive Diagrams, In M. Oudshoorn (Ed.): Proceedings of the 24th Australasian Conference on Computer science, Gold Coast, Queensland, Australian Computer Science Communications **23**, IEEE Computer Society (2001) 47-53.

[LDW]   Freese, R.: LatDrawWin, a lattice drawing applet, available at http://www.math.hawaii.edu/~ralph/LatDraw.

[Fre]   R. Freese: Automated Lattice Drawing, Second International Conference on Formal Concept Analysis, ICFCA 2004, Sydney, Australia, Lecture Notes on Artificial Intelligence (LNAI) **2961**, Springer-Verlag, P. Eklund (ed) (2004).

[GaL]   University of Montreal, Galicia - Galois Lattice Interactive Constructor. http://www.iro.umontreal.ca/~galicia/visualization.html.

[FCA]   Ganter, B. and Wille, R.: Formal Concept Analysis: Mathematical Foundations, Springer, NY (1999) 68-79.

[UCI]   Hettich, S. and Bay, S. D., The UCI KDD Archive http://kdd.ics.uci.edu. Irvine, CA: University of California, Department of Information and Computer Science, 1999.

[PHM]   Pogel, A., Hannan, T., and Miller, L.: Visualization of Concept Lattices Using Weight Functions, Supplementary Proceedings of ICCS04, Shaker, 2004.

[Lea]   Langsdorf, R., Skorsky, M., Wille, R. and Wolf, A.: An Approach to Automated Drawing of Concept Lattices. Technical Report 1874, Technical University of Darmstadt, Schlossgatenstasse, Darmstadt, Germany (1996).

[OPH]   Ozonoff, D., Pogel, A., and Hannan, T.: Generalized Contingency Tables and Concept Lattices, to appear in AMS-DIMACS Special Volume: Discrete Methods in Epidemiology, Eds. J. Abello and G. Cormode, AMS, 2006.

[PSL]   For higher resolution images of various Figures in this paper and additional images of the results of the drawing algorithm, go to http://www.psl.nmsu.edu/~apogel/SpringBasedLatticeDrawing.

[St1]   Stumme, G., Taouil, R., Bastide, Y., Pasquier, N. and Lakhal, L.: Computing iceberg concept lattices with Titanic, Data and Knowledge Engineering (Elsevier) **42** (2002) 189-222.

[StW]   Stumme, G. and Wille, R.: A Geometrical Heuristic for Drawing Concept Lattices, Graph Drawing (Springer-Verlag) (1995) 452-460.

[Yev]   Yevtushenko, S., et al: Concept Explorer, Open source java software available at http://sourceforge.net/projects/conexp, Release 1.2 (2003).

Characterizing Planar Lattices Using Left-Relations

Christian Zschalig

Institut für Algebra, TU Dresden, Germany
zschalig@math.tu-dresden.de

Abstract. With the help of the *left-relation on lattices* [11] we give two characterizations for planar lattices. They can be used to decide already in a context, whether the associated concept lattice is planar. With the help of these results we hope to find a quick algorithm to recognize planar lattices and draw them in the plane in the near future.

1 Introduction

We assume that all sets in this paper are finite.

1.1 Motivation

In order to draw "nice" diagrams of lattices, it is helpful to minimize the number of edge crossings. In particular we want every planar lattice to be displayed plane. Most lattice drawing algorithms ignore this issue or find plane diagrams heuristically. There exist characterizations of planar lattices ([5], [7], [1]). However, efficient algorithms for embedding planar lattices into the plane are not developed yet.

The aim of our work is to provide tools to create such an algorithm.

1.2 Diagrams of Lattices

A lattice $\mathfrak{V} = (\mathfrak{V}, \leq)$ is often represented by a diagram which we denote by[1] pos($\mathfrak{V}$). We draw a small circle for each lattice element and a line for each pair v, w of lattice elements in covering relation. Lattice diagrams are drawn upward. For a formal definition, see [7], [11]. Line diagrams are very common, here the diagram edges are straight line segments.

When looking at a diagram, one may intuitively think of nodes being left or right of others. Obviously this relation should affect only nodes representing incomparable elements since the comparable ones can be understood to be situated below or above each other. We define for two nodes v and w that v is *left of w* and denote this with $v\lambda w$ if there exists an upward polyline from bottom to top element in the lattice containing w and if v is left of this line.

---

[1] In the same manner we write pos(X) for the diagram representation of an arbitrary part X of the lattice $\mathfrak{V}$.

R. Missaoui and J. Schmid (Eds.): ICFCA 2006, LNAI 3874, pp. 280–290, 2006.

Definition 1. *[7], [11] Let $\mathfrak{V} = (\mathfrak{V}, \leq)$ be a finite lattice and* $\mathrm{pos}(\mathfrak{V})$ *a diagram of it. A maximal chain C is a sequence[2] $0_{\mathfrak{V}} = z_0 \prec z_1 \prec \ldots \prec z_n = 1_{\mathfrak{V}}$ of lattice elements z_i. In a diagram $\mathrm{pos}(\mathfrak{V})$ let $x_C(y)$ denote the unique[3] x-coordinate of the polyline $\mathrm{pos}(C)$ at height y (where y is in between the y-coordinates $y(0_{\mathfrak{V}})$ and $y(1_{\mathfrak{V}})$ of the points corresponding to the bottom and the top element of $\mathfrak{V}$). For a maximal chain C,*

$$F_l(C) := \{(x,y) \in \mathbb{R}^2 \mid y \in [y(0_{\mathfrak{V}}), y(1_{\mathfrak{V}})], x < x_C(y)\}$$

is the area left *of $\mathrm{pos}(C)$ and dually $F_r(p)$ the* area right *of $\mathrm{pos}(C)$. We define the* left- *and the* right-relation *λ and ϱ induced by $\mathrm{pos}(\mathfrak{V})$ by*

$$v \lambda w : \Longleftrightarrow (\exists C \ni w : \mathrm{pos}(v) \in F_l(C)) \wedge (v \parallel w)$$
$$v \varrho w : \Longleftrightarrow (\exists C \ni w : \mathrm{pos}(v) \in F_r(C)) \wedge (v \parallel w)$$

for all elements $v, w \in \mathfrak{V}$.

1.3   Conjugate Orders

A *conjugate order* is a (strict) order on the incomparable elements of an ordered set.

Definition 2. *[5] A conjugate order L_c on an ordered set $\underline{P} = (P, \leq)$ is a relation meeting the following conditions (where $\parallel$ denotes the incomparability relation in $\underline{P}$).*

1. *$L_c \cup L_c^{-1} = \parallel$.*
2. *L_c is a strict order*

If only the first condition holds, we call L_c a conjugate relation.

The existence of a conjugate order on a lattice characterizes its plane diagrams. In Figure 1 we provide two examples for left-relations of diagrams, the second being a conjugate order.

Theorem 1. *[11] Let $\underline{\mathfrak{V}}$ be a finite lattice. The following statements are equivalent.*

1. *There exists a plane diagram $\mathrm{pos}(\mathfrak{V})$ with the induced left-relation λ.*
2. *λ is a conjugate order on $\underline{\mathfrak{V}}$.*

How can we find conjugate orders on a lattice $\underline{\mathfrak{V}}$? One can try to test all conjugate relations on $\underline{\mathfrak{V}}$ for being a strict order. Of course this is by far too tedious, since we have at most $2^{|\mathfrak{V}|^2 - |\mathfrak{V}|}$ of those relations.

Our approach uses an observation when considering lattice diagrams drawn with the convention of *attribute additivity*.

---

[2] The symbol $\prec$ denotes the covering relation.
[3] Since $\mathrm{pos}(\underline{\mathfrak{V}})$ is drawn upward.

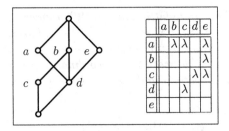

	a	b	c	d	e
a		λ	λ		λ
b					λ
c				λ	λ
d		λ			
e					

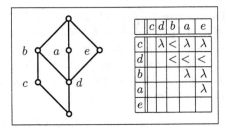

	c	d	b	a	e
c		λ	<	λ	λ
d			<	<	<
b				λ	λ
a					λ
e					

Fig. 1. Two diagrams of the same lattice with their respective left-relations. The left one is not plane and its left-relation is not antisymmetric. The right one is plane and its left-relation is a strict order.

Definition 3. *[8, 11] Let $\mathfrak{B}(G, M, I)$ be a concept lattice. A line diagram of $\mathfrak{B}(G, M, I)$ is attribute additive if there is a map* vec $: M \mapsto \mathbb{R}^2$, *such that the equation*

$$\text{pos}(A, B) = \sum_{m \in B} \text{vec}(m)$$

holds for all concepts $(A, B) \in \mathfrak{B}(G, M, I)$.

The layout of the resulting diagrams is defined by the vectors assigned to attribute concepts or $\bigwedge$-irreducibles for concept lattices or arbitrary lattices respectively. Therefore, we assert that the relationship of these vectors in the plane determines already, whether the respective diagram is plane. This assertion turns out to be true, we even can restrict our attention to $\bigwedge$-irreducibles with common upper neighbour. We call this relationship *sorting relation*. On this base we will introduce *left-relations on lattices*.

2   Left-Relations on Lattices

2.1   Definition

In this section we will introduce a possibility to characterize conjugate orders on a lattice. As already mentioned we consider in a first step *sorting relations* on the set of $\bigwedge$-irreducibles. With m^* we denote the unique upper cover of an $\bigwedge$-irreducible m.

Definition 4. *[11] Let $\mathfrak{V}$ be a finite lattice and $M = M(\mathfrak{V})$ be the set of its $\bigwedge$-irreducible elements. A strict order $L_a \subseteq M \times M$ is called a sorting relation if the following condition holds for all elements $m, n \in M$:*

$$m^* = n^* \iff m \, L_a \, n \text{ or } n \, L_a \, m.$$

The sorting relation just gives a relationship of $\bigwedge$-irreducibles with common upper neighbour. We extend it to the set of all pairs of incomparable elements.

Definition 5. *[11] Let $\mathfrak{V}$ be a finite lattice with a given sorting relation L_a. For arbitrary lattice elements v and w, we define*

$$M(v,w) = \{(v',w') \subseteq M \times M \mid v \le v', w \le w', v \parallel w', w \parallel v'\}.$$

We define the relation $L \subseteq \mathfrak{V} \times \mathfrak{V}$ according to:

$$v\,L\,w : \Longleftrightarrow \begin{cases} v\,L_a\,w, & v,w \in M, v^* = w^* \\ \exists(m,n) \in M(v,w) : m\,L\,n, & else \end{cases}$$

L is called left-relation *and $R := L^{-1}$ is called* right-relation *on the lattice $\mathfrak{V}$.*

Consider the picture on the right for an example of calculating the left-relation on the depicted lattice for a given sorting relation. Notice that we are interested just in the underlying lattice, not in the particular diagram used. We assume $m_1\,L_a\,m_2$, i.e. $m_1\,L\,m_2$. Consider now the pair (m_3,v_1). We observe $(m_3,m_2) \in M(m_3,v_1)$ and $(m_1,m_2) \in M(m_3,m_2)$ and conclude $m_3\,L\,v_1$.

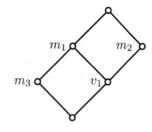

2.2   Left-Relations and Conjugate Orders

The left-relation gives indeed a possibility to find all conjugate orders, as we will show here.

Lemma 1. *[11] For every left-relation from Definition 5, the identity $L \cup R = \parallel$ holds.*

Lemma 2. *[11] A conjugate order L_c on a finite lattice $\mathfrak{V}$ is a left-relation on $\mathfrak{V}$.*

We notice that for a left-relation L, which is a strict order, exactly one of the five statements

$$v_1 < v_2, \quad v_1 > v_2, \quad v_1 = v_2, \quad v_1\,L\,v_2 \text{ or } v_1\,R\,v_2$$

holds for all pairs v_1, v_2 of lattice elements.

Proposition 1. *[11] Let L be a relation on a finite lattice $\mathfrak{V}$. Then the below-mentioned statements are equivalent:*

1. *L is a conjugate order.*
2. *L is a left-relation and a strict order.*

Proposition 1 provides a possibility to calculate all conjugate orders:

Compute the left-relations from all possible sorting relations (at most $|M|!$) and check whether they are strict orders. This strategy is already much quicker than the already described, most naïve way, but obviously still too time consuming.

2.3   Properties of Asymmetric Left-Relations

In this section we prove two lemmas that will help to prove the Propositions 2 and 3. The first lemma is used in later proofs to decide, whether two lattice elements are in left-relation.

Lemma 3. *Let $\mathfrak{V} = (\mathfrak{V}, \leq)$ be a finite lattice. If L is an asymmetric left-relation on $\mathfrak{V}$ then the equivalence*

$$v_1 \ L \ v_3 \iff v_2 \ L \ v_3$$

holds for all elements $v_1, v_2, v_3 \in \mathfrak{V}$ fulfilling the conditions $v_1 \leq v_2$ and $v_1 \parallel v_3 \parallel v_2$.

Proof.
Since v_2 and v_3 are incomparable we can find a $\bigwedge$-irreducible m_2 satisfying $m_2 \geq v_2$ and $m_2 \not\geq v_3$. Similarly we find an $\bigwedge$-irreducible m_3 such that $m_3 \geq v_3$ and $m_3 \not\geq v_1$ hold. We observe $(m_2, m_3) \in M(v_1, v_3) \cap M(v_2, v_3)$. Since L is asymmetric, we conclude

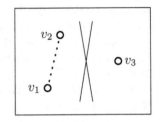

$$v_1 \ L \ v_3 \iff m_2 \ L \ m_3 \iff v_2 \ L \ v_3.$$

$\square$

Corollary 1. *Let $\mathfrak{V} = (\mathfrak{V}, \leq)$ be a finite lattice. Let L be an asymmetric left-relation on $\mathfrak{V}$. Let v_1, v_2, v_3 be lattice elements such that $v_1 \ L \ v_2 \ L \ v_3$ holds. Then v_1 and v_3 are incomparable.*

Proof.
If v_1 and v_3 are comparable then we find by applying Lemma 3 $v_1 \ L \ v_2 \implies v_3 \ L \ v_2$. This contradicts the fact that L is asymmetric. $\square$

The next lemma shows that asymmetric left-relations are already transitive. It is useful when proving that a left-relation is a strict order.

Lemma 4. *Let $\mathfrak{V} = (\mathfrak{V}, \leq)$ be a finite lattice. Let L be an asymmetric left-relation on $\mathfrak{V}$. Then L is transitive.*

Proof.
- We assume L to be not transitive. Then we have lattice elements v_1, v_2 and v_3 such that $v_1 \ L \ v_2 \ L \ v_3$ holds, but not $v_1 \ L \ v_3$. With Corollary 1 we find $v_3 \ L \ v_1$.
- We show first that the suprema of two of these elements are all equal and equal to the supremum of all three elements. Let w.l.o.g.

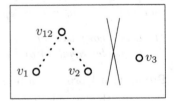

$$v_{12} := v_1 \vee v_2 < v_1 \vee v_2 \vee v_3 =: v_{123}.$$

It follows $v_{12} \parallel v_3$. Since $v_2 \ L \ v_3$ holds, we conclude with Lemma 3 $v_{12} \ L \ v_3$ and $v_1 \ L \ v_3$. This contradicts our assumption that L is asymmetric.

- 1. We assume that there exists an element $v_4 \in \mathfrak{V}$ meeting the conditions $v_4 \parallel v_{123}$ and $v_4 > v_1$. Then we notice $v_2 \parallel v_4 \parallel v_3$ since $v_{12} = v_{123} = v_{13}$ holds. With Lemma 3 we conclude

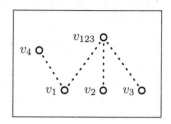

$$v_1 \ L \ v_2 \implies v_4 \ L \ v_2 \implies v_4 \ L \ v_{123} \text{ and}$$
$$v_3 \ L \ v_1 \implies v_3 \ L \ v_4 \implies v_{123} \ L \ v_4$$

  contradicting the asymmetry of L.
- 2. If there exists an element v_4 with $v_4 \parallel v_{123}$ and either $v_4 > v_2$ or $v_4 > v_3$ respectively, our argumentation is analogous, Use that $v_1 \ L \ v_2 \ L \ v_3 \ L \ v_1$.
- 3. Since v_4 as described in the previous cases does not exist, we can find $\bigwedge$-irreducibles $m_1, m_2, m_3 \prec v_{123}$ satisfying $m_1 \geq v_1, m_2 \geq v_2$ and $m_3 \geq v_3$. With lemma 3 we conclude $m_1 \ L_a \ m_2 \ L_a \ m_3 \ L_a \ m_1$. This is a contradiction, since a sorting relation L_a is a strict order by definition. $\qquad\square$

3   Two Conditions to Characterize Planarity

In this section we want to give two conditions for a left-relation L which are necessary and sufficient for the planarity of the underlying lattice $\mathfrak{V}$. The first planarity condition acts on $\bigwedge$-irreducibles only. It can be understood as a way to calculate, which $\bigwedge$-irreducibles are allowed to be "in between" others (see Figure 2 for an intuitive explanation).

Definition 6. *A conjugate relation R on a lattice $\mathfrak{V}$ fulfills the* first planarity condition (FPC) *if*

$$m_i \ R \ m_k \ R \ m_j \implies m_k > (m_i \wedge m_j)$$

holds for all $\bigwedge$-irreducibles $m_i, m_k, m_j \in M$.

Fig. 2. When considering a diagram of a lattice, the necessity of the FPC is obvious for its planarity: If $m_i \ L \ m_k \ L \ m_j$ or $m_j \ L \ m_k \ L \ m_i$ holds then also $m_k > (m_i \wedge m_j)$. Otherwise every chain of diagram edges from m_k to the bottom element of the lattice intersects with a chain of edges from either m_i or m_j to $m_i \wedge m_j$.

Proposition 2. *Let L be a left-relation on a lattice $\mathfrak{V}$, then the following equivalence holds:*

$$L \text{ satisfies the FPC} \iff L \text{ is a conjugate order.}$$

Proof.

$\Rightarrow$: We show that L is asymmetric. With Lemma 4 we can then conclude that L is a strict order, i.e. a conjugate order.

We assume L not to be asymmetric. Then we find two lattice elements v_1 and v_2 with $v_1 \ L \ v_2 \ L \ v_1$. Applying Definition 5 we know that there exist $\bigwedge$-irreducibles

$$(m_{11}, m_{21}) \in M(v_1, v_2) : m_{11} \ L \ m_{21}$$
$$(m_{12}, m_{22}) \in M(v_1, v_2) : m_{22} \ L \ m_{12}.$$

Obviously m_{11} and m_{22} are incomparable. The first case $m_{11} \ L \ m_{22}$ implies $m_{11} \ L \ m_{22} \ L \ m_{12}$ and further with the FPC $m_{22} \geq (m_{11} \wedge m_{12}) \geq v_1$. This is a contradiction, since m_{22} and v_1 are incomparable. The second case $m_{22} \ L \ m_{11}$ leads to a similar contradiction, namely $m_{11} \geq (m_{22} \wedge m_{21}) \geq v_2$.

$\Leftarrow$: We assume L not to meet the FPC. Then we find $\bigwedge$-irreducibles m_i, m_k, m_j with

$$m_i \ L \ m_k \ L \ m_j; \quad m_k \not\geq m_i \wedge m_j =: v.$$

In this case m_k and v are incomparable. W.l.o.g. let $m_k \ L \ v$. We observe then $m_i \ L \ m_k \ L \ v$. Since L is transitive, we conclude $m_i \ L \ v$. This is a contradiction, since we know $m_i \geq v$. $\qquad \square$

The FPC provides a possibility to introduce a ternary relation T by

$$(m_i, m_k, m_j) \in T : \iff m_k > (m_i \wedge m_j).$$

This relation can be understood as "in-betweenness", i.e. m_k can be drawn in between (in terms of the left relation of the diagram) m_i and m_j in a plane diagram of $\mathfrak{V}$.

The second planarity condition clusters the $\bigwedge$-irreducibles above some specified lattice elements. See Figure 3 for an intuitive understanding.

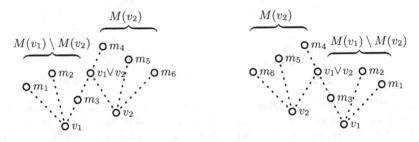

Fig. 3. The SPC holds, if for each two incomparable lattice elements v_1 and v_2, the $\bigwedge$-irreducibles above v_2 are either right or greater *or* left or greater than the ones which are above v_1 but not above v_2

We introduce the notation $M(v) := \{m \in M \mid m \geq v\}$ for the set of all $\bigwedge$-irreducibles above one lattice element v. Furthermore, for an arbitrary relation R we define $M_1 \; R \; M_2 : \iff M_1 \times M_2 \subseteq R$, i.e. a set M_1 is in relation R to another set M_2, if every element of M_1 is in relation to any of M_2. Finally, if R is a relation on a lattice $\mathfrak{V}$, we write $R_<$ and $R_>$ for the union of a R with the strict lattice order $<$ and its inverse $>$ respectively.

Definition 7. *A conjugate relation R on a lattice $\mathfrak{V}$ fulfills the* second planarity condition (SPC) *if the requirements stated below are satisfied:*

1. *R is a strict order on $M \times M$.*
2. *For all lattice elements $v_1 \parallel v_2 \in \mathfrak{V}$ holds*

$$(M(v_1) \setminus M(v_2)) \; R_< \; M(v_2) \quad or \quad M(v_2) \; R_> \; (M(v_1) \setminus M(v_2)).$$

Lemma 5. *Let L be a left-relation on a lattice $\mathfrak{V}$, then the following equivalence holds:*

$$L \text{ satisfies the SPC} \iff L \text{ is a conjugate order.}$$

Proof.
$\Leftarrow$: Since $\mathfrak{V}$ is planar, L is a strict order, in particular on $M \times M$.

Let v_1 and v_2 be arbitrary incomparable lattice elements. With m_1 we denote an element of $M(v_1) \setminus M(v_2)$, with m_2 one of $M(v_2)$. If m_1 and m_2 are comparable then we notice $m_1 < m_2$. Otherwise we would have $m_1 \geq m_2 \geq v_2$ in contradiction to $m_1 \ngeq v_2$.

Let w.l.o.g. $\tilde{m}_1 \; L \; \tilde{m}_2$ for two $\bigwedge$-irreducibles .
We must show that all $m_1 \parallel m_2$ satisfy $m_1 \; L \; m_2$.
With Lemma 3 we conclude

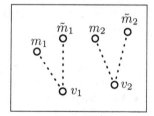

$$\tilde{m}_1 \; L \; \tilde{m}_2 \implies \tilde{m}_1 \; L \; v_2 \implies v_1 \; L \; v_2$$
$$\implies m_1 \; L \; v_2 \implies m_1 \; L \; m_2.$$

$\Rightarrow$: We show that the SPC implies the FPC. Let $m_i, m_j, m_k \in M$ be arbitrary $\bigwedge$-irreducibles such that $m_i \; L \; m_k \; L \; m_j$ holds. Let $v := m_i \wedge m_j$.

- $v > m_k$ implies $m_i > m_k$ contradicting the fact that L is a conjugate relation.
- From $v \parallel m_k$, w.l.o.g. $m_k \; L \; (m_i \wedge m_j)$ we conclude

$$m_k \in (M(m_k) \setminus M(v)) \; L \; M(v) \ni m_i, m_j,$$

in particular $m_k \; L \; m_i$. This contradicts our precondition that L is asymmetric on $M \times M$.
- The remaining case $m_k > v$ is our assertion. $\square$

In the following we will give a more efficient version of the SPC. Instead of clustering attributes of arbitrary lattice elements, we only consider $\bigvee$-irreducibles. It turns out that this is already sufficient to characterize planarity.

Definition 8. *A conjugate relation R on a lattice $\mathfrak{V}$ fulfills the* reduced second planarity condition (rSPC) *if the succeeding requirements are satisfied:*

1. *R is a strict order on $M \times M$.*
2. *For all $\bigvee$-irreducibles $g_1 \parallel g_2 \in \mathfrak{V}$ holds*

$$(M(g_1) \setminus M(g_2)) \; R_< \; M(g_2) \quad or \quad M(g_2) \; R_> \; (M(g_1) \setminus M(g_2)).$$

Proposition 3. *Let L be a left-relation on a lattice $\mathfrak{V}$, then the following equivalence holds:*

$$L \; satisfies \; the \; rSPC \quad \Longleftrightarrow \quad L \; is \; a \; conjugate \; order.$$

Proof.
$\Leftarrow$: Since the rSPC is implied by the SPC, this proof follows immediately from Lemma 5.
$\Rightarrow$: We prove that the rSPC implies the FPC. Let $m_i, m_j, m_k \in M$ be arbitrary $\bigwedge$-irreducibles satisfying $m_i \; L \; m_k \; L \; m_j$. Let $v := m_i \wedge m_j$. We assume $v \parallel m_k$, w.l.o.g. let $v \; L \; m_k$. We search for $\bigvee$-irreducibles g_1 and g_2 not meeting the requirements of the rSPC. This will contradict our assumption and prove the claim $m_k > v$.

- We will use formal concept analysis notation for the proof. We consider the reduced context $\mathbb{K} = (G, M, I)$. Thereby G is the set of $\bigvee$-irreducibles, M is the set of $\bigwedge$-irreducibles and $gIm : \Longleftrightarrow \gamma g \leq \mu m$. We know $\mathfrak{B}(\mathbb{K}) \cong \mathfrak{V}$.
- Since v is incomparable to m_k, we notice that v is not the bottom element of the lattice, i.e. $m_i' \cap m_j' \neq \emptyset$. If gIm_k holds for all $g \in m_i' \cap m_j'$ then it follows $\mu m_k \geq v$ contradicting our assumption. Hence there exists a $g_2 \in G$ with

$$g_2 I m_i, \quad g_2 I m_j \; \text{und} \; g_2 \not I m_k.$$

  In analogy we find a $g_1 \in m_k'$ not possessing both m_i and m_j. In particular, g_i and g_j are incomparable.
- We conclude $m_k \in (g_1' \setminus g_2') \; L \; g_2' \ni m_i, m_j$ contradicting the asymmetry of L on $M \times M$. $\square$

4   Planar Contexts

In this section we want to apply the previous results to contexts in order to characterize, whether the appropriate lattice is planar.

Theorem 2. *Let $\mathbb{K} = (G, M, I)$ be a column reduced context and $\mathfrak{B}(\mathbb{K})$ its corresponding concept lattice. The following statements are equivalent:*

1. *$\mathfrak{B}(\mathbb{K})$ is planar.*
2. *There exists an enumeration of the attributes, s.t. the condition*

$$gIm_i, g \not I m_j, gIm_k \implies \mu m_j < \mu m_k$$

holds for all objects $g \in G$ and all attributes $m_i, m_j, m_k \in M$ with $1 \leq i < j < k \leq |M|$.

Proof.

$\Rightarrow$: Since $\mathfrak{B}(\mathbb{K})$ is planar, we can find a conjugate order L on $B(\mathbb{K})$. The relation $L_<:=L \cup <$ is a strict linear order since it is connex and both L and $<$ are strict orders. If we enumerate the attributes according to $L_<$, we have for all $g \in G, m_i, m_j, m_k \in M$ with $1 \le i < j < k \le |M|$:

$$gIm_i, g \nmid m_j, gIm_k \implies \mu m_j \not\ge \mu m_i \wedge \mu m_k$$
$$\stackrel{FPC}{\implies} \mu m_i \not\mathrel{L} \mu m_j \text{ or } \mu m_j \not\mathrel{L} \mu m_k$$
$$\stackrel{L_<=L\cup<}{\implies} \mu m_i < \mu m_j \text{ or } \mu m_j < \mu m_k$$
$$\stackrel{\mu m_i \not\le \mu m_j}{\implies} \mu m_j < \mu m_k.$$

$\Leftarrow$: Define $\mu m_i \mathrel{R} \mu m_j :\iff \mu m_i \parallel \mu m_j$ and $i < j$. Obviously R is asymmetric and irreflexive.

1. Let m_i, m_j, m_k be arbitrary attributes satisfying $\mu m_i \mathrel{R} \mu m_j \mathrel{R} \mu m_k$. We conclude $i < j < k$ and with the precondition we find that the implication $gIm_i, gIm_k \implies gIm_j$ holds for all objects $g \in G$. We observe that neither $\mu m_i < \mu m_k$ nor $\mu m_i > \mu m_k$ hold since this would imply $\mu m_i < \mu m_j$ or $\mu m_j > \mu m_k$ respectively. Therefore we have $\mu m_i \parallel \mu m_k$, i.e. $\mu m_i \mathrel{R} \mu m_k$. Hence R is transitive.

2. Therefore we gain a sorting relation L_a if we restrict R to pairs of attributes whose attribute concepts have a common upper cover.

3. Let $\mu m_i \mathrel{R} \mu m_j$ hold for $m_i, m_j \in M$ and let $(\mu m_k, \mu m_l) \in M(\mu m_i, \mu m_j)$. The case $\mu m_l \mathrel{R} \mu m_k$ leads to either

$$i < l \implies \mu m_i \mathrel{R} \mu m_l \mathrel{R} \mu m_k \implies \mu m_i \mathrel{R} \mu m_k \text{ or}$$
$$l < i \implies \mu m_l \mathrel{R} \mu m_k \mathrel{R} \mu m_j \implies \mu m_l \mathrel{R} \mu m_j,$$

contradicting $\mu m_i < \mu m_k$ and $\mu m_l < \mu m_j$ respectively Therefore we conclude $\mu m_k \mathrel{R} \mu m_l$. That means that R is a subset of the left-relation L induced by L_a, in particular we have $L_a \subseteq R \subseteq L$ and $R = L \cap (M \times M)$.

4. Let m_i, m_j and m_k be attributes satisfying $\mu m_i \mathrel{R} \mu m_j \mathrel{R} \mu m_k$. From the precondition we derive $gIm_i, gIm_k \implies gIm_j$, i.e. $\mu m_j > (\mu m_i \wedge \mu m_k)$. Since we found $R = L \cap (M \times M)$ we conclude that L satisfies the FPC. Hence $\mathfrak{B}(\mathbb{K})$ is planar. $\square$

5 Results and Further Work

Left-relations are a useful tool to characterize planar lattices. The FPC and the SPC give a new view on the topic. There necessity is intuitively clear, the sufficiency could be proven. With the help of the FPC we could give a condition to decide already in a context, whether the associated concept lattice is planar.

Unfortunately we could not yet reach our aim to find a quick algorithm to recognize a planar lattice and draw it plane. The problem, how to create conjugate orders on a lattice remains unsolved.

A quick algorithm for drawing planar lattices without edge crossings in the plane will help us to reach our main goal: we want to design an algorithm for lattice drawing. Based on the left-relation and the tools supplied so far, we hope to find strategies to minimize the number of edge crossings for non-planar lattices. Of course, considering just this esthetic criterion would not be sufficient. Additionally, we will improve the diagram's quality by maximizing the conflict distance ([9], [10]). This can be done for instance by an optimization process called *force directed placement* ([4], [6]). Finally, diagrams shall be drawn with the attribute-additive convention or with similar drawing rules, e.g.[3].

References

1. K. A. Baker, P. Fishburn, F. S. Roberts: *Partial Orders of Dimension 2*. Networks, 2, 11-28, 1971.
2. G. Birkhoff: *Lattice Theory*. Amer. Math. Soc., Third Edition, 1967.
3. R. Cole, *Automated Layout of Concept Lattice Using Layer Diagrams and Additive Diagrams*. Austr. Comp. Sc. Conf., 2000.
4. G. DiBattista, P. Eades, R. Tamassia, I. G. Tollis, *Graph Drawing*. Prentice Hall, 1999.
5. B. Dushnik, E.W. Miller: *Partially Ordered Sets*. Amer. J. Math. 63, 1941, pp. 600-610.
6. P. Eades: *A Heuristic for Graph Drawing*. Congressus Numerantium 42, pp. 149-160, 1984.
7. D. Kelly, I. Rival: *Planar Lattices*. Can. J. Math. Vol. 27, No. 3, pp. 636-665, 1975.
8. B. Ganter, R. Wille: *Formal Concept Analysis*. Springer, 1999.
9. B. Ganter: *Conflict Avoidance in Order Diagrams*. preprint, TU Dresden, 2003.
10. B. Schmidt, *Ein Optimierungsalgorithmus für additive Liniendiagramme*. Diploma Thesis, TU Dresden, 2002.
11. C. Zschalig: *Planarity of Lattices - An approach based on attribute additivity*. Proc. of ICFCA 05, LNAI 3403, pp. 391-402, 2005.

Automated Layout of Small Lattices Using Layer Diagrams

Richard Cole[1], Jon Ducrou[2], and Peter Eklund[2]

[1] School of Information Technology and Electrical Engineering,
University of Queensland, St. Lucia, Australia
`richard.j.cole@gmail.com`
[2] School of Economics and Information Systems, University of Wollongong,
Woolongong, Australia
{`jrd990, peklund`}`@uow.edu.au`

Abstract. Good quality concept lattice drawings are required to effectively communicate logical structure in Formal Concept Analysis. Data analysis frameworks such as the Toscana System use manually arranged concept lattices to avoid the problem of automatically producing high quality lattices. This limits Toscana systems to a finite number of concept lattices that have been prepared a priori. To extend the use of formal concept analysis, automated techniques are required that can produce high quality concept lattice drawings on demand. This paper proposes and evaluates an adaption of layer diagrams to improve automated lattice drawing.

1 Introduction

The automatic production of high quality concept lattice diagrams in applications of formal concept analysis remains a challenge even for comparatively small lattices [1, 2, 3, 4, 5]. The Toscana data analysis framework relies on the use of concept lattices that have been drawn by human experts prior to the data analysis phase. This is possible within the Toscana system because the diagrams produced by the system are composed of one or more conceptual scales. The conceptual scales are often one of a standard set of scales which have known diagrams. For the remainder, scale diagrams are designed and drawn offline using a program called ANACONDA (or in the case of ToscanaJ, a program called Sienna). However, in a growing number of applications of FCA concept lattice diagrams need to be produced on demand. Examples include: CEM [6], CASS [7], SurfMachine [8] and D–Sift [9].

Concept lattices express a logical structure and to make this logical structure more easily identified by human readers, diagrams of concept lattices are usually drawn as additive diagrams. This paper details and evaluates a method to derive good quality concept lattice diagrams for small lattices by adapting layer diagrams [10]. The method involves the incremental construction of diagrams for a concept lattice by finding satisfactory diagrams for successively larger sublattices. A number of diagram metrics are then calculated for these satisfactory diagrams and used to rank them (according to a fairly simple rule explained in

R. Missaoui and J. Schmid (Eds.): ICFCA 2006, LNAI 3874, pp. 291–305, 2006.
© Springer-Verlag Berlin Heidelberg 2006

Section 4.4). The satisfactory lattice diagrams were also independently assessed by the authors and divided into "good" and "bad" diagrams. The automated layout algorithm is then assessed in terms of the rank of the highest ranked good diagram for each of a set of test lattices.

The work reported in this paper builds on earlier work conducted by one of the authors and reported in [11, 12, 13]. The approach in this paper extends earlier work by introducing a large number of diagram metrics, a formula for combining them to achieve a diagram ranking, and a systematic evaluation of the results.

The structure of the paper is as follows: Sections 2, 3 and 4 set the scene for the construction of concept lattice diagrams, and Section 5 explains details of our algorithm which is then evaluated in Sections 6 and 7.

2   Additive Diagrams

This section introduces definitions for lattice diagrams that help to delineate the space of diagrams that we are concerned with. We seek to be consistent with the notation established by Ganter and Wille in [14] wherein additive diagrams and the mathematical theory of formal concept analysis are defined. In this paper we are primarily concerned with small finite lattices (generally with fewer than 50 concepts) and so the theory in this paper only considers the finite case.

A *diagram* of a lattice, L, is defined by a mapping pos $: L \to \mathbb{R}^n$ called the position assignment function of the diagram. When $n = 2$ the diagram is called a 2D-diagram and when $n = 3$ the diagram is called a 3D-diagram. In this paper we will focus on 2D-diagrams.

The *Hasse diagram constraint* requires that if an element a is less than an element b then a occurs further down the page than b. For a 2D-diagram pos $: L \to \mathbb{R}^2$, the constraint says that for all $a, b \in L$ with $a < b$, pos$(a) = (x_a, y_a)$ and pos$(b) = (x_b, y_b)$ it is the case that $y_a < y_b$. When drawing concept lattices we follow the convention that the x-dimension extends from the left of the page to the right, and that the y-dimension extends from the bottom of the page towards the top.

An *additive diagram* is a diagram defined by a position assignment function, $pos : L \to \mathbb{R}^n$, calculated from a representation set X, a representation function rep $: L \to \mathcal{P}(X)$, and a vector assignment function vec $: X \to \mathbb{R}^n$ according to the following equation:

$$\text{pos}(a) = \sum_{x \in \text{rep}(a)} \text{vec}(x) \tag{1}$$

The upset of $A \subseteq L$ with respect to an ordered set L, is defined as:

$$\uparrow_L(A) := \{\, x \in L \mid x \geq a \text{ for some } a \in A \,\} \tag{2}$$

As a shorthand we write $\uparrow_L(a)$ instead of the more verbose $\uparrow_L(\{a\})$.

Every diagram of a concept lattice that obeys the Hasse diagram constraint is equal to an additive diagram. For a diagram with position assignment function $pos : L \to \mathbb{R}^n$ we construct an additive diagram from rep$(a) = \uparrow_L(a)$, and

$$\text{vec}(a) = \text{pos}(a) - \sum_{x \in X \,|\, x > a} \text{vec}(x) \qquad (3)$$

An *attribute-additive diagram* of a lattice L is an additive diagram whose representation set is composed of the meet irreducibles of $\mathcal{M}(L)$. The representation function is given by: $\text{rep}(a) = \uparrow_L(a) \cap \mathcal{M}(L)$ and the vector assignment function is some function $\mathcal{M}(L) \to R^n$. The reason such a diagram is called attribute-additive is that $\mathcal{M}(L)$ are the attributes of the formal concept $(\mathcal{J}(L), \mathcal{M}(L), \leq)$ which is a reduced formal context whose concept lattice is isomorphic to L. Two finite concept lattices are isomorphic if and only if their two reduced formal contexts are isomorphic. Not all additive diagrams are equal to an attribute additive diagram.

An *x-dimensional additive diagram* is an additive diagram in which only the x-component of the concept positions is given by Equation 1. The y-component is allowed to vary so long as it preserves the Hasse diagram constraint. An x-dimensional attribute-additive diagram is defined similarly.

3   Layout Objectives

In conceptual data analysis the purpose of a concept lattice diagram is to convey information to the human reader of the diagram. A diagram of a concept lattice can convey the following:

1. Intents: Which attributes are possessed by an object or group of objects.
2. Extents: Which objects possess an attribute or a group of attributes.
3. Attribute implications: Which groups of attributes imply other groups of attributes.
4. Object implications: Which groups of objects imply other groups of objects.
5. Partial implications: what proportion of objects with attributes X additionally have attributes Y.

The layout of concept lattices therefore is concerned primarily with optimizing the rate at which this information can be conveyed. A layout that optimizes these tasks is called a *functional* layout. A secondary task is also important: the speed at which a concept lattice can be memorized.

Principles guiding the automated layout of concept lattices are often derived from aesthetic criteria because there is a correlation between aesthetic layout and functional layout. It is often the case that an unaesthetic layout is also a non-functional layout.

3.1   Overlaps

Within a diagram, concepts (represented as circles) and the lines between them have a thickness, and thus can overlap. Overlapping concepts and lines can be misleading to the reader of the concept lattice. We therefore seek diagrams in which (i) the concepts (modelled as circles with radius r) do not overlap and

(ii) the concepts do not overlap with lines they are not connected to. Lines which overlap to a large extent with other lines may be detected by concept-line overlaps. If two lines are parallel and overlap they will also have a concept-line overlap. For these reasons, in discarding diagrams we only look for concept-concept and concept-line overlaps.

3.2   Symmetry

Symmetry in the form of reflection about a horizontal axis seems to be important in making a layout functional. Readers seem to be able to compare the left and right hand side of a diagram and recognize symmetries and departures from symmetry. This seems to aid in the memorizing of diagrams — an indicator of the functionality of a diagram. A number of our proposed metrics seek to select symmetrical structures.

3.3   Background Theory

In some cases the layout of a concept lattice is performed in the presence of some background theory. A classic case is in the program CEM where attributes of the concept lattice come from a background theory in which they are organized within a partial order. In CEM a concept lattice constructed from the data was embedded in a lattice constructed from the background theory as a way to help the reader recognized places where the data deviated from the background theory. In this paper, however, we are concerned with the layout of concept lattices in the absence of a background theory providing clues to good layout.

3.4   Distributive Embedding

Implications can be emphasized by drawing the concept lattice embedded within a distributive lattice. Distributive lattices have the advantage that they have regular layouts since in an attribute-additive diagram of a distributive lattice every edge has a direction and length equal to one of the attribute vectors. It is also the case that the logic of a distributive lattice is relatively simple: all reduced implications have only a single element in their premise. For a more expansive discussion see [15].

Let L be a concept lattice, then the distributive completion of L, denoted $DC(L)$ is the concept lattice $\mathfrak{B}(\mathcal{M}(L), \mathcal{M}(L), \not\leq)$. There is a meet-preserving order embedding from L to the intents of $DC(L)$ given by $a \mapsto \uparrow_L(a) \cap \mathcal{M}(L)$.

Unless L is a distributive lattice this order embedding is not surjective. We distinguish two types of concepts in $DC(L)$, realized and unrealized concepts, corresponding to concepts that are mapped to, and concepts that are not, respectively.

An unrealized concept is evidence of an implication. When a concept is not realized its intent implies the intent of the concept below it. Since L is a lattice there is always a maximal realized concept below any unrealized concept.

We have observed in conceptual data analysis applications that the existence of implications are often discussed by humans readers with reference to missing

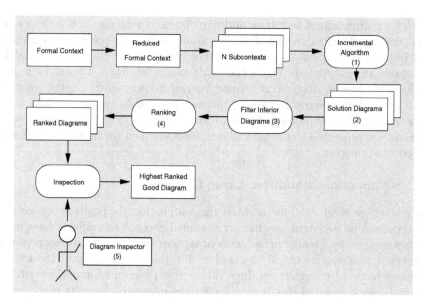

Fig. 1. Flow chart for generating and ranking diagrams

concepts. By embedding L visually in $DC(L)$ the human reader can physically point to unrealized concepts that visually reinforce the presence of an implication.

In many cases the lattice $DC(L)$ is very much larger than L. In such cases two solutions present themselves: the first is to cut out unrealized concepts that have no realized concept below them other than the bottom concept. The second is to automatically decide not to draw $DC(L)$ but instead just draw L.

3.5   Artificial Emphasis

Many layouts emphasize some properties of a concept lattice at the expense of others [16]. For example a concept lattice may contain two attribute implications, one of which is easier to identify, and another which is harder. Other properties such as mutual exclusion of attributes and objects, or the doubling of a particular sublattice may also be emphasized in a diagram.

Artificial emphasis can be potentially misleading to human readers in the task of data analysis, and thus requires caution, as an aesthetic diagram can be potentially non-functional if it over-emphasizes some aspect of the logical structure to the detriment of another aspect.

4   Layout Algorithm

Fig. 1 presents an overview of our algorithm for automated layout of concept lattices. The steps are explained in detail in the following sections. Briefly stated however the process is as follows. A formal context is reduced and then split into n subcontexts. The n subcontexts are then used by the incremental algorithm to search for x-dimensional attribute-additive diagrams with no node-node or edge-node overlaps. The first N such diagrams found by the incremental algorithm are collected and termed *solution* diagrams. For each of these diagrams a collection of metrics are calculated. The solution diagrams are then filtered to remove inferior diagrams. A diagram is termed inferior if there exists another diagram which gets a better score for one of the metrics, and is better or equal for all other metrics. The remaining diagrams are then ranked and the top n ranked diagrams are presented to the user who chooses between them, or selects the top ranked good diagram.

4.1 x-Dimensional Additive Layer Diagrams

The advantage of an attribute-additive diagram is that the position of a concept is determined by its intent, so that an identified concept can quickly have its intent determined by a reader of the concept lattices, and vice-versa a concept with a specific intent can be quickly located within the diagram. Attribute-additive diagrams however generally produce rather bad diagrams for non-distributive lattices because in a non-distributive lattice the intents of concepts in a covering relation (i.e. concepts having an edge between them) may differ from each other by more than a single reduced attribute. This has two consequences: (i) the number of distinct vectors formed by considering edges as vectors is large, and (ii) the variance in vertical displacement between concepts in the covering relation can be large, leading to too much white space in the interior of the diagram.

The idea behind an x-dimensional attribute-additive layout is to keep a correspondence between the concept position and its intent while reducing the variance in the vertical displacement between concepts in a covering relation.

The *uprank* of an element p of a lattice L is the length of the longest path from the element to the top of the lattice. So the top concept has uprank 0, its children have uprank 1 and so on.

$$\text{uprank}(\top) = 0$$
$$\text{uprank}(p) = 1 + \max_{q \in L \,|\, q \prec p} \text{uprank}(q)$$

In an x-dimensional additive diagram the x-position of a concept is determined by an attribute-additive diagram while the y-position is given by the uprank. For distributive lattices rank assignment produces diagrams that are attribute-additive in the y-dimension. An alternative that we do not discuss, apart from this mention in passing, is to make the y-position equal to either the *downrank*, or the average of the *downrank* and *uprank*.

4.2   Incremental Vector Assignment

Let us consider that we have a sequence of vectors, $\mathcal{V} = v_1, \ldots, v_n$, with $v_i \in \mathbb{R}^2$ and we want to incrementally derive an attribute-additive diagram for L using those vectors. The attribute-additive representation function for a lattice L is $\text{rep}(x) = \uparrow_L(x) \cap \mathcal{M}(L)$. Now set $\mathcal{M}(L) := \{m_1, \ldots, m_N\}$ such that $m_i \neq m_j$

for $i \neq j$ and if $m_i \leq m_j$ then $i > j$. Next, define $M_k = \{m_1, \ldots, m_k\}$ for $k = 1 \ldots N$, $m_k \in \mathcal{M}(L)$, and $\text{vec}_k : M_k \to \mathcal{V}$. Now, we introduce the formal contexts $\mathbb{K}_k := (\mathcal{J}(L), M_k, \leq)$ and the associated lattices $L_k := \mathfrak{B}(\mathbb{K}_k)$. Then $\text{rep}_k : \mathfrak{B}(\mathbb{K}_k) \to \mathcal{P}(M_k)$ with $\text{rep}_k(x) = \text{Intent}_L(x) \cap M_k$ is equivalent to the attribute-additive representation for L_k because $(\mathcal{J}(L_k), M_k, \leq)$ is a reduced context.

Now let $\text{pos}_k : L_k \to \mathbb{R}^n$ be the position of points derived from vec_k and rep_k. A position function pos_k is *unsatisfactory* if there exists $x, y \in L_k$ with $x \neq y$ and $\text{pos}(x) = \text{pos}(y)$.

It can be shown (for details see [11]), by considering the natural embedding of L_k into L_{k+1} that if pos_k is *unsatisfactory* then it is also the case that pos_{k+1} will be unsatisfactory as a collision in L_k will be duplicated in L_{k+1}. This gives us a way to incrementally search for satisfactory diagrams.

The incremental algorithm is outlined in Fig. 2. Each diagram is represented by an integer array. The i'th element of the array selects an attribute vector for the i'th attribute. The algorithm starts with an integer array of length 0 (first argument to the function next in Fig. 2). The function next then recursively assigns vectors to attributes each time checking that the result is a satisfactory diagram. If the result is an unsatisfactory diagram then the next function returns, because according to the previous paragraph once an unsatisfactory diagram is produced it cannot be elaborated into a satisfactory diagram. The terminating condition for the recursion is either that a satisfactory diagram for L_N has been produced (line 2). Solution diagrams are only stored if they have no vertex-line overlaps. It is already the case, due to the test on line 10, that they have no vertex line overlaps. Line 11 causes the algorithm to terminate if it has produced a number of solutions equal to max_solutions.

The essential characteristic of the algorithm in Fig. 2 is that it prunes the search space of diagrams by not pursuing a vector assignment v_k that has produced a diagram for L_k that is unsatisfactory.

The solutions function makes available candidate attribute vectors in blocks. If n1=5 and n2=3 then the first call by the solutions function will look for diagrams using just the first 5 canidate attribute-vectors. In the subsequent call 10 candidate vectors will be available and in the final call all 15 candidate vectors will be available. Since the y-coordinate is assigned by layer assignment only the x-component of the candidate vectors are relevent and these are given by the expression $(-1)^i \lfloor i/2 \rfloor$ for $i = 1 \ldots n_1 n_2$.

The store_solution function (not shown, but called on line 4) only stores a solution in the case that the solution is not dominated, or equal in its metrics, to an already recorded solution. Also, if the new solution dominates any recorded solution, that recorded solution is discarded. One solution dominates another if it improves on at least one of the metrics, and improves upon (or does as well as) w.r.t. each of the other metrics. In addition a planar diagram always dominates a non-planar diagram. This process of discarding dominated diagrams and diagrams that are equal in their metric values to an already recorded solution has the effect of eliminating diagrams that are symmetries of one another. Consider a diagram of the b3 lattice (see Section 5) with attributes m_1, m_2, m_3, then a

```
1.    def next(v: Int Array, num_attr: Int, base: Int, max_solutions: Int)
2.        if v.length == num_attr then
3.            if not has_line_node_overlaps(v) then
4.                store_solution(v)
5.                self.solution_count += 1
6.            end
7.        else
8.            for i in 1..base do
9.                v.push(i)
10.               if sat(v) then next(v, num_attr, base) end
11.               if self.solution_count >= max_solutions then break; end
12.               v.pop
13.           end
14.       end
15.   end
16.
17.   def solutions(num_attr: Int, n1: Int, n2: Int, max_solutions: Int)
18.       self.output_count = 0
19.       for i in 1..n2 do
20.           next([], num_attr, i*n1, max_solutions)
21.       end
22.   end
```

Fig. 2. Algorithm to iterate though all satisfactory diagrams. The algorithm is incremental in the sense that is seeks a satisfactory solution for $[v_1, \ldots, v_k]$ before extending it to obtain a solution for $[v_1, \ldots, v_{k+1}]$.

diagram resulting from the vector assignments, (1,2,3), (1,3,2), (2, 1, 3), (2, 3, 1), etc. all look the same. Since they all score the same values on the metrics, only the first solution will be stored.

The output of the incremental algorithm is dependent on the ordering of the meet irreducibles $m_1, \ldots, m_n$ which, while constrained to respect the lattice ordering, is by no means fixed. For example when the irreducibles are unordered there are $n!$ potential orderings. A good ordering will mean that good diagrams are found quickly while a bad ordering will mean that good diagrams occur later in the sequence of diagrams output by the algorithm. To make our experiments more deterministic we ordered the attributes by their extent size in the reduced context so that attributes with the largest extents come first. If two attributes have the same extent size then we ordered based on the order of the attributes in the input context.

4.3   Diagram Metrics

To automatically distinguish good diagrams from bad ones automatically, we calculated the following metrics[1]

---

[1] See http://cvs.sourceforge.net/viewcvs.py/griff/cass_browser/metrics.rb?rev=1.7&view=markup for exact details.

- Number of edge crossings.
- Number of edge vectors.
- Number of edge vectors applied to meet irreducibles.
- Number of edge gradients.
- Number of absolute edge gradients.
- Average path width. For each path from the top of the lattice to the bottom calculate the maximum x-displacement minus the minimum x-displacement of concepts on the path. Average these values.
- Total Edge Length.
- Horizontal Shift. Difference between the top-most and bottom-most concepts x dimension.
- Child Balance. Counts the number of unbalanced children around the parent's x dimension.
- Number of symmetric siblings. Counts the number of sibling pairs whose x-position relative to their parent is the negative of each other.
- Number of symmetric siblings (Non Zero). As above, but excludes all child concepts without siblings.
- Sum of logs of number of elements at average points

$$\sum_{a,b \in L} \text{logth}(\text{count\_ave\_points}(a,b))$$

where

$$\text{logth}(x) = \begin{cases} log(x) & \text{if } x \geq 1 \\ 0 & \text{otherwise} \end{cases}$$

and count_ave_points(a, b) is the number of concepts whose x-position is the average of the x-positions of a and b.
- Count of two chains. A two chain is a child, parent, and grandparent in which the displacement from the child to the parent is the same as the displacement from the parent to the grandparent.
- Count of three chains. A three chain is like a two chain but involves a great grandparent.
- Sum of *well–placed* children. *Well–placed* children are defined for a parent with 1 to 3 children, where the children have parent-relative positions considered well–placed (See Fig. 3(a)).

(a)   (b)

Fig. 3. (a) The three possible child arrangements accepted by the *well–placed* children metric. (b) The solid children show the required positions for acceptance by the *ok–placed* children metric; the dashed children show that other children can be placed in any other configuration.

– Sum of *ok–placed* children. *Ok–placed* children are defined by a parent with 2 or more children where 2 of the children are placed at an x-position relative to the parent by -1 and $+1$ (See in Fig. 3(b)).

These metrics were chosen by a process of iterative refinement by looking at diagrams and trying to determine metrics that would be selective of them. Some experimentation was conducted as part of this design process. Diagrams were produced and ranked and some metrics discarded and new ones produced.

Many of the metrics re-enforce one another. The choice of metrics is linked to the system for ranking diagrams is outlined in the following section.

4.4   Ranking Diagrams

The best value achieved by any generated diagram for each metric is stored. An overall score is then assigned to each diagram equal to the number of metrics for which the diagram has a value equal to the best value for that metric. For example, if a diagram got the best value for 8 out of the 17 metrics then it would get an overall score of 8.

This overall score was used to rank diagrams. In the case that two diagrams have the same overall score then the diagram that was produced earlier is given a higher rank.

For measurement purposes explained in the following section, the top n diagrams by rank are chosen. When n is greater than 1 we presume that the user can be offered n diagrams and choose the one they like the best.

5   Experiments

5.1   Data

We generated diagrams and metrics for a collection of 28 lattices, 10 of which are distributive. The lattices were taken from a number of sources including: (i) applications of FCA, e.g. 2mutex, call-graph1, second and surfmachine, (ii) papers on automated lattice drawing, e.g. fig8, fig16, freese, freese2 and rjthesis, and (iii) standard FCA examples, e.g. b3 and b4, chain-4, mesh and n5. Formal contexts for each of the lattices used in our experiments are available on line at: http://griff.sf.net/test_lattices.html.

Some of the lattices are the distributive completion $DC(L)$ of another lattice in the collection. Specifically any lattice named contra_L is the distributive completion of L. This was done with a view to applications that draw diagrams of L embedded within $DC(L)$ (see Section 3.4).

The lattices in our collection are all small, including less than 50 concepts, and are all known to have good x-dimensional attribute-additive diagrams employing uprank layer assignment. There are some small lattices for which no good layouts have been found either automatically or manually and we purposely did not include any of these in our test set.

5.2   Evaluation

Our algorithm generates n diagrams that are presented to the user to choose between. In order to evaluate the performance of our algorithm we measured the rank of the first *good* diagram. A diagram was termed *good* if there was no

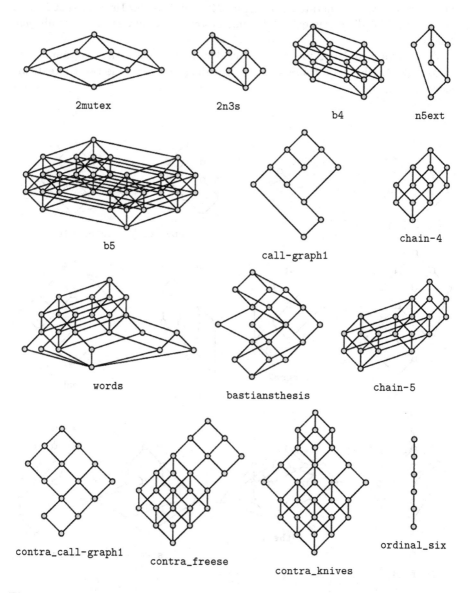

Fig. 4. The top ranked diagram for the test contexts (Part 1). Each diagram was judged to be a *good* diagram, meaning there was no attribute-additive diagram using uprank vector assignment produced either manually or automatically that was significantly better.

other x-dimensional attribute additive diagram using uprank layer assignment that was considered by the authors to be significantly better with respect to the layout objectives outlined in Section 3. The judgement of whether or not a diagram is a good diagram is somewhat subjective.

The top ranked diagrams for the test lattices are shown in Figs. 4 and 5. The diagrams were generated with n1=5, n2=3 and max_solutions=5000 (refer to Fig. 2). The diagrams except for n5, summersurf and second are all *good* diagrams.

The diagram ranked 2 for n5 was considered good, while the diagram ranked 3 was considered good for summersurf. The diagrams for these two contexts are shown in Fig. 7.

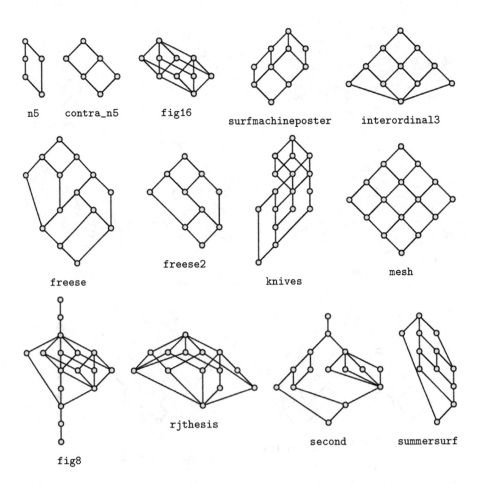

Fig. 5. Top ranked diagrams for the test contexts (Part 2). The top ranked diagram for the test contexts. The top ranked diagrams for n5 and summersurf and second were judged (by us) to be poor diagrams.

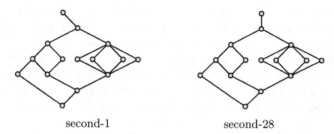

second-1   second-28

Fig. 6. A *good* diagram for `second` was not produced until `max_solutions` was increased to 62,000. The *good* diagram was then ranked 28.

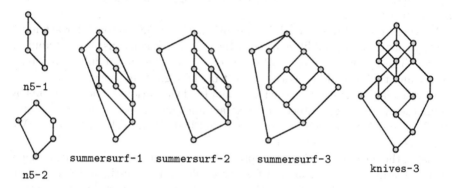

n5-1

summersurf-1   summersurf-2   summersurf-3

n5-2   knives-3

Fig. 7. The top ranked diagrams up to the first *good* diagram for the contexts whose top ranked diagram was judged to be poor (by us). For `n5` the second to top diagram was judged to be *good*. For `summersurf` the third to top ranked diagram was judged to be *good*. The third ranked `knives` diagram could be considered better than the top ranked `knives` diagram.

With the parameter `max_solutions` set to 5,000 a good diagram for `second` was not produced. In order to obtain a good diagram for `second` it was necessary to increase the number of solutions to 62,000. When this was done the highest ranked good diagram was ranked 28 (see Fig. 6).

6   Conclusions

This paper has presented and evaluated an adaptation of layer diagrams for the layout of concept lattices. A mechanism to incrementally search for diagrams that avoid vertex collisions is presented and used to generate a collection of 5,000 diagrams for each lattice. Diagram metrics are then used to prune these diagram collections before a simple classifier ranks the small set of pruned diagrams.

We choose a simple classifier because with such a small number of training points, just 28 lattices, there is a significant danger of over fitting.

In 25 out of the 28 test lattices the first ranked diagram was a good diagram. For two of the remaining cases a good diagram was found in the top 3. In the final remaining case a good diagram was poorly ranked.

We tried to be representative in the selection of test lattices, however this was under the condition that the lattices are small (less than 50 concepts) and also be known to have a good x-dimensional attribute additive diagrams. The algorithm presented does not handle large lattices well (for example the layout of a b6 lattice).

The algorithm presented is restricted to x-dimensional attribute additive diagrams using uprank layer assignment. There are some small lattices for which there is no known good diagram that uses uprank layer assignment. Our algorithm could be adapted to include other rank assignment strategies, but we leave that for future work.

Another area of further work is to consider the same approach for attribute additive diagrams (i.e. additive in both the x and y dimensions) for distributive lattices. This would require a selection of a good set of candidate attribute vectors and thus would be another aspect of the algorithm to tune.

A principled comparison of the functionality of the various lattice diagrams by timing the rate at which subjects can extract information as detailed in Section 3 would make the distinction between good and bad diagrams less subjective. We leave this too as an area of further work.

Lastly, the current trend in microprocessors seems to be towards multiprocessor architectures. Another area of further would be to investigate the extent to which the incremental layout algorithm and the calculation of diagram metrics can be parallelized.

References

1. Wille, R.: Lattices in Data Analysis: How to draw them with a computer. Technical Report 1067, University of Darmstadt (1987)
2. Vogt, F., Wille, R.: TOSCANA - a graphical tool for analyzing and exploring data. In Tamassia, R., Tollis, I., eds.: Proceedings of the DIMACS International Workshop on Graph Drawing (GD'94). Lecture Notes in Computer Science 894, Berlin-Heidelberg, Springer-Verlag (1995) 226–233
3. Becker, P., Hereth, J., Stumme, G.: TOSCANAJ - an open source tool for qualitative data analysis. In: Advances in Formal Concept Analysis for Knowledge Discovery in Databases, FCAKDD 2002. (2002) 1–2
4. Freese, R.: Automated lattice drawing. In: Proc. of the 2nd Int. Conference on Formal Concept Analysis. LNAI 2961, Springer (2004) 112–127
5. Ganter, B.: Conflict Avoidance in Additive Order Diagrams. Journal of Universal Computer Science **10**(8) (2004) 955–966
6. Cole, R., Eklund, P., Stumme, G.: CEM — a program for visualization and discovery in email. In D. A. Zighed, J. Komorowski, J.Z., ed.: Proc. of the European Conf. on Knowledge and Data Discovery, PKDD'00. LNAI 1910, Springer-Verlag (2000) 367–374
7. Cole, R., Becker, P.: Navigation spaces for the analysis of software structure. In: Proceedings of the 3rd International Conference on Formal Concept Analysis. LNCS 3403. Springer (2005) 113–128

8. Ducrou, J., Eklund, P.W.: Combining spatial and lattice-based information land-scapes. In Ganter, B., Godin, R., eds.: Proc. of the 3rd Int. Conference on Formal Concept Analysis. LNAI 3403, Springer-Verlag (2005) 64–78

9. Ducrou, J., Wormuth, B., Eklund, P.: D-SIFT: A dynamic simple intuitive FCA tool. In: Conceptual Structures: Common Semantics for Sharing Knowledge: Proceedings of the 13th International Conference on Conceptual Structures. LNAI 2595, Springer-Verlag (2005) 295–306

10. Battista, G., Eades, P., Tamassia, R., Tollis, I.: Graph Drawing. Algorithms for the Visualisation of Graphs. Prentice Hall, New Jersey (1999)

11. Cole, R.: The Management and Visualisation of Document Collections Using Formal Concept Analysis. PhD thesis, Griffith University, School of Information Technology, Parklands Drive, Southport, QLD (2000)

12. Cole, R.J.: Automatic layout of concept lattices using force directed placement and genetic algorithms. In Edwards, J., ed.: 23th Australiasian Computer Science Conference. Volume 22 of Australian Computer Science Communications., IEEE Computer Society (2000) 47–53

13. Cole, R.J.: Automatic layout of concept lattices using layer diagrams and additive diagrams. In Oudshoorn, M., ed.: 24th Australiasian Computer Science Conference. Volume 23 of Australian Computer Science Communications., IEEE Computer Society (2001) 47–53

14. Ganter, B., Wille, R.: Formal Concept Analysis: Mathematical Foundations. Springer-Verlag, Berlin (1999)

15. Wille, R.: Truncated distributive lattices: Conceptual structures of simple-implicational theories. Order **20**(3) (2003) 229 – 238

16. Becker, P.: Using intermediate representation systems to interact with concept lattices. In: Proc. of the 3rd Int. Conference on Formal Concept Analysis. Volume 3403 of LNAI., Springer (2005) 265 – 268

Counting Pseudo-intents and #P-completeness

Sergei O. Kuznetsov[1] and Sergei Obiedkov[2]

[1] VINITI Institute, ul. Usievicha 20, Moscow, 125190 Russia
serge@viniti.ru
[2] University of Pretoria, Pretoria 0002, South Africa
sergei.obj@gmail.com

Abstract. Implications of a formal context (G, M, I) have a minimal implication basis, called Duquenne-Guigues basis or stem base. It is shown that the problem of deciding whether a set of attributes is a premise of the stem base is in coNP and determining the size of the stem base is polynomially Turing equivalent to a #P-complete problem.

1   Introduction

Since the introduction of the Duquenne-Guigues basis of implications [4, 5] (called also the stem base in [2]), a long standing problem was that concerning the upper bound of its size: whether the size of the basis can be exponential in the size of the input. In [6] we proposed a general form of a context where the number of implications in the basis is exponential in the size of the context. Moreover, in [6] it was shown that the problem of counting pseudo-intents, which serve premises for the implications in the basis, is a #P-hard problem.

A closely related question is that posed by Bernhard Ganter at ICFCA 2005: what is the complexity class of the problem of determining if an attribute set is a pseudo-intent? There was also a conjecture that this problem is PSPACE-complete. This paper provides a proof that this problem is just in coNP. Then, the polynomial Turing equivalence to a #P-complete counting problem is a direct consequence of this fact and the previous #P-hardness result from [6].

2   Definitions and Main Results

We assume that the reader is familiar with basic definitions and notation of formal concept analysis [2]. Recall that, given a context (G, M, I) with derivation operator $(\cdot)'$ and $B, D \subseteq M$, an *implication* $D \to B$ holds if $D' \subseteq B'$.

A minimal (in the number of implications) subset of implications from which all other implications of a context follow semantically [2] was characterized in [4, 5]. This subset is called Duquenne-Guigues basis or stem base in the literature. The premises of implications of the stem base can be given by pseudo-intents [1, 2]: a set $P \subseteq M$ is a *pseudo-intent* if $P \neq P''$ and $Q'' \subsetneq P$ for every pseudo-intent $Q \subsetneq P$.

The notions of quasi-closed and pseudo-closed sets used below have first been formulated in [4] under the name of saturated gaps (*noeuds de non-redondance*

R. Missaoui and J. Schmid (Eds.): ICFCA 2006, LNAI 3874, pp. 306–308, 2006.

in [5]) and minimal saturated gaps (*noeuds minimaux* in [5]), respectively. The terms *quasi-closed* and *pseudo-closed* have been introduced in [1]. The corresponding definitions in [5] and [1] are different but equivalent (except that saturated gaps are not closed by definition). We use notation from [1].

A set $Q \subseteq M$ is *quasi-closed* if for any $R \subseteq Q$ one has $R'' \subseteq Q$ or $R'' = Q''$. For example, closed sets are quasi-closed.

Below we will use the following properties of quasi-closed sets:

Proposition 1. [1] *A set $Q \subseteq M$ is quasi-closed iff $Q \cap C$ is closed for every closed set C with $Q \not\subseteq C$. Intersection of quasi-closed sets is quasi-closed.*

A set P is called *pseudo-closed* if it is quasi-closed, not closed, and for any quasi-closed set $Q \subsetneq P$ one has $Q'' \subsetneq P$. It can be shown that a set P is pseudo-closed if and only if $P \neq P''$ and $Q'' \subsetneq P$ for every pseudo-closed $Q \subsetneq P$. Hence, a pseudo-closed subset of M is a pseudo-intent and vice versa, and we use these terms interchangeably. By the above, a pseudo-intent is a minimal quasi-closed set in its closure class, i.e., among quasi-closed sets with the same closure. In some closure classes there can be several minimal quasi-closed elements.

Proposition 2. *A set S is quasi-closed iff for any object $g \in G$ either $S \cap \{g\}'$ is closed or $S \cap \{g\}' = S$.*

Proof. By Proposition 1, to test quasi-closedness of $S \subseteq M$, one should verify that for all $R \subseteq M$ the set $S \cap R''$ is closed or coincides with S. Any closed set of attributes R'' can be represented as the intersection of some object intents:

$$R'' = \bigcap_{g \in R'} \{g\}' \text{ and } S \cap R'' = \bigcap_{g \in R'} (S \cap \{g\}').$$

If $S \cap \{g\}' = S$ for all $g \in R'$, then $S \cap R'' = S$. Thus, if intersection of S with each object intent is either closed or coincides with S, then this also holds for the intersection of S with any R''. If $S \cap \{g\}'$ is not closed and $S \cap \{g\}' \neq S$ for some g, then this suffices to say that S is not quasi-closed.   □

Corollary 1. *Testing whether $S \subseteq M$ is quasi-closed in the context (G, M, I) may be performed in $O(|G|^2 \cdot |M|)$ time.*

Proof. By Proposition 2, to test whether S is quasi-closed, it suffices to compute intersection of S with intents of all objects from G and check whether these intersections are closed or equal to S. Testing closedness of intersection of S with an object intent takes $O(|G| \cdot |M|)$ time, testing this for all $|G|$ objects takes $O(|G|^2 \cdot |M|)$ time.   □

Proposition 3. *The following problem is in NP:*

> INSTANCE: *A context (G, M, I) and a set $S \subseteq M$*
> QUESTION: *Is S not a pseudo-intent of (G, M, I)?*

Proof. First, we test if S is closed. If it is, then it is not pseudo-closed and the answer to our problem is positive. Otherwise, note that a nonclosed set S is pseudo-closed if and only if there is no pseudo-closed set $P \subsetneq S$ with $P'' = S''$. However, such P exists if and only if there is a quasi-closed set $Q \subsetneq S$ with the same property. Therefore, we nondeterministically obtain for S such a set Q and verify if Q is indeed a quasi-closed subset of S such that $Q'' = S''$. By the corollary of Proposition 2, this test can be done in polynomial time.   □

Corollary 2. *The following problem is in coNP:*

   INSTANCE: A context (G, M, I) and a set $S \subseteq M$
   QUESTION: Is S a pseudo-intent of (G, M, I)?

Consider the problem of counting the number of all pseudo-intents. #P [7] is the class of problems of the form "compute $f(x)$", where f is the number of accepting paths of an NP machine [3]. A problem is #P-hard if any problem in #P can be reduced by Turing to it in polynomial time. A problem is #P-complete if it is in #P and is #P-hard. #P-completeness of a problem in #P, can be proved by reducing a #P-complete problem to it in polynomial time.

   Since the problem of checking whether a set is nonpseudo-closed is in NP, the problem of counting such sets is in #P. Since the number of pseudo-intents is $2^{|M|} - k$ if the number of sets that are not pseudo-intents is k, the #P-hardness of the problem of counting pseudo-intents [6] implies #P-hardness of the problem of counitng the sets that are not pseudo-intents. Hence, we proved

Proposition 4. *The following problem is #P-complete:*

   INSTANCE: A context (G, M, I)
   QUESTION: What is the number of sets that are not pseudo-intents?

Hence, the problem of counting pseudo-intents is polynomially Turing equivalent to a #P-complete problem. It remains still open if deciding that a set is a pseudo-intent can be done in polynomial time.

References

1. B. Ganter: Two Basic Algorithms in Concept Analysis, *Preprint Nr. 831, Technische Hochschule Darmstadt* (1984).
2. B. Ganter, R. Wille: *Formal Concept Analysis: Mathematical Foundations*; Springer, Berlin (1999).
3. M. Garey, D. Johnson: *Computers and Intractability: A Guide to the Theory of NP-Completeness*; Freeman, San Francisco (1979).
4. J.-L. Guigues, V. Duquenne: Informative implications derived from a table of binary data. *Preprint, Groupe Mathématiques et Psychologie, Université René Descartes, Paris* (1984).
5. J.-L. Guigues, V. Duquenne: Familles minimales d'implications informatives résultant d'un tableau de données binaires; *Math. Sci. Hum.* **24**, 95 (1986), 5-18.
6. S.O. Kuznetsov: On the Intractability of Computing the Duquenne-Guigues Base, *Journal of Universal Computer Science*, **10**, no. 8 (2004), 927-933.
7. L. G. Valiant: The Complexity of Enumeration and Reliability Problems; *SIAM J. Comput.* **8**, 3 (1979), 410–421.

Author Index

Lecture Notes in Artificial Intelligence (LNAI)